ANAEROBIC BIOTECHNOLOGY FOR BIOENERGY PRODUCTION

Principles and Applications

ANAEROBIC BIOTECHNOLOGY FOR BIOENERGY PRODUCTION

Principles and Applications

SAMIR KUMAR KHANAL

University of Hawai'i at Mānoa

A John Wiley & Sons, Ltd., Publication

Edition first published 2008

Blackwell Publishing was acquired by John Wiley & Sons in February 2007. Blackwell's publishing program has been merged with Wiley's global Scientific, Technical, and Medical business to form Wiley-Blackwell.

Editorial Office
2121 State Avenue, Ames, Iowa 50014-8300, USA

For details of our global editorial offices, for customer services, and for information about how to apply for permission to reuse the copyright material in this book, please see our website at www.wiley.com/wiley-blackwell.

Library of Congress Cataloging-in-Publication Data
Anaerobic biotechnology for bioenergy production : principles and applications / Samir Kumar Khanal [contributor and editor].
p. cm.
Includes bibliographical references and index.
ISBN-13: 978-0-8138-2346-1 (alk. paper)
ISBN-10: 0-8138-2346-3 (alk. paper)
1. Alcohol–Synthesis. 2. Biomass energy. 3. Anaerobic bacteria. 4. Industrial microbiology.
5. Sewage sludge digestion. 6. Refuse and refuse disposal–Biodegredation. I. Khanal, Samir Kumar.
TP358.A55 2008
662′.88–dc22

2008013128

A catalogue record for this book is available from the U.S. Library of Congress.
Set in 11.5/13.5pt Adobe Garmound by Aptara Inc., New Delhi, India
Printed in Singapore by Fabulous Printers Pte Ltd

Contents

Contributors

Srisuda Dhamwichukorn is President and Chief Scientist of Pacific Time Co. Ltd. and Universe and Science Co. Ltd., Bangkok, Thailand. She received her Ph.D. in environmental biotechnology from Michigan Technological University, Michigan, and was Postdoctoral Research Associate at the University of Notre Dame, Indiana. Dr Dhamwichukorn's research interests are in microbial community analysis using molecular techniques, bioremediation, applications of DNA microarrays, and bioinformatics.

Santha Harikishan is a Process Engineer at Black and Veatch, Kansas City, Missouri. He obtained his M.S. in environmental engineering from Iowa State University. Santha has extensive practical experience with biosolids processing and bioenergy production. He is actively involved in process and facility design for wastewater treatment plants, and evaluation and development of residuals management schemes for treatment facilities.

Hong Liu is an Assistant Professor of Biological and Ecological Engineering at Oregon State University. She obtained her Ph.D. in environmental engineering from the University of Hong Kong and received postdoctoral training at Pennsylvania State University. Dr Liu has published over 30 research papers, mainly in the areas of microbial fuel cell and biohydrogen production. She and her coworkers at Pennsylvania State University won *Popular Mechanics 2005 Breakthrough Award* for their invention of a bioelectrochemically assisted process for hydrogen generation. Her current research interests are in the areas of environmental biotechnology, microbial fuel cell, and bioenergy production.

Keshab Raj Sharma is a Research Fellow in Advanced Water Management Center at the University of Queensland, Australia. He received Ph.D. in environmental engineering from the Hong Kong University of Science and Technology, Hong Kong. Dr Sharma's research focuses on biofilm processes, and process modeling and simulation of biological systems.

Illustrator

Prachand Shrestha is a Ph.D. candidate in Biorenewable Resources Program at Iowa State University. He obtained his M.S. in environmental engineering from Iowa State University. His research interests are in the area of renewable energy and biobased products.

Technical Editor

Christopher S. Jones is Laboratory Supervisor at Des Moines Water Works, Des Moines, Iowa. He holds Ph.D. in analytical chemistry from Montana State University, Bozeman, Montana. He has over a decade of technical writing experience.

Preface

Technological advances have improved the quality of life throughout the twentieth century. Although we have been quick to enjoy the benefits of our technological prowess, we have been slow to acknowledge its negative consequences. Increasingly, we are forced to confront these negative consequences: climate change, increased global demand for energy, growing energy insecurity, and continuous exploitation of limited natural resources. World energy demand is expected to grow by as much as 50% by 2025, mainly due to increasing demand from rapidly growing Asian countries such as China and India.

Sustainability must be the foundation for economic growth in the twenty-first century. We need to redirect our efforts toward bioenergy production from renewable, low-cost, and locally available feedstocks such as biowastes and agri-residues. Such efforts will not only alleviate environmental pollution, but also reduce energy insecurity and demand for declining natural resources. The most cost-effective and sustainable approach is to employ a biotechnology option. Anaerobic biotechnology is a sustainable technology that generates renewable bioenergy and biofuels and helps us achieve our environmental and energy objectives. This book is a result of fine contribution made by several distinguished researchers, and inspiration by King *Bhumibol Adulyadej*'s (Thailand) self-sufficiency economy concept.

Information on this subject is limited, and this textbook will be a first reference for both undergraduate and graduate students, researchers, instructors, consulting engineers, and others interested in bioenergy. The book is intended to be a useful resource to both engineering and science students in agricultural, biological, chemical, and environmental engineering, renewable energy, and bioresource technology.

The book does not assume a previous background in anaerobic biotechnology, although most readers should have a good working knowledge of science or engineering. The first six chapters cover the fundamental aspects of anaerobic processes. The remaining six chapters focus on applications with an emphasis on bioenergy production from wastes and agri-residues. Pertinent calculations and design examples are included in each chapter.

Chapter 1 presents an overview of anaerobic fermentation, including definitions, biochemical reactions, major considerations in an anaerobic system, benefits, limitations, and calculations of the energy generation from various feedstocks.

Chapter 2 covers the common metabolic stages of the anaerobic fermentation of organics and microbiological processes, and aims to provide readers with the necessary basics of microbiology, biochemistry, and stoichiometry involved in an anaerobic system. Chapter 3 focuses on the effect of environmental factors such as temperature, pH, nutrients, and toxicity on the growth of key microbial groups involved in bioenergy production. Chapter 4 describes the biokinetics of anaerobic systems and application of mathematical modeling (e.g., anaerobic digestion model 1 (ADM1)) as a tool in design, operation, and optimization of anaerobic processes for bioenergy production. Chapter 5 covers bioreactor configurations and growth systems (e.g., attached, granular, and suspended) used in anaerobic processes. Appropriate reactor selection and design for bioenergy production are also addressed.

The modern molecular techniques in anaerobic fermentation and their application for the generation of methane, hydrogen, ethanol, and butanol are presented in Chapter 6. Chapter 7 outlines the selection of a suitable reactor design and operating conditions for bioenergy production from a sulfate-rich feedstock without sulfide inhibition. Strategies for sulfide control by converting aqueous and gaseous sulfides to elemental sulfur are also discussed. The next chapter covers bioenergy production from residues of emerging biofuel industries, including feedstocks, biofuel production processes from these feedstocks, stillage and glycerin generation, and anaerobic digestion of these residues. Also covered are water reclamation/reuse and biosolids disposal issues in biofuel industries. Chapter 9 describes the fundamentals of fermentative hydrogen production, including the hydrogen production pathway, strategies of obtaining enriched cultures, factors affecting hydrogen yield, process engineering, and microbiology. In addition, the concept of a bioelectrochemical-based microbial reactor for hydrogen production is also covered. The focus of Chapter 10 is development of the microbial fuel cell (MFC), with emphasis on principles, stoichiometry, energetics, microbiology, design, and operation. Different pretreatment technologies to enhance hydrolysis of high-solids feedstocks are discussed in Chapter 11. The last chapter provides an overview of digester gas production from various feedstocks along with a discussion of the cleaning requirements and energy use options.

I acknowledge several people whose selfless guidance and inspiration positively impacted my career path and preparation of this book. Bhoj Raj Bhattarai, Dhruba Narayan Pathak, and Ram Charan Chauhan were my early mentors who introduced me to science. My advisors Akhilendra Bhusan Gupta, Heinz Eckhardt, and Ju-Chang Huang taught me the art of waste treatment. Shihwu Sung, J. (Hans) van Leeuwen, Robert Brown, and Anthony L. Pometto III at Iowa State University provided an opportunity to work on bioenergy-related research projects for nearly 6 years. Chettiyapan Visvanathan, Chongrak Polprasert, and Preeda Parkpian at Asian Institute of Technology, Thailand, graciously offered me a home to work on the book project during the past three summers. Several of my graduate students at Iowa State University assisted me with workout examples and suggestions.

I am thankful to my contributing authors Harikishan Santha, Hong Liu, Keshab Sharma, and Srisuda Dhamwichukorn for their timely help in providing excellent write-ups in their respective areas of expertise. I am indebted to Christopher Jones for critically reviewing the manuscript. I am grateful to my graduate student Prachand Shrestha at Iowa State University for transforming my poor handwritten sketches into beautiful drawings for the completed book. I also acknowledge his help in formatting chapters. I am thankful to Justin Jeffryes of Wiley-Blackwell for his encouragement in preparing a book in the bioenergy area. Shelby Hayes at Wiley-Blackwell provided administrative assistance and kept me on schedule with her frequent reminder e-mails. Funds from United States Department of Agriculture (USDA), through the Biotechnology Bioproducts Consortium and Subvention Grant at Iowa State University, and the University of Hawai'i at Mānoa partially supported the book project.

Last but not the least, I extend my sincere gratitude, love, and appreciation to my family members, especially my mom, Uma Devi, my three loving sisters, Mala, Malati, and Parbati, and my brother Barma for their support throughout the years. I am also indebted to Saranya for her support and encouragement. And especially my young son Sivarat (Irwin), who missed my company in the many evenings, holidays, and weekends I was working on this book. I hope he will appreciate this effort when he grows up. This book is dedicated to my late father, Shri Lok Bahadur Khanal.

Samir Kumar Khanal, Ph.D., P.E.
University of Hawai'i at Mānoa

CHAPTER

1

Overview of Anaerobic Biotechnology

Samir Kumar Khanal

> We are convinced . . . that socially compatible and environmentally sound economic development is possible only by charting a course that makes full use of environmentally advantageous technologies. By this, we mean technologies that utilize resources as efficiently as possible and minimize environmental harm while increasing industrial productivity and improving quality of life (United States National Research Council Committee, 1995).

1.1 Anaerobic Biotechnology and Bioenergy Recovery

Environmental pollution is one of the greatest challenges human beings face in the twenty-first century. We are also faced with the consequences of climate change, increased global demand on fossil fuels, energy insecurity, and continuous exploitation of limited natural resources. The traditional approach of pollution control, which focuses on ridding pollutants from a single medium, that is, transformation of pollutants from liquid to solid or gas phases and vice versa, is no longer a desirable option. It has become enormously important to direct research efforts toward sustainable methods that not only alleviate environmental pollution, but also ease the stress on depleted natural resources and growing energy insecurity. The most cost-effective and sustainable approach is to employ a biotechnology option. Anaerobic biotechnology is a sustainable approach that combines waste treatment with the recovery of useful byproducts and renewable biofuels. Widespread application of anaerobic technology could ease increasing energy insecurity and limit the emission of toxic air pollutants, including green house gases to the atmosphere.

Figure 1.1 illustrates the potentials of anaerobic biotechnology in recovery of value-added products and biofuels from waste streams. Carbon, nitrogen, hydrogen, and sulfur from municipal, industrial, and agricultural solid and liquid wastes are converted into value-added resources. These include biofuels (hydrogen, butanol, and methane), electricity from microbial fuel cells (MFCs), fertilizers (biosolids), and useful chemicals (sulfur, organic acids, etc.). The sulfur can be used

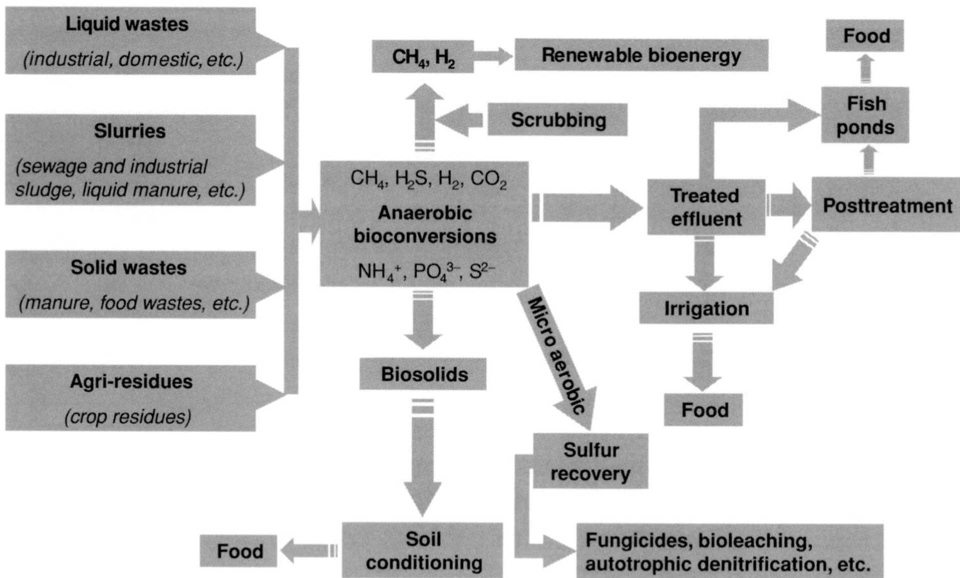

FIG. 1.1. Integrated anaerobic bioconversion processes in recovery of resources from wastes.

as an electron donor for bioleaching of heavy metals or removal of nitrate through autotrophic denitrification. Posttreatment effluent can be lagooned or reused for fish farming, algal production, and irrigation (see Box 1.1).

Box 1.1

Research Need

Due to the concern of endocrine disrupting chemicals (EDCs), e.g., natural steroidal hormones, pharmaceuticals, and personal care products in human/livestock wastes, growing fish, and algae for protein in effluent for human consumption could become a major heath issue. More research is needed to examine the residual levels of EDCs in the effluent and their potential impact on aquatic species.

From the perspective of developing and underdeveloped nations, a wider application of anaerobic biotechnology has even larger implications, as it would fulfill three basic needs: (a) improvement in health and sanitation through pollution control; (b) generation of renewable energy for household activities, such as cooking, lighting, and heating, and running small-scale businesses, for example, poultry farming and silkworm raising; and (c) supply of digested materials (biosolids) as a biofertilizer for crop production. Thus, anaerobic biotechnology plays a significantly greater role not only in controlling pollution but also in supplementing valuable resources: energy and value-added products. This chapter presents a general overview of anaerobic biotechnology and builds up a foundation for the subsequent chapters.

1.2 Historical Development

The chronological development of anaerobic biotechnology is presented in Table 1.1. The application of anaerobic biotechnology dates back to at least the tenth century, when the Assyrians used it for heating bath water (Ostrem 2004). In 1776, Volta recognized that the anaerobic process results in conversion of organic matter to methane gas (McCarty 2001). The French journal *Cosmos* cited the first full-scale anaerobic treatment of domestic wastewater in an airtight chamber known as "Mouras Automatic Scavenger" in 1881. A septic tank modeled on the Mouras Automatic Scavenger was built in the city of Exeter, England, in 1895 by Donald Cameron. Cameron recognized the importance of methane gas, and the septic tank at Exeter was designed to collect methane for heating and lighting. In 1897, waste disposal tanks at a leper colony in Matunga, Bombay, India, were reported to have been designed with a biogas collection system, and the gas used to drive gas engines (Bushwell and Hatfield 1938).

With the development of a two-stage system known variously as the Travis tank (1904) and the Imhoff tank (1905), the focus shifted from wastewater treatment to settled sludge treatment. With the installation of the first sludge heating apparatus, separate digestion of sludge was reported at the Essen-Rellinghausen Plant, Germany, in 1927 (Imhoff 1938). The separate sludge digestion became immensely popular in larger cities, and the importance of methane gas generation was widely recognized. Methane gas was used for digester heating; it was collected and delivered to municipal gas systems, and it was used for power generation for operating biological wastewater treatment systems. Today, anaerobic digestion is widely adopted for the stabilization of municipal sludge and animal manure, and recovery of useful renewable energy—methane and biosolids.

Due to a failure to understand the fundamental of the process, application of anaerobic biotechnology was limited until 1950. Stander (1950) was the first to recognize the importance of solids retention time (SRT) for successful anaerobic treatment of different wastewaters. This has been the basis for the development of the so-called high-rate anaerobic reactor in which SRT and hydraulic retention time (HRT) were uncoupled. This development led to a wider application of anaerobic biotechnology, particularly for industrial wastewater treatment and biogas recovery.

Some of the widely used high-rate anaerobic treatment processes for industrial wastewater treatment include upflow anaerobic sludge blanket (UASB) reactor, expanded granular sludge bed (EGSB), anaerobic filter, fluidized bed, and hybrid

Table 1.1. Historical development of anaerobic biotechnology.

Anaerobic Technologies	Investigator(s) and Place	Developments in Chronological Order
Discovery of combustible air—methane	A. Volta, Italy	Recognized that anaerobic decomposition of organic matters produces methane (1776)
Mouras Automatic Scavenger	M. L. Mouras, France	Patented in 1881; the system had been installed in the 1860s
Anaerobic filter	Massachusetts Experimental Station, United States	Began operation in the 1880s
A hybrid system—a digester and an anaerobic filter	W. D. Scott Moncrieff, England	Constructed around 1890 or 1891
Septic tank	D. Cameron, Exeter, England	Designed in 1895 with provision for recovery of biogas for heating and lighting
	A. L. Talbot, United States	Designed in 1894 (Urbana); 1897 (Champaign)
Waste disposal tank	Leper colony, Matunga, Bombay, India	Digestion tank with gas collection system (1897)
Travis tank	W. O. Travis	Development of a two-stage system for a separate solid digestion (1904)
Imhoff tank	K. Imhoff, Germany	Modified the Travis tank (1905)
Sludge heating system	Essen-Rellinghausen Plant, Germany	Development of first separate sludge digestion system (1927)
Digester seeding and pH control	Fair and More	Realized the importance of seeding and pH control (1930)
High-rate anaerobic digestion	Morgan and Torpey	Developed digester mixing system (1950)
Clarigester (high-rate anaerobic processes)	G. J. Stander, South Africa	Realized the importance of SRT (1950)
Anaerobic contact process (ACP)	G. J. Schroepfer, United States	Developed ACP similar to aerobic-activated sludge process (1955)
Anaerobic filter (AF)	J. C. Young and P. L. McCarty, United States	Reexamined AF for the treatment of soluble wastewater (1969)
Anaerobic membrane bioreactor (AnMBR)	H. E. Grethlein, United States	An external cross-flow membrane coupled with anaerobic reactor (1978)
	Dorr-Oliver, United States	Developed commercial-scale AnMBR in early 1980s
Upflow anaerobic sludge blanket reactor	G. Lettinga, The Netherlands	Based on his first observation of granular sludge in Clarigester in South Africa (1979)
Expanded-bed reactor	M. S. Switzenbaum and W. J. Jewell, United States	Developed fixed-film expanded-bed reactor (1980)
Anaerobic baffled reactor	P. L. McCarty, United States	Retention of biomass within the baffles (1981)
Trace elements for methanogens	R. Speece, United States	Reported the importance of trace elements for methanogenic activity (1983)
Anaerobic sequential batch reactor (ASBR)	R. Dague and S. R. Pidaparti, United States	Developed ASBR for the treatment of swine manure (1992)

Sources: Lettinga (2001), Liao et al. (2006), McCarty (2001), Pidaparti (1991).

Table 1.2. Applications of anaerobic biotechnology in industrial wastewater treatment.

Types of Industries	Numbers of Plants
Breweries and beverages	329
Distilleries and fermentation	208
Chemicals	63
Pulp and paper	130
Food	389
Landfill leachate	20
Undefined/unknown	76
Total in database	1,215

Source: Franklin (2001). Reprinted with permission.

systems. Table 1.2 shows the different applications of high-rate anaerobic reactors in industrial wastewater treatment worldwide. There will be continued efforts to obtain improved bioreactor design to meet the future needs of environmental protection and resource recovery.

1.3 Importance of Anaerobic Biotechnology in Overall Waste Treatment

Although aerobic processes are widely used worldwide for municipal wastewater treatment, anaerobic processes still play a significant role in overall waste treatment as illustrated in Fig. 1.2.

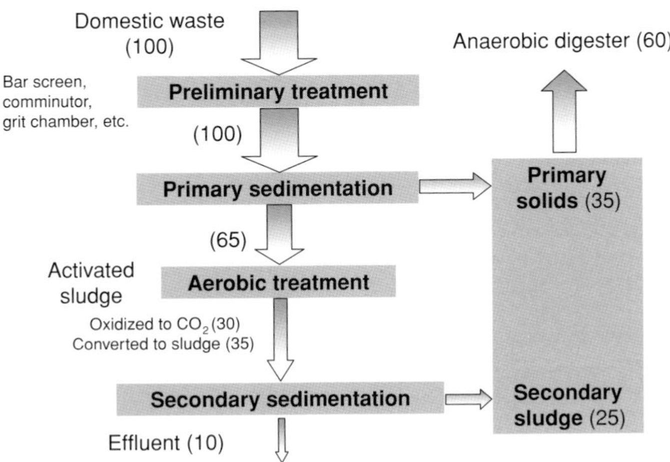

FIG. 1.2. Role of anaerobic biotechnology in overall waste treatment.

During conventional biological wastewater treatment process, preliminary treatment does not reduce biochemical oxygen demand (BOD). This is because particle sizes resulting from preliminary treatment are too large to be measured during a conventional BOD or chemical oxygen demand (COD) analysis. In a typical aerobic biological waste treatment system such as an activated sludge process, the organic pollutants (soluble, colloidal, and/or suspended) are merely transferred from the liquid phase to the solid phase. The solids (primary solids and secondary sludge) account for about 60% of the total influent waste strength, which requires further treatment before final disposal. The fate of the solids and sludge is the anaerobic digester, which reduces their mass and putricibility.

1.4 Definition and Principle of Anaerobic Processes

Anaerobic processes are defined as biological processes in which organic matter is metabolized in an environment free of dissolved oxygen or its precursors (e.g., H_2O_2). Anaerobic process is classified as either anaerobic fermentation or anaerobic respiration depending on the type of electron acceptors.

1.4.1 Anaerobic Fermentation

In an anaerobic fermentation, organic matter is catabolized in the absence of an external electron acceptor by strict or facultative anaerobes through internally balanced oxidation–reduction reactions under dark conditions. The product generated during the process accepts the electrons released during the breakdown of organic matter. Thus, organic matter acts as both electron donor and acceptor. In fermentation the substrate is only partially oxidized, and therefore, only a small amount of the energy stored in the substrate is conserved. The major portion of the adenosine triphosphate (ATP) or energy is generated by substrate-level phosphorylation. Figure 1.3 shows the anaerobic fermentation of glucose to ethanol. It is important to point out that the major portion (two-thirds) of methane is produced through anaerobic fermentation in which acetate acts as both electron

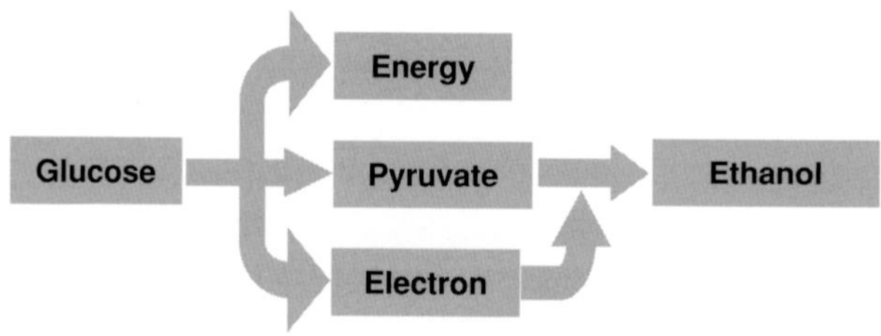

FIG. 1.3. Anaerobic fermentation of glucose to ethanol.

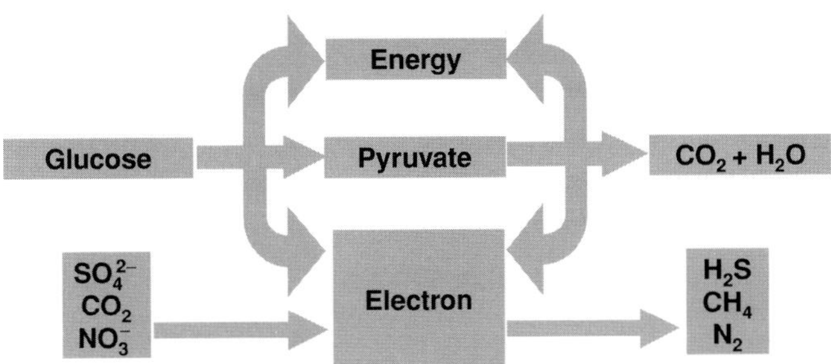

FIG. 1.4. Anaerobic respiration of glucose.

donor and electron acceptor. Methane production through this route is commonly known as acetotrophic (or acetoclastic) methanogenesis. Anaerobic fermentation can be applied for the recovery of both biofuels (e.g., hydrogen and butanol) and biochemicals (nisin and lactic acid) from low-value feedstock.

1.4.2 Anaerobic Respiration

Anaerobic respiration on the other hand requires external electron acceptors for the disposal of electrons released during the degradation of organic matter (Fig. 1.4). The electron acceptors in this case could be CO_2, $SO_4{}^{2-}$, or $NO_3{}^-$. Both substrate-level phosphorylation and oxidative phosphorylation generate energy (or ATP). The energy released under such a condition is much greater than anaerobic fermentation.

When CO_2 accepts the electrons released by the organic matter, it is reduced to CH_4 gas. Methane production through this route is known as hydrogenotrophic methanogenesis and accounts for about one-third of total methane production. Some anaerobes such as homoacetogens also use CO_2 as an electron acceptor and reduce hydrogen to acetic acid (Müller 2001). The presence of sulfate in an anaerobic environment diverts part of organic matter toward sulfate reduction by a specialized group of anaerobic bacteria known as sulfate-reducing bacteria (SRB). The release of odorous hydrogen sulfide (H_2S) gas is a characteristic of anaerobic environment in which sulfate acts as an electron acceptor. SRB are mostly obligate anaerobes, although studies have shown that some species of SRB are capable of aerobic respiration.

When $NO_3{}^-$ acts as an electron acceptor, it is reduced to nitrogen gas. This is a standard biological process for the removal of nitrogenous compounds from wastewater. The process is commonly referred as denitrification or anoxic denitrification. The group of bacteria involved in the process is known as nitrate-reducing bacteria (NRB) or denitrifiers. NRB are usually facultative bacteria, which are capable of aerobic respiration and/or nitrate respiration. The anaerobic environment

Table 1.3. Microbial groups and their preferred electron acceptors and donors, and carbon sources.

Microbes	Electron Acceptor	Electron Donor	Carbon Source
Methane-producing bacteria			
Acetotrophic (or acetoclastic)	Acetate	Acetate	Acetate
Hydrogenotrophic	CO_2	H_2	CO_2
Nitrate/nitrite-reducing bacteria			
Heterotrophic denitrifiers	NO_3^-, NO_2^-	Organic carbon	Organic carbon
Autotrophic denitrifiers	NO_3^-, NO_2^-	S° or H_2	CO_2
Sulfate-reducing bacteria			
Acetotrophic (or acetoclastic)	SO_4^{2-}	Acetate	Acetate
Hydrogenotrophic	SO_4^{2-}	H_2	CO_2
Anaerobic ammonia-oxidizing bacteria	NO_2^-	NH_4^+	CO_2

in which denitrification occurs is sometime known as *anoxic*. Major anaerobic microbes involved in carbon, nitrogen, and sulfur pollution control and the respective electron donors, electron acceptors, and carbon sources are presented in Table 1.3.

From energetic standpoint, oxygen is the most favorable electron acceptor as it releases the greater Gibb's free energy change ($\Delta G^{\circ\prime}$), and hence is favored by microorganisms. In an environment devoid of oxygen, the next best electron acceptor is NO_3^- followed by MnO_2, FeOH, SO_4^{2-}, and CO_2. Findings, however, suggest that fermentation reactions and reductions of SO_4^{2-} and CO_2 may occur almost simultaneously. The affinity of microorganism for the electron acceptor is in the following order (Kiene 1991):

$$O_2 > NO_3^- > MnO_2 > FeOH > SO_4^{2-} > CO_2$$

1.5 Important Considerations in Anaerobic Biotechnology

From both the waste treatment and resource recovery perspectives, it is important to examine some of the important factors that govern the anaerobic bioconversion process. These include organic loading rate, biomass yield, substrate utilization rate, HRT and SRT, start-up time, microbiology, environmental factors, and reactor configuration. The following sections elaborate on these factors.

1.5.1 Volumetric Organic Loading Rate

Anaerobic processes are characterized by high volumetric organic loading rates (VOLRs). High-rate anaerobic reactors such as UASB, EGSB, anaerobic filter, and fluidized bed reactors are capable of treating wastewater at VOLR of 10–40 kg

COD/m^3· day, and on occasion can exceed 100 kg COD/m^3· day in fluidized bed reactors. A high VOLR indicates that more wastewater can be treated per unit of reactor volume. VOLR is one of the most important factors in designing or sizing an anaerobic bioreactor. VOLR is given by the following expression:

$$\text{VOLR} = \frac{C_i Q}{V} \tag{1.1}$$

where C_i is influent wastewater biodegradable COD concentration (mg/L), Q is wastewater flow rate (m^3/day), and V is anaerobic bioreactor volume (m^3).

C_i and Q are known parameters, and VOLR is determined based on on-site pilot-scale testing. For a biological process, the VOLR to the reactor is dependent on several factors, such as the kinetics of pollutant degradation, biomass level in the bioreactor, and types of bioreactor.

1.5.2 Biomass Yield

Biomass yield is a quantitative measure of cell growth in a system for a given substrate. The commonly used term to represent biomass yield is yield coefficient (Y), which is mathematically expressed as:

$$Y = \frac{\Delta X}{\Delta S} \tag{1.2}$$

where ΔX is increase in biomass concentration (mg VSS/L), and ΔS is decrease (consumed) in substrate concentration (mg COD/L).

Of note is the biomass yield per mole of ATP, which totals 10.5 g volatile suspended solids (VSS) for both aerobic and anaerobic processes (Henze and Harremöes 1983). However, when considering the metabolic processes of microorganism, the total aerobic ATP generation is 38 mol, while the anaerobic ATP generation is only 4 mol ATP/mol glucose. This results in a significantly lower biomass yield for the anaerobic treatment process compared to the aerobic process.

Anaerobic degradation of organic matter is accomplished through a number of metabolic stages in a sequence by several groups of microorganisms. This differs from the aerobic treatment process, in which such synergistic relation does not exist. The yield coefficient of acid-producing bacteria is significantly different from that of methane-producing bacteria. The aerobic treatment process gives a fairly constant yield coefficient for biodegradable COD irrespective of the type of substrates. Some common yield coefficients for different processes are presented in Table 1.4.

For an anaerobic system, the yield coefficient depends not only on COD removed but also on the types of substrates being metabolized. Table 1.5 shows the yield coefficients of anaerobic systems under different substrate conditions.

Table 1.4. Yield coefficients.

Process	Yield Coefficient (kg VSS/kg COD)	References
Acidogenesis	0.15	Henze and Harremöes (1983)
Methanogenesis	0.03	
Overall	0.18	
Anaerobic filter (mixed culture) (carbohydrate + protein as substrate)	0.115–0.121	Young and McCarty (1969)
Anaerobic treatment process	0.05–0.15	van Haandel and Lettinga (1994)

Carbohydrate and protein have relatively high yield coefficients, as the two groups of microorganisms (acidogens and methanogens) are involved in the metabolism of the substrates to methane. The overall yield coefficients for these substrates are the sum of individual yield coefficient of acidogens and methanogens. Acetate and hydrogen on the other hand have relatively low yield coefficients as only methanogens are involved in the metabolism of these substrates.

1.5.3 Specific Biological Activity

Specific biological activity indicates the ability of biomass to utilize the substrate. It is usually reported as:

$$\text{Specific substrate utilization rate} = \frac{\text{kg COD}_{\text{removed}}}{(\text{kg VSS} \cdot \text{day})} \tag{1.3}$$

Anaerobic processes have a substrate utilization rate of 0.75–1.5 kg COD/kg VSS · day, which is more than double that of the aerobic treatment process. Henze and Harremöes (1983) also reported substrate removal rate of 1.0 kg COD/kg VSS · day, assuming 50% of the VSS is active. These are quite reasonable rates, as O_2 transfer/diffusion limitation is not an issue in an anaerobic process, unlike an

Table 1.5. Yield coefficients with different substrates.

Types of Substrates	Yield Coefficient (Y) (kg VSS/kg COD)
Carbohydrate	0.350
Protein	0.205
Fat	0.038
Butyrate	0.058
Propionate	0.037
Acetate	0.032
Hydrogen	0.038

Source: Pavlostathis and Giraldo-Gomez (1991).

aerobic system. Furthermore, by maintaining a high concentration of diversified group of biomass in close proximity through biomass immobilization or granulation, a good balance of syntrophic relation between acidogens and methanogens can be achieved. The significant improvement in specific activity of anaerobic system has been the result of studies conducted by Speece in early 1980s, who reported on the specific nutrient requirements of methanogens.

1.5.4 Hydraulic Retention Time and Solids Retention Time

HRT and SRT are two important design parameters in biological treatment processes. HRT indicates the time the waste remains in the reactor in contact with the biomass. The time required to achieve a given degree of treatment depends on the rate of microbial metabolism. Waste containing simple compounds such as sugar is readily degradable, requiring low HRT, whereas complex wastes, for example, chlorinated organic compounds, are slowly degradable and need longer HRT for their metabolism. SRT, on the other hand, controls the microbial mass (biomass) in the reactor to achieve a given degree of waste stabilization. SRT is a measure of the biological system's capability to achieve specific effluent standards and/or to maintain a satisfactory biodegradation rate of pollutants. Maintaining a high SRT produces a more stable operation, better toxic or shock load tolerance, and a quick recovery from toxicity. The permissible organic loading rate in the anaerobic process is also determined by the SRT. Speece (1996) indicated that HRT is a deciding factor in process design for complex and slowly degradable organic pollutants, whereas SRT is the controlling design parameter for easily degradable organics.

For the slow-growing microorganisms such as methanogens, care must be exercised to prevent their washout from the reactor in order to achieve a longer SRT. Continuous stirred tank reactors (CSTRs) without solid separation and recycling are often prone to failure due to excessive biomass washout unless long HRTs (or SRTs) are maintained. Elevated HRTs require a bigger reactor volume (volume = flow rate × HRT), which is costly. An early attempt to maintain a long SRT irrespective of HRT was the use of the clarigester or anaerobic contact process, where the anaerobic sludge was allowed to settle in the settling tank and was then returned back to the reactor.

A wide variety of high-rate anaerobic reactors have been able to maintain extremely high SRTs due to biomass immobilization or agglomeration. Such systems operate under short HRTs without any fear of biomass washout. The first full-scale installation of a UASB reactor in the Netherlands has adequately demonstrated that anaerobic treatment is possible with an HRT as short as 4 h, up to an organic loading rate of 16 kg COD/m^3· day (Lettinga et al. 1980). The empirical HRTs for different anaerobic systems to achieve the same degree of treatment are presented in Table 1.6.

Table 1.6. HRTs of anaerobic systems needed to achieve 80% COD removal efficiency at temperature >20°C.

Anaerobic System	HRT (h)
UASB	5.5
Fluidized/expanded bed	5.5
Anaerobic filter	20
Anaerobic pond[a]	144 (6 days)

Source: van Haandel and Lettinga (1994). Reprinted with permission.
[a] BOD removal efficiency.

1.5.5 Start-Up Time

Start-up is the initial commissioning period during which the process is brought to a point where normal performances of the biological treatment system can be achieved with continuous substrate feeding. Start-up time is one of the major considerations in anaerobic processes because of the slow growth rate of anaerobic microorganisms, especially methanogens, and their susceptibility to changes in environmental factors. Anaerobic treatment systems often need quite a long start-up time, which may weaken their competitiveness with aerobic treatment systems that have a relatively short start-up time of 1–2 weeks. The start-up time could be reduced considerably if the exact microbial culture for the waste in question is used as a seed. Under such a situation the generation time of the microorganisms is greatly reduced. A start-up time of 2–4 months is quite common at a mesophilic temperature range (37°C). Periods exceeding a year may be needed under thermophilic conditions (55°C), due to the high decay rate of biomass. The start-up time also depends on the initial biomass inventory (i.e., the initial amount of seed placed in the reactor). The more seed used, the shorter the start-up time. Loading rates and environmental factors such as pH, nutrient availability, temperature and oxidation–reduction potential (ORP) should be maintained within the limits of microbial comfort during the start-up.

1.5.6 Microbiology

The microbiology of the anaerobic treatment system is much more complicated than that of the aerobic one. An anaerobic process is a multistep process in which a diverse group of microorganisms degrades the organic matter in a sequential order resulting a synergistic action (see Fig. 2.1). The stability of an anaerobic treatment system is often debated, mainly due to the fragile nature of microorganisms especially methanogens to the changes in environmental conditions such as pH, temperature, ORP, nutrients/trace metals availability, and toxicity. When an anaerobic treatment system fails because of lack of proper environmental factors or biomass washout from the reactor, it may take several months for the system to return to a normal operating condition because of an extremely slow growth rate of methanogens.

1.5.7 Environmental Factors

It has been pointed out earlier that anaerobic processes are severely affected by the changes in environmental conditions. Anaerobic treatment system is much more susceptible than the aerobic one for the same degree to deviation from the optimum environmental conditions. The successful operation of anaerobic reactors, therefore, demands a meticulous control of environmental factors close to the comfort of the microorganisms involved in the process. The effect of environmental factors on treatment efficiency is usually evaluated by the methane yield because methanogenesis is a rate-limiting step in anaerobic treatment of wastewater. Hence, the major environmental factors are usually governed by the methanogenesis. Brief descriptions of the important environmental factors are outlined here.

1.5.7.1 Temperature

Anaerobic processes, like other biological processes, strongly depend on temperature. The anaerobic conversion of organic matter has its highest efficiency at a temperature 35–40°C for mesophilic conditions and at about 55°C for the thermophilic conditions (van Haandel and Lettinga 1994). Anaerobic processes, however, can still operate in a temperature range of 10–45°C without major changes in the microbial ecosystem. Generally, anaerobic treatment processes are more sensitive to temperature changes than the aerobic treatment process.

1.5.7.2 Operating pH

There are two groups of bacteria in terms of pH optima, namely acid-producing bacteria (acidogens) and methane-producing bacteria (methanogens). The acidogens prefer a pH of 5.5–6.5, while methanogens prefer a range of 7.8–8.2. In an environment where both cultures coexist, the optimal pH range is 6.8–7.4. Since methanogenesis is considered as the rate-limiting step, where both groups of bacteria are present, it is necessary to maintain the reactor pH close to neutral.

1.5.7.3 Oxidation–reduction potentials

Morris (1975) reported that to obtain the growth of obligate anaerobes in any medium, the culture ORP value should be maintained from −200 to −350 mV at pH 7. It is well established that methanogens require an extremely reducing environment, with redox potentials as low as −400 mV (Archer and Harris 1986; Hungate 1969).

1.5.7.4 Nutrients and Trace Metals

All microbial-mediated processes require nutrients and trace elements during waste stabilization. A question may arise how nutrients and trace elements are involved in waste stabilization. In fact nutrients and trace metals are not directly involved in waste stabilization; but they are the essential components of a microbial cell and are thus required for the growth of an existing microbial cell and synthesis of new cell. Besides, nutrients and trace metals also provide a suitable physicochemical condition for optimum growth of microorganisms. It is important to note that if the waste stream in question does not have one or more of the important nutrients and trace elements, the waste degradability is severely affected. This is because of inability of microbial cell to grow at optimum rate and to produce new cells.

1.5.7.5 Toxicity and Inhibition

Anaerobic microorganisms are inhibited by the substances present in the influent waste stream and by the metabolic byproducts of microorganisms. Ammonia, heavy metals, halogenated compounds, and cyanide are examples of the former, while ammonia, sulfide, and volatile fatty acids are examples of the latter. It is interesting to point out that many anaerobic microorganisms are also capable of degrading refractory organics (Stronach et al. 1986) that otherwise might be considered toxic. In some cases, toleration is manifested by acclimation to toxicants. These observations provide a considerable cause for optimism about the feasibility of anaerobic treatment of industrial wastewaters that contain significant concentrations of toxic compounds (Parkin and Speece 1982).

1.5.8 Reactor Configuration

Selection of a proper reactor configuration is of prime importance in anaerobic processes. The relatively low biosynthesis rate of methanogens in an anaerobic system demands special consideration for reactor design. The selection of reactor types is based on the requirement of a high SRT/HRT ratio, so as to prevent the washout of slow-growing methanogens.

The treatment performance of the selected reactors is, therefore, mainly dependent on their capability to retain biomass, thus maintaining a high SRT/HRT ratio.

Another approach for reactor configuration selection is based on required effluent quality. Because of relatively high half-saturation constants (K_s) for anaerobic microorganisms, CSTRs may not be suitable, as immediate dilution of the waste leads to low concentrations of organic matters, but still too high to meet the effluent discharge standards, which are below the range of anaerobic degradation. Under such circumstances, a staging or plug flow type reactor would be more beneficial.

1.6 Merits of Anaerobic Biotechnology

Anaerobic biotechnology is becoming immensely popular due to its potential to produce renewable biofuels and value-added products from low-value feedstock such as waste streams. In addition, it provides an opportunity for the removal of pollutants from liquid and solid wastes more economically than the aerobic processes. These merits are illustrated in the following sections.

1.6.1 Recovery of Bioenergy and Biofuels

1.6.1.1 Biomethane Production

Methane gas is a major byproduct of anaerobic degradation of organic solid and liquid wastes. Methane gas has an energy content of 55,525 kJ/kg at 25°C and 1 atm (CRC Handbook of Chemistry and Physics 1996). With one-third conversion efficiency of heat energy into electrical energy, the electricity generation is 5.14 kWh/kg CH_4 [55,525 × (1/3) × (1/3,600)]. The methane energy generation is calculated as follows:

Stoichiometrically, 1 kg of COD releases about 15.625 mol (or 0.35 m^3 at standard temperature and pressure (STP)) of methane gas (see Example 1.1). Thus, 1 kg COD is needed to produce 15.625 mol (or 0.25 kg) of methane. The electrical energy generated from methane is 1.29 kWh/kg $COD_{removed}$ ((5.14 kWh/kg CH_4) × (0.25 kg CH_4/kg COD)).

Example 1.1

How much methane gas could be generated through complete anaerobic degradation of 1 kg COD at STP?

Solution

Step 1: Calculation of COD equivalent of CH_4

It is necessary to calculate the COD equivalent of methane by considering its complete oxidation, as shown in the following chemical equations:

$$\begin{array}{lll} CH_4 & + 2O_2 & \rightarrow CO_2 + 2H_2O \\ 16\text{ g} & 64\text{ g} & \\ \rightarrow 16\text{ g } CH_4 & \sim 64\text{ g } O_2\text{ (COD)} & \\ \rightarrow 1\text{ g } CH_4 & \sim 64/16 & = 4\text{ g COD} \end{array} \quad (1)$$

Step 2: Conversion of CH_4 mass to equivalent volume

Based on ideal gas law, 1 mol of any gas at STP occupies a volume of 22.4 L.

$$\begin{aligned}
&\rightarrow 1 \text{ mol } CH_4 \sim 22.4 \text{ L } CH_4\\
&\rightarrow 16 \text{ g } CH_4 \sim 22.4 \text{ L } CH_4\\
&\rightarrow 1 \text{ g } CH_4 \sim 22.4/16 = 1.4 \text{ L } CH_4
\end{aligned} \quad (2)$$

Step 3: CH_4 generation rate per unit of COD removed

From Eqs (1) and (2), we have:

$$\begin{aligned}
&\rightarrow 1 \text{ g } CH_4 \sim 4 \text{ g COD} \sim 1.4 \text{ L } CH_4\\
&\rightarrow 1 \text{ g COD} \sim 1.4/4 = 0.35 \text{ L } CH_4\\
&\text{or } 1 \text{ kg COD} \sim 0.35 \text{ m}^3 \text{ } CH_4
\end{aligned} \quad (3)$$

So, complete anaerobic degradation of 1 kg COD produces 0.35 m^3 CH_4 at STP.

1.6.1.2 Biohydrogen Production

In anaerobic fermentation, hydrogen is produced during acidogenic phase. The consumption of hydrogen by hydrogenotrophic methanogens is prevented by proper process control, such as pH and heat treatment (Khanal et al. 2006). From a global environmental perspective, production of hydrogen from renewable organic wastes represents an important area of bioenergy production. Using glucose as a model substrate, the hydrogen production can be represented by the equations (Miyake et al. 1984):

$$C_6H_{12}O_6 + 2H_2O \rightarrow 2CH_3COOH + 2CO_2 + 4H_2 \quad \Delta G^\circ = -184 \text{ kJ}$$

$$C_6H_{12}O_6 \rightarrow CH_3CH_2CH_2COOH + 2CO_2 + 2H_2 \quad \Delta G^\circ = -255 \text{ kJ}$$

Most of the studies on hydrogen production have been primarily confined to the laboratory scale. The low yield of hydrogen through anaerobic fermentation alone has long been a major challenge to engineers and scientists. Liu and Fang (2003) reported a maximum hydrogen yield of about 3.76 mol H_2/mol sucrose using acidogenic granular sludge at an HRT of 13.7 h, temperature of 26°C, and pH of 5.5 in a CSTR. Kataoka et al. (1997) studied continuous hydrogen production in a chemostat using a pure culture of *Clostridium butyricum* SC-E1 with glucose as an organic substrate at an HRT of 8 h, temperature of 30°C, and pH of 6.7. The authors reported the maximum hydrogen yield of 1.3–2.2 mol H_2/mol glucose. The hydrogen production potential of cellulose was investigated using two types of natural inocula: anaerobically digested sludge and sludge compost in batch cultures at 60°C (Ueno et al. 1995). The authors reported a hydrogen yield of

0.9 and 2.4 mol/mol hexose for anaerobic digested sludge and sludge compost as inocula, respectively. More research is needed to improve the hydrogen yield for commercial viability.

1.6.1.3 Butanol Production

Butanol is also a potential substitute for fossil fuel and is considered a superior fuel to ethanol for several reasons: more favorable physical properties, better economics, and safety. In addition, the butanol eliminates the need for engine modification that has been running on gasoline. Butanol is produced by fermentative bacteria including *Clostridium acetobutylicum* (Qureshi et al. 2006) and *Clostridium beijerinkii* (Formanek et al. 1997). The ratio of acetone, butanol, and ethanol (ABE) is 3:6:1, with butanol being the major fermentation byproduct. The ABE fermentation consists of two distinct phases: acidogenesis and solventogenesis. The solvent production particularly, butanol, takes place during the solventogenesis and is directly correlated to the spore-forming ability of the culture (Long et al. 1984). Low butanol yield through fermentation coupled with cheap petroleum feedstock is the major impediment to the widespread development of butanol fuel. Environmental Energy Inc., Blacklick, OH, claimed that the use of fibrous bed bioreactor along with their patent process could produce 2.5 gal butanol per bushel of corn (http://www.butanol.com/). Carbohydrate-rich waste stream could serve as an ideal feedstock for butanol production. Hydrogen gas is another byproduct of butanol fermentation and can also be recovered as a renewable energy.

1.6.1.4 Biodiesel Production from Biogas

The biogas generated during anaerobic digestion of organic waste can be converted into liquid fuel—biodiesel. The biogas is first converted into liquid methanol using a thermal catalytic process. Biodiesel or methyl ester is then produced by transesterification of fats or oil with methanol in the presence of a base catalysis (e.g., sodium or potassium hydroxide). Smithfield's Circle Four Swine farm in Southwestern, Utah, USA, is running two full-scale anaerobic digesters under mesophilic condition. The biogas produced will be used for in situ methanol production, which is currently under construction. The produced methanol will be shipped to Texas for biodiesel production. Some of the important features of biogas/biodiesel process are outlined in Box 1.2.

1.6.1.5 Electricity Generation Using Microbial Fuel Cell

An MFC is a device that directly converts biochemical energy stored in the carbohydrate and other organic matter in wastewater into electricity. An MFC contains two chambers, consisting of an anode and cathode similar to hydrogen fuel cell,

Box 1.2

Number of hogs: 144,000 head
Total feed flow rate: ~872 m^3/day (230,400 gal/day)
Hydraulic retention time: ~30 days
Feed total solids content: ~3–4%
Volatile solids destruction rate: 55–64%
Biogas production: ~11,300–14,100 m^3/day (400,000–500,000 ft^3/day)
Methanol production potential: ~15–19 m^3/day (4,000–5,000 gal/day)

separated by a proton (cation) exchange membrane (PEM). The organic matter is oxidized by anaerobic microbes in the anode chamber and electrons are released. These electrons are then transferred to the anode (positively charged terminal) and flowed to the cathode (negatively charged terminal) through a conductive material such as a resistor or to an external load. The electrons in the cathode combine with protons that diffuse through the PEM and oxygen (from air). The oxygen is reduced to water. In the MFC, the driving force is the redox reaction of substrates (wastewater) mediated by anaerobic microorganisms. Thus, MFC research has a potential to treat the wastewater and produce electricity. MFC studies, however, have been mainly confined to the laboratory-scale level, and their full-scale application is still on the far horizon.

1.6.2 Recovery of Value-Added Products

1.6.2.1 Recovery of Acetic Acid

Miller and Wolin (1995) reported production of high concentration of acetate (0.33 M) from cellulose substrate by a coculture of cellulolytic bacterium and a reductive acetogen that yields acetate from H_2 and CO_2. In a conventional anaerobic process, acetate is produced by homoacetogens from H_2 and CO_2, and with proper control of environmental conditions, acetate production can be enhanced. The control strategies include pH, redox potential, periodic depletion of hydrogen via nitrate addition, and addition of other substances, such as protein, bile salts, or by varying feed composition variation (Verstraete and Vandevivere 1999).

1.6.2.2 Production of Nisin and Lactic Acid

Nisin is a bacteriocin produced commercially by fermentation using lactic acid bacteria (LAB), primarily *Lactococcus lactis.* Nisin is of considerable interest because of its increasing use as a natural food preservative against a wide range of gram-positive pathogens. Waste streams from soy-processing (soy whey), cheese-processing (cheese whey), corn-processing, and other food-processing industries serve as an ideal feedstock for nisin production. These waste streams are nutritionally rich, containing protein, carbohydrate, phosphorus, and numerous trace elements needed for the growth of *L. lactis*. The LAB are also able to produce lactic acid during anaerobic fermentation.

1.6.3 Waste Treatment

1.6.3.1 Less Energy Requirement

Aerobic treatments are energy-intensive processes for the removal of organic matter, requiring 0.5–0.75 kWh of aeration energy for every 1 kg of COD removed (van Haandel and Lettinga 1994). Anaerobic treatments need no air/O_2 supply. The aeration energy requirement is calculated based on the following consideration:

For the removal of 1 kg COD, 0.5–0.75 kg O_2 is required during a conventional aerobic treatment process. The higher end of the range can be explained by the O_2 requirement for endogenous respiration. The energy input for the transfer of O_2 into liquid for most aerators is in the order of 1 kWh/kg O_2.

The aeration energy requirement is:

$$= \frac{1\ \text{kWh}}{\text{kg O}_2} \times \frac{0.5\text{–}0.75\ \text{kg O}_2}{\text{kg COD}}$$
$$= 0.5\text{–}0.75\ \text{kWh/kg COD}$$

The reader should bear in mind that the use of anaerobic treatment provides a net financial gain through energy generation from methane gas, as well as savings realized through the elimination of energy inputs required for aeration. The energy balance between anaerobic and aerobic treatment processes is shown in Box 1.3.

1.6.3.2 Less Biomass (Sludge) Generation

Aerobic wastewater treatment process, especially activated sludge process, generates considerable amounts of sludge. Biological oxidation of every kilogram of soluble BOD produces 0.5 kg of sludge as depicted in Fig. 1.5. The cost of treatment and disposal of sludge accounts for 30–60% of the total operational costs in a conventional activated sludge process.

Anaerobic treatment processes, on the other hand, utilize more than 90% of the biodegradable organic matter (COD) for methane production, with only 10% or less converted to biomass, as illustrated in Fig. 1.6.

Sludge production in an anaerobic process, as depicted in Figs 1.5 and 1.6, is <20% of the aerobic treatment process. Furthermore, the anaerobic sludge is well stabilized and needs no further treatment other than dewatering for final disposal.

Box 1.3

Compare the energy balance between aerobic and anaerobic processes for treating a food-processing wastewater with the following characteristics:

Wastewater flow rate: 10 MGD (~37.85 m^3/day)
Wastewater soluble chemical oxygen demand: 10,000 mg/L
Influent temperature: 20°C

The anaerobic reactor will be operated under mesophilic condition (35°C).

Anaerobic process:

(a) Energy generation from methane gas

Methane yield = 0.35 m^3/kg COD at STP
COD loading rate = 10,000 mg/L (10^{-6} kg/10^{-3} m^3) × 37.85 m^3/day
= 378.5 kg COD/day
Total methane generation = 0.35 m^3/kg COD × 378.5 kg COD/day
= 132.5 m^3/day

The net heating energy content of methane = 35,846 kJ/m^3 (at STP)
Thus, the total net energy content of methane = 35,862 kJ/m^3
× 132.5 m^3/day
= 4.75 × 10^6 kJ/day

(b) Energy need for temperature increase from 20 to 35°C

Heat energy needed = 37,850 kg/day × ((35–20)°C) × (4,200 J/kg °C)
= 2.38 × 10^6 kJ/day

Aerobic process:

Aeration energy requirement = (0.75 kWh/kg COD) × (3,600 s/h)
× (378.5 kg COD/day) = 1.02 × 10^6 kJ/day

Energy	Anaerobic Treatment	Aerobic Treatment
Methane gas (kJ/day)	4.75 × 10^6	—
Energy for reactor heating (kJ/day)	−2.38 × 10^6	—
Aeration energy (kJ/day)	—	−1.02 × 10^6

Note: Anaerobic treatment provides a net energy gain, whereas aerobic process requires energy input. If the costs of sludge handling, treatment, and disposal are included in this calculation, anaerobic process will result even higher net energy gain.

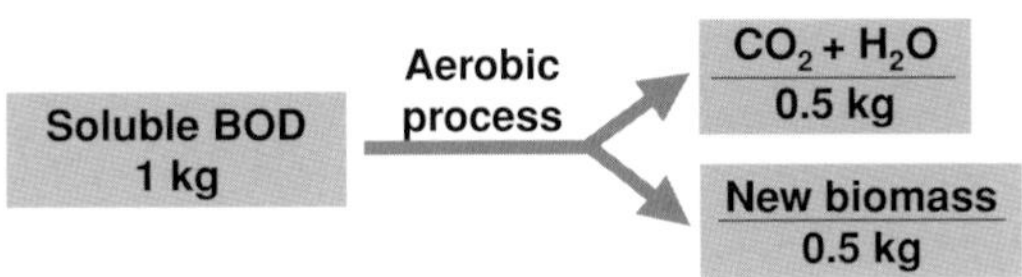

FIG. 1.5. Fate of organic matter in an aerobic process.

1.6.3.3 Less Nutrients (N and P) Requirement

Owing to the lower biomass synthesis rate during the anaerobic process, the nutrient requirements are considerably lower, with the anaerobic process requiring just 20% of the nutrients required for the aerobic process.

1.6.3.4 Higher Volumetric Organic Loading Rate

A higher organic loading rate is not recommended for aerobic treatment processes primarily due to the following:

1. Limited O_2 supply/transfer rate, especially in fixed-film reactors, such as a trickling filter and a rotating biological contactor.
2. Limitation related to the maintenance of high biomass concentrations due to poor settleability, especially in the activated sludge process.

Anaerobic treatment processes are not limited by O_2 transfer capability, and extremely high concentrations of biomass can be maintained in high-rate reactors such as UASB, anaerobic filters, and expanded/fluidized bed reactors. Therefore, loading rates 10–20 times higher for anaerobic treatment processes are possible. The completely mixed anaerobic digesters are the exception in this case, where a maximum concentration of solid/biomass in the reactor is governed by the adequate mixing requirement.

1.6.3.5 Space Considerations

Since a relatively high biomass concentration is maintained in an anaerobic system compared to an aerobic one, large volumetric organic loading rates can be applied.

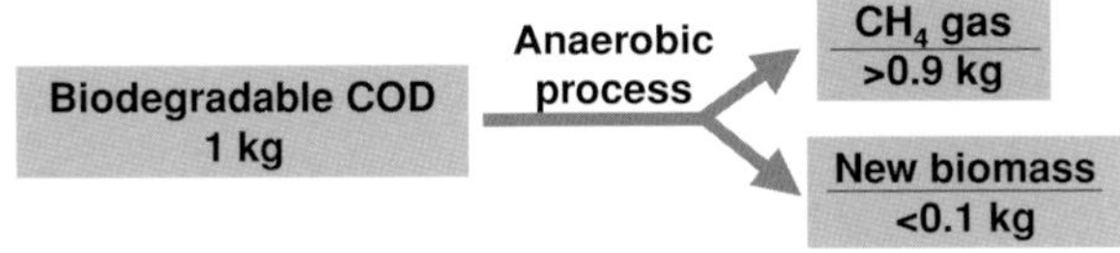

FIG. 1.6. Fate of organic matter in an anaerobic process.

The application of a higher loading rate, therefore, requires a smaller reactor volume, reducing land requirements for the anaerobic treatment units.

1.6.3.6 Ability to Reduce Concentrations of Refractory Organics

With proper acclimation, many of the previously identified refractory organics such as carbon tetrachloride, chloroform, trichloroethane, tetrachloroethane, and polychlorinated biphenyl have been successfully transformed to a lower chlorine functionality by anaerobic microorganisms. These byproducts can then be further degraded by aerobic bacteria to nontoxic end products (Petersen and Samual 1998).

1.6.3.7 Odor Control

Anaerobic treatment largely proceeds in a closed reactor to avoid oxygen contact with the anaerobic biomass and to collect the produced biogas. This prevents the emanation of malodorous compounds, especially hydrogen sulfide.

1.7 Limitations of Anaerobic Process

Although the anaerobic process has many inherent benefits as reported earlier, it is not a panacea for the treatment of all types of wastewaters. Some of the limitations of anaerobic treatment system are outlined here.

1.7.1 Long Start-Up Time

Low sludge yield is deemed one of the major advantages of anaerobic treatment systems. The flip side is that low sludge yields require longer start-up times to attain a given biomass concentration. Start-up times can be reduced by maintaining a higher biomass inventory during the reactor start-up.

1.7.2 Long Recovery Time

If an anaerobic treatment system is subjected to disturbances, due to either biomass washout, toxic substances, or shock loading, it may take a longer time for the system to return to the normal operating condition. However, the extent of such effect could be alleviated by using high-rate anaerobic reactors, such as UASB, anaerobic filter, anaerobic membrane bioreactor (AnMBR), and anaerobic sequential batch reactor, which maintain relatively high SRTs.

1.7.3 Specific Nutrients and Trace Metal Requirements

Anaerobic microorganisms have very specific nutrient requirements. Trace amounts of iron, nickel, and cobalt are essential for optimum growth of methanogens. Municipal wastewater usually contains sufficient amounts of micronutrients and trace metals. However, industrial wastewater often lacks such micronutrients and trace metals, and requires external supplementation. Speece (1996) reported that failure of many anaerobic reactors prior to the 1970s may have been due to lack of understanding of the micronutrient requirement of methanogens.

1.7.4 More Susceptible to Changes in Environmental Conditions

Anaerobic microorganisms, especially methanogens, are more prone to changes in environmental factors such as temperature, pH, and redox potentials. Thus, treatment of low-temperature wastewater requires heating to bring the temperature to an optimum level. Wastewater with a low pH or low alkalinity generation potential, such as dilute wastewater or carbohydrate-rich wastewater, may require alkalinity supplementation to maintain optimum pH. Moreover, an anaerobic reactor operating at the thermophilic temperature is more likely to fail due to changes in environmental conditions than one operating at a mesophilic condition. It is important to note that the degree of susceptibility could be reduced by maintaining a high biomass concentration through the use of high-rate anaerobic reactors.

1.7.5 Treatment of High-Sulfate Wastewater

Anaerobic treatment of high-sulfate wastewater poses considerable challenges to engineers. The presence of sulfate reduces the methane yield due to substrates (such as hydrogen and acetate) diversion to sulfate reduction. In addition, methanogens are inhibited by the presence of sulfide produced by sulfate reducers. The hydrogen sulfide also lowers the quality of the biogas as fuel. Finally, hydrogen sulfide is extremely corrosive gas and produces an objectionable odor. The author has successfully developed an online sulfide control method for the treatment of such wastewaters (Khanal and Huang 2006).

1.7.6 Effluent Quality of Treated Wastewater

The minimum substrate concentration (S_{min}) from which microorganisms are able to generate energy for their growth and maintenance is much higher for an anaerobic treatment system than the aerobic one. Owing to this fact, the anaerobic process may not be able to degrade the organic matter to a level meeting discharge limits required by many environmental agencies for ultimate disposal. Thus, in many cases, the anaerobically treated effluent may require posttreatment before final disposal.

1.7.7 High Protein- and Nitrogen-Containing Wastewater

Proteins are not completely degraded during anaerobic treatment. The partial degradation of proteins produces amines that impart a foul smell. Little information exists on anaerobic degradation of amines (Verstraete and Vandevivere 1999). Similarly, nitrogen concentrations remain unchanged during anaerobic treatment, as reducing equivalents necessary for denitrification are removed. Thus, in anaerobic treatment, only the forms of nitrogen are changed; that is, organic nitrogen is simply transformed to inorganic ammonia or ammonium, depending on pH. However, recent findings suggest that NH_4^+ can be anaerobically oxidized to N_2 in the presence of NO_2^-, as shown by the following biochemical reaction:

$$NH_4^+ + HNO_2 \rightarrow N_2 + 2H_2O$$

The above process is commonly referred as *an*aerobic *amm*onia *ox*idation (ANAMMOX).

1.7.8 Meticulous Attention

A successful operation of an anaerobic treatment system requires careful attention. Attention is needed especially on the availability of trace metals, nutrients, and alkalinity; avoidance of toxic chemicals, volatile fatty acids accumulation, shock loadings, air exposure, and sludge washout; and maintenance of proper environmental conditions, for example, temperature, pH, and ORP. Such attention is quite often crucial during the start-up phase. Poor attention to these details may lead to complete failure of anaerobic reactors.

Example 1.2

A UASB reactor has been employed to treat leachate from an acidogenic fermentation unit in a two-phase anaerobic digestion of food waste at 20°C. The leachate flow rate is 2,000 L/day with mean soluble COD of 7,000 mg/L. Calculate the maximum methane generation rate in m^3/day. What would be the biogas generation rate at 85% COD removal efficiency with 10% of the COD removed diverted to biomass? The mean methane content of the biogas is 80%.

Solution

Maximum methane generation rate:

The complete degradation of organic matter in the waste could only lead to maximum methane generation, which is also regarded as theoretical methane generation rate.

$$\therefore \text{Total COD removed} = \frac{(7,000 \times 10^{-6})}{(10^{-3})} \times (2,000 \times 10^{-3}) \text{ kg/day}$$
$$= 14 \text{ kg/day}$$

From Eq. (3) in Example 1.1, we have:

1 kg COD produces 0.35 m^3 CH_4 at STP
14 kg COD produces ~ 0.35 × 14 = 4.9 m^3 CH_4/day at STP

$$\text{Total COD removed} = \frac{(7,000 \times 10^{-6})}{(10^{-3})} \times (2,000 \times 10^{-3}) \times 0.85 \text{ kg/day}$$
$$= 11.9 \text{ kg/day}$$

As 10% of the removed COD has been utilized for biomass synthesis, the remaining 90% of the removed COD has thus been converted to CH_4 gas.

COD utilized for CH_4 generation = 11.9 × 0.9 kg/day = 10.71 kg/day

From Eq. (3) in Example 1.1, we have:

1 kg COD produces 0.35 m^3 CH_4 at STP
10.71 kg COD produces 0.35 × 10.71 = 3.75 m^3 CH_4/day at STP
At 20°C, the CH_4 gas generation = 3.75 × (293/273) = 4.02 m^3/day

The biogas generation rate = 4.02/0.80 = 5.03 m^3/day

References

Archer, D. B., and Harris, J. E. 1986. Methanogenic bacteria and methane production in various habitats. In *Anaerobic Bacteria in Habitats Other Than Man*, edited by E. M. Barnes and G. C. Mead, pp. 185–223. Blackwell Scientific Publications.

Bushwell, A. M., and Hatfield, W. D. 1938. *Anaerobic Fermentation.* Bulletin No. 32, State Water Supply.

CRC Handbook of Chemistry and Physics, 76th edn (1995–1996), edited by D. R. Lide. CRC Press, Boca Raton, FL, USA.

Formanek, J., Mackie, R., and Blaschek, H. P. 1997. Enhanced butanol production by *Clostridium beijerinckii* BA101 grown in semidefined P2 medium containing 6 percent maltodextrin or glucose. *Appl. Environ. Microbiol.* 63:2306–2310.

Franklin, R. J. 2001. Full-scale experience with anaerobic treatment of industrial wastewater. *Water Sci. Technol.* 44(8):1–6.

Henze, M., and Harremöes, P. 1983. Anaerobic treatment of wastewater in fixed film reactors—a literature review. *Water Sci. Technol.* 15:1–101.

Hungate, R. E. 1967. A roll tube method for cultivation of strict anaerobes. In *Methods in Microbiology*, edited by J. R. Norris and D. W. Ribbons, pp. 117–132. Academic Press, New York, USA.

Imhoff, K. 1938. Sedimentation and digestion in Germany. In *Modern Sewage Disposal*, edited by L. Pearse, p. 47. Lancaster Press, Lancaster, PA, USA.

Kataoka, N., Miya, K., and Kiriyama, K. 1997. Studies on hydrogen production by continuous culture system of hydrogen-producing anaerobic bacteria. *Water Sci. Technol.* 36(6–7):41–47.

Khanal, S. K., Chen, W.-H., Li, L., and Sung, S. 2006. Biohydrogen production in continuous flow reactor using mixed microbial culture. *Water Environ. Res.* 78(2):110–117.

Khanal, S. K., and Huang, J.-C. 2006. Online oxygen control for sulfide oxidation in anaerobic treatment of high sulfate wastewater. *Water Environ. Res.* 78(4):397–408.

Kiene, R. P. 1991. Production and consumption of methane in aquatic systems. In *Microbial Production and Consumption of Greenhouse Gases: Methane, Nitrogen Oxides, and Halomethane*, edited by J. E. Rogers and W. B. Wiliams, pp. 111–146. American Society of Microbiology.

Lettinga, G. 2001. Digestion and degradation, air for life. *Water Sci. Technol.* 44(8):1567–1576.

Lettinga. G., Velsen, A. F. M., Hobma, S. W., De Zeeuw, and Klapwijk, A. 1980. Use of the upflow sludge blanket (USB) reactor concept for biological wastewater treatment, especially for anaerobic treatment. *Biotechnol. Bioeng.* XXII:699–734.

Liao, B. Q., Kraemer, J. T., and Bagley, D. M. 2006. Anaerobic membrane bioreactors: Applications and research directions. *Crit. Rev. Environ. Sci. Technol.* 36:489–530.

Liu, H., and Fang, H. H. P. 2003. Hydrogen production from wastewater by acidogenic granular sludge. *Water Sci. Technol.* 47(1):153–158.

Long, S., Jones, D. T., and Woods, D. R. 1984. The relationship between sporulation and solvent production in *Clostridium acetobutylicum* P262. *Biotechnol. Lett.* 6(8):529–534.

McCarty, P. L. 2001. The development of anaerobic treatment and its future. *Water Sci. Technol.* 44(8):149–156.

Miller, T. L., and Wolin, M. J. 1995. Bioconversion of cellulose to acetate with pure cultures of ruminococcus albus and a hydrogen-using acetogens. *Appl. Environ. Microbiol.* 61:3832–3835.

Miyake, J., Mao, X. Y., and Kawamura, S. 1984. Photoproduction of hydrogen by a co-culture of a photosynthetic bacterium and *Clostridium butyricum*. *J. Ferment. Technol.* 62:531–535.

Morris, J. G. 1975. The physiology of obligate anaerobiosis. *Adv. Microb. Physiol.* 12:169–246.

Müller, V. 2001. Bacterial fermentation. In *Encyclopedia of Life Sciences*, pp. 1–7. Nature Publishing.

Ostrem, K. 2004. *Greening Waste: Anaerobic Digestion for Treating the Organic Fraction of Municipal Solid Wastes.* M.S. thesis, Earth Resources Engineering, Columbia University.

Parkin, G. F., and Speece, R. E. 1982. Attached versus suspended growth anaerobic reactors: Response to toxic substances. *Water Sci. Technol.* 15(8–9):261–289.

Pavlostathis, S. G., and Girolda-Gomez, E. 1991. Kinetics of anaerobic treatment: A critical review. *Crit. Rev. Environ. Sci. Technol.* 21(5–6):411–490.

Petersen, J. N., and Samual, Y. B. 1998. The effect of oxygen exposure on the methanogenic activity of an anaerobic bacterial consortium. *Environ. Prog.* 17(2):104–110.

Pidaparti, S. 1991. *Sequential Batch Reactor Treatment of Swine Manure at 35° C and 25° C.* Master's thesis, Iowa State University, Ames, IA, USA.

Qureshi, N., Li, X.-L., Hughes, S., Saha, B. C., and Cotta, M. A. 2006. Butanol production from corn fiber xylan using *Clostridium acetobutylicum*. *Biotechnol. Prog.* 22(3):673–680.

Speece, R. E. 1996. *Anaerobic Biotechnology for Industrial Wastewater Treatments.* Archae Press, Nashvillee, TN, USA.

Stander, G. J. 1950. Effluents from fermentation industries. Part IV. A new method for increasing and maintaining efficiency in the anaerobic digestion of fermentation effluents. *Journal of the Institute of Sewage Purification*, Part 4, 447.

Stronach, S. M., Rudd, T., and Lester, J. N. 1986. *Anaerobic Digestion Processes in Industrial Wastewater Treatment.* Springer-Verlag, Berlin, Germany.

Ueno, Y., Kawai, T., Sato, S., Otsuka, S., and Morimoto, M. 1995. Biological production of hydrogen from cellulose by nature anaerobic microflora. *J. Ferment. Bioeng.* 79:395–397.

van Haandel, A. C., and Lettinga, G. 1994. *Anaerobic Sewage Treatment: A Practical Guide for Regions with a Hot Climate.* John Wiley & Sons, Chichester, England.

Verstraete, W., and Vandevivere, P. 1999. New and broader applications of anaerobic digestion. *Crit. Rev. Environ. Sci. Technol.* 28(2):151–173.

Young, J. C., and McCarty, P. L. 1969. The anaerobic filter for waste treatment. *J. Water Pollut. Control Fed.* 41(5):R160–R173.

CHAPTER 2

Microbiology and Biochemistry of Anaerobic Biotechnology

Samir Kumar Khanal

2.1 Background

One of the key factors in the success of microbial-mediated processes is an adequate understanding of process microbiology, more specifically the study of microscopic organisms involved in waste degradation and byproduct formation. An anaerobic process is much more complex than an aerobic process due to the involvement of a diverse group of microorganisms and a series of interdependent metabolic stages. The low growth rate and the specific nutrient and trace mineral requirements of methanogens, coupled with their susceptibility to changes in environmental conditions, demand meticulous process control for stable operation. The biochemistry mainly involves enzyme-mediated chemical changes (the chemical activities of microorganisms), types of substrates (wastes/residues) microorganism can destroy or transform to new compounds, and the step-by-step pathway of degradation. This chapter illustrates the common metabolic stages and process microbiology of anaerobic processes.

2.2 Organics Conversion in Anaerobic Systems

The transformation of complex macromolecules, for example, proteins, carbohydrates (polysaccharides), and lipids present in wastewater, or solids into end products such as methane and carbon dioxide is accomplished through a number of metabolic stages mediated by several groups of microorganisms. Figure 2.1 illustrates the schematics of the various steps and the bacterial groups involved in anaerobic digestion of complex wastes (Gujer and Zehnder 1983).

Complex organic compounds such as proteins, carbohydrates, and lipids are transformed into simple soluble products such as amino acids, sugars, and

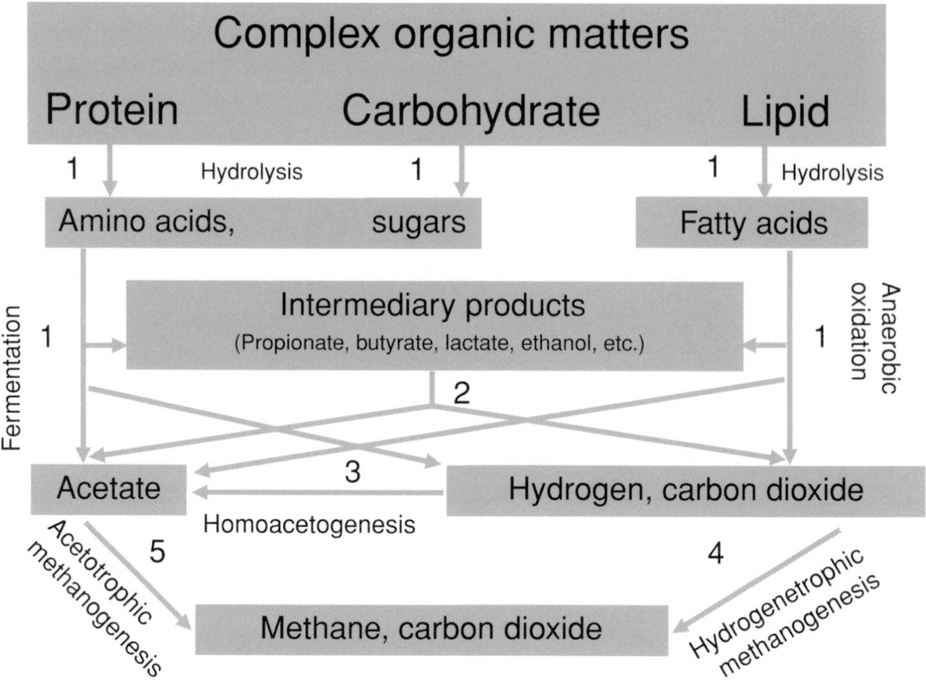

FIG. 2.1. Conversion steps in anaerobic digestion of complex organic matter. (The number indicates the group of bacteria involved in the process.)

long-chain fatty acids and glycerine, by the action of extracellular enzymes excreted by the fermentative bacteria (group 1). This step is commonly known as hydrolysis or liquefaction. Hydrolysis can be a rate-limiting step in the overall anaerobic treatment processes for waste containing lipids and/or a significant amount of particulate matter (e.g., sewage sludge, animal manure, and food waste) (Henze and Harremöes 1983; van Haandel and Lettinga 1994). For such wastes, the rate of methane production in a mature digester is proportional to the net rate of particle solubilization (Gujer and Zehnder 1983).

The fermentative bacteria ferment the soluble products of the first step to a mixture of organic acids, hydrogen, and carbon dioxide. Acidogenesis is the generation of volatile fatty acids (VFAs) ($C > 2$), such as propionic and butyric acid. These VFAs along with ethanol are converted to acetic acid, hydrogen, and carbon dioxide by another group of bacteria known as hydrogen-producing acetogenic bacteria (group 2). The acetic acid–producing step is known as acetogenesis. Acetogenesis is regarded as thermodynamically unfavorable unless the hydrogen partial pressure is kept below 10^{-3} atm via efficient removal of hydrogen by the hydrogen-consuming organisms such as hydrogenotrophic methanogens (Zinder

1988) and/or homoacetogens. Elevated hydrogen partial pressure is reported to inhibit propionate degradation in particular (Speece 1996).

Acetate, H_2, and CO_2 are the primary substrates for methanogenesis. On chemical oxygen demand (COD) basis about 72% of methane production comes from the decarboxylation of acetate, while the remainder is from CO_2 reduction (McCarty 1964). The groups of microorganisms involved in the generation of methane from acetate are known as acetotrophic or aceticlastic methanogens (group 3). The remaining methane is generated from H_2 and CO_2 by the hydrogenotrophic methanogens (group 4). Since methane is largely generated from acetate, acetotrophic methanogenesis is the rate-limiting step in anaerobic wastewater treatment. The synthesis of acetate from H_2 and CO_2 by homoacetogens (group 5) has not been widely studied. Mackie and Bryant (1981) reported that acetate synthesis through this pathway accounts for only 1–2% of total acetate formation at 40°C and 3–4% TS at 60°C in a cattle waste digester.

In anaerobic waste stabilization, there exists a symbiotic (or syntrophic) relationship between acetogens and methanogens. The symbiotic relationship keeps the anaerobic system well balanced. One of the most important tests to judge this balance is the determination of individual VFAs in the effluent. VFAs are short-chain organic acids and are the intermediates formed during anaerobic fermentation of complex organic materials. For a normal operating anaerobic system, the effluent VFA concentration ranges from 50 to 250 mg/L as acetic acid (Sawyer et al. 2003). When the symbiotic relationship is disturbed, due to either overloading, toxicity, nutrient deficiency, or biomass washout, there is an accumulation of VFAs and their levels continue to increase. This may cause an abrupt drop in pH and subsequent souring of the anaerobic reactor. If corrective measures are not taken in a timely manner, the reactor may eventually fail (see Box 2.1).

Box 2.1

Upflow Anaerobic Sludge Blanket Granules: Distribution of Acidogens, Acetogens, and Methanogens

An upflow anaerobic sludge blanket (UASB) granule is an aggregate of densely packed diverse microbial communities. Based on cell morphologies and ultrastructures, Fang (2000) reported a three-layered structure of a UASB granule treating carbohydrate-rich wastewater. The outer layer comprises hydrolytic/fermentative acidogens that convert the complex organic matter into VFAs. The middle layer is mainly populated with syntrophic microbes consisting of acetogens (which convert the diffused VFAs from outer layer into acetic acid and hydrogen) and hydrogenotrophic methanogens (which eventually convert hydrogen into methane). The inner layer consists of acetoclastic (acetotrophic) methanogens, primarily *Methanosaeta*. The author further reported that the granule did not exhibit layered microbial distribution for substrates (e.g., propionate, peptone, ethanol, and glutamate) with rate-limiting hydrolytic/fermentative step. For such granules, the microbes of different morphologies intertwined randomly throughout the cross section.

2.3 Process Microbiology

The anaerobic degradation of complex organic matter is carried out by different groups of bacteria as indicated in Fig. 2.1. There exists a coordinated interaction among these bacteria. The process may fail if one group is inhibited.

2.3.1 Fermentative Bacteria (1)

This group of bacteria is responsible for the first stage of anaerobic digestion—hydrolysis and acidogenesis. The anaerobic species belonging to the family of Streptococcaceae and Enterobacteriaceae and the genera of *Bacteroides*, *Clostridium*, *Butyrivibrio*, *Eubacterium*, *Bifidobacterium*, and *Lactobacillus* are most commonly involved in this process (Novaes 1986). Bacillaceae, Lactobacillaceae and Enterobacteriaceae are all present in digesting sludge with Bacillaceae predominating (Novaes 1986). The hydrolyzed products of proteins, for example, peptides and amino acids, are fermented to VFAs, CO_2, H_2, NH_4^+, and S^{2-} by fermentative bacteria such as the *Clostridia.*

The various types of substrates and environmental conditions determine the end products of the metabolism. Especially important is regulating the presence of H_2. Only at a low hydrogen partial pressure, the formation of acetate, CO_2, and H_2—the major substrates for methanogens—is thermodynamically favorable. If the partial pressure of H_2 is high, the formation of propionate and some other organic acids occurs (Novaes 1986).

In recent years, there has been considerable interest in recovery of value-added products from waste streams. The incomplete breakdown of organic matter by fermentative bacteria results in the formation of useful byproducts including organic acids, solvents, nisin, hydrogen, and others. For example, lactic acid and nisin can be produced by lactic acid bacteria. Hydrogen is produced by fermentative bacteria, especially the genera of *Clostridia*, during acidogenesis. A detailed discussion of hydrogen production is given in Chapter 9.

2.3.2 Hydrogen-Producing Acetogenic Bacteria (2)

This group of bacteria metabolizes C3 or higher organic acids (propionate, butyrate, etc.), ethanol, and certain aromatic compounds (i.e., benzoate) into acetate, H_2, and CO_2. The anaerobic oxidation of these compounds is not favorable

Table 2.1. Free energy changes for anaerobic oxidation of propionate, butyrate, benzoate, and ethanol by hydrogen-producing acetogenic bacteria in pure cultures under standard conditions.[a]

Reactions	$\Delta G^{\circ\prime}$ (kJ/reaction)
Propionate → acetate (i) $CH_3CH_2COO^- + 3H_2O \rightarrow CH_3COO^- + H^+ + HCO_3^- + 3H_2$	+76.1
Butyrate → acetate (ii) $CH_3CH_2CH_2COO^- + 2H_2O \rightarrow 2CH_3COO^- + H^+ + 2H_2$	+48.1
Benzoate → acetate (iii) $C_7H_5CO_2^- + 7H_2O \rightarrow 3CH_3COO^- + 3H^+ + HCO_3^- + 3H_2$	+53
Ethanol → acetate (iv) $CH_3CH_2OH + H_2O \rightarrow CH_3COO^- + H^+ + 2H_2$	+9.6

Source: Dolfing (1988). Reprinted with permission.
[a] H_2 in gaseous form and other compounds in aqueous solution at 1 mol/kg activity; 25°C.

thermodynamically by hydrogen-producing bacteria in a pure culture, as illustrated by positive Gibb's free energy changes (Table 2.1). However, in a coculture of hydrogen-producing acetogenic bacteria and hydrogen-consuming methanogenic bacteria, there exists a symbiotic relationship between these two groups of bacteria. Hungate was the first researcher to realize the importance of hydrogen production and its consumption in an anaerobic system (Hungate 1967). The hydrogen-consuming methanogenic bacteria rapidly scavenge the hydrogen and keep the level of hydrogen partial pressure extremely low. This provides a thermodynamically favorable condition for the hydrogen-producing acetogenic bacteria to break down the aforementioned organic compounds into acetate, H_2, and CO_2 as evident by negative Gibb's free energy changes (Table 2.2). This phenomenon is commonly known as *interspecies hydrogen transfer*. The propionic acid oxidation to acetate becomes thermodynamically favorable only at hydrogen partial pressures below 10^{-4} atm, and for butyrate and ethanol oxidation below 10^{-3} and 1 atm, respectively (Pohland 1992). It is important to point out that during anaerobic treatment of complex wastes such as sewage sludge, as many as 30% of the electrons are

Table 2.2. Free energy changes for anaerobic oxidation of propionate, butyrate, benzoate, and ethanol by coculture of hydrogen-producing acetogenic and hydrogen-consuming methanogenic bacteria under standard conditions.

	$\Delta G^{\circ\prime}$			
	H_2-Consuming Methanogenic Bacteria		H_2-Consuming, Sulfate-Reducing Bacteria	
Substrate	Substrate (kJ/mol)	CH_4 (kJ/mol)	Substrate (kJ/mol)	CH_4 (kJ/mol)
Propionate	−25.6	−34.1	−37.8	−50.4
Butyrate	−19.7	−39.4	−27.9	−55.7
Benzoate	−10.7	−14.3	−22.9	−30.6
Ethanol	−58.2	−116.4	−66.4	−132.7

Source: Dolfing (1988). Reprinted with permission.

associated with propionate oxidation (McCarty and Smith 1986). Thus, oxidation of propionate appears to be more critical than oxidation of other organic acids and solvents.

Under thermophilic conditions, propionic acid accumulation is even more severe than that occurring under mesophilic conditions. Reactor configuration, nutrient supplementation, substrate characteristics, and microbial proximity are equally important to enhance propionic acid metabolism (Speece et al. 2006) (see Box 2.2).

Box 2.2

Interspecies Hydrogen Transfer: High-Rate Anaerobic Reactor
McCarty and Smith (1986) described how high-rate anaerobic reactors could maintain low hydrogen partial pressure to efficiently oxidize intermediates (e.g., propionate, butyrate, and ethanol). Hydrogen produced during anaerobic oxidation of intermediates must be consumed rapidly by hydrogen-consuming microbes such as methanogens, sulfate reducers, or homoacetogens for efficient oxidation of these intermediates to take place. Fixed-film bioreactors (such as an anaerobic filter or fluidized bed), upflow anaerobic sludge blanket, static granular bed reactor, expanded granular sludge bed, and membrane bioreactor provide an excellent opportunity for the proximate growth of a diverse microbial community. Thus, high-rate anaerobic systems are considered to be highly efficient for maintaining lower hydrogen partial pressure.

2.3.3 Homoacetogens (3)

Homoacetogenesis has attracted much attention in recent years because of its final product, acetate, an important precursor to methane generation. The responsible bacteria are either autotrophs or heterotrophs. The autotrophic homoacetogens utilize a mixture of hydrogen and carbon dioxide, with CO_2 serving as the carbon source for cell synthesis. Some homoacetogens can use carbon monoxide as a carbon source. The heterotrophic homoacetogens, on the other hand, use organic substrates such as formate and methanol as a carbon source while producing acetate as the end product.

$$CO_2 + H_2 \rightarrow CH_3COOH + 2H_2O \quad (2.1)$$

$$4CO + 2H_2O \rightarrow CH_3COOH + 2CO_2 \quad (2.2)$$

$$4HCOOH \rightarrow CH_3COOH + 2CO_2 + 2H_2O \quad (2.3)$$

$$4CH_3OH + 2CO_2 \rightarrow 3CH_3COOH + 2CO_2 \quad (2.4)$$

Clostridium aceticum and *Acetobacterium woodii* are the two mesophilic homoacetogenic bacteria isolated from sewage sludge (Novaes 1986). Homoacetogenic bacteria have a high thermodynamic efficiency; as a result there is no accumulation of H_2 and CO_2 during growth on multicarbon compounds (Zeikus 1981).

The kinetic parameters for mesophilic homoacetogens growing on H_2/CO_2 are a specific growth rate (μ) of 0.4–1.9/day; growth yield (Y) of 0.35–0.85 g dry weight/g hydrogen; and specific substrate utilization rate (q) of (ca) 1.4 mg H_2/g·min (Dolfing 1988). These ranges are comparable to those obtained with hydrogenotrophic methanogens. In addition, the Gibb's free energy changes are comparable (see Eqs (2.5) and (2.6)), suggesting a close competition for available hydrogen as an electron donor. More thorough research is needed to understand the competition between homoacetogens and methanogens.

$$4H_2 + HCO_3^- + H^+ \rightarrow CH_4 + 3H_2O, \quad \Delta G^{\circ\prime} = -135.6 \text{ kJ/reaction} \tag{2.5}$$

$$4H_2 + 2HCO_3^- + H^+ \rightarrow CH_3COO^- + 4H_2O, \quad \Delta G^{\circ\prime} = -104.6 \text{ kJ/reaction} \tag{2.6}$$

2.3.4 Methanogenic Bacteria (4 and 5)

Before molecular identification techniques were developed, microorganisms were classified based on their taxonomy, that is, observable physical properties. In the present day, however, microorganisms are classified using genetic characteristics known as phylogeny, a classification system based on an organism's evolutionary history. Taxonomic classification of microorganisms is based on observable cell characteristics. Methanogens, which were previously classified as bacteria, are now classified as archaea. Archaea are unique group of microbes that are distinguished from the true bacteria by the presence of membrane lipids, absence of basic cellular characteristics (e.g., peptidoglycan), and distinctive ribosomal RNA (Boone et al. 1993). Methanogens are obligate anaerobes and considered as a rate-limiting species in anaerobic treatment of wastewater. Abundant methanogens are found in anaerobic environments rich in organic matter such as swamps, marshes, ponds, lake and marine sediments, and the rumen of cattle.

2.3.4.1 Metabolism

Methanogenesis occurs through three major pathways: CO_2-reducing or hydrogenotrophic methanogenesis, acetotrophic or aceticlastic methanogenesis, and methylotrophic pathways. The CO_2-reducing pathways use a series of four two-electron reductions to convert CO_2 or HCO_3^- to methane (Boone et al. 1993).

Table 2.3. Selected substrates and energy yields by methanogens.

Reactions	ΔG (kJ/mol)
(A) $4H_2 + CO_2 \rightarrow CH_4 + 2H_2O$	−139
(B) $4HCOO^- + 2H^+ \rightarrow CH_4 + CO_2 + 2HCO_3^-$	−127
(C) $CH_3COO^- + H_2O \rightarrow CH_4 + 2HCO_3^-$	−28
(D) $4CH_3OH \rightarrow 3CH_4 + CO_2 + 2H_2O$	−103
(E) $4CH_3NH_2 + 2H_2O + 4H^+ \rightarrow 3CH_4 + CO_2 + NH_4^+$	−102
(F) $(CH_3)_2S + H_2O \rightarrow 1.5CH_4 + 0.5CO_2 + H_2S$	−74

Source: Oremland (1988). Reprinted with permission.

Most methanogens can grow by using H_2 as a source of electrons via hydrogenase as shown in Reaction (A) of Table 2.3. The source of H_2 is the catabolic product of other bacteria in the system, such as hydrogen-producing fermentative bacteria, especially *Clostridia* (group 1) and hydrogen-producing acetogenic bacteria (group 2). The hydrogenotrophic pathway contributes up to 28% of the methane generation in an anaerobic treatment system. It bears mentioning that there are many H_2-using methanogens that can use formate as a source of electrons for the reduction of CO_2 to methane (Reaction (B)). A limited number of methanogens can even oxidize primary or secondary alcohols for CO_2 reduction to methane.

The acetotrophic (sometimes also known as acetogenic or aceticlastic) pathway is a major catabolic process contributing up to 72% of the total methane generation (Gujer and Zehnder 1983). In an acetotrophic pathway, acetate is converted to methane (Reaction (C)). The two important genera of acetotrophic methanogens are *Methanosarcina* and *Methanosaeta* (previously known as *Methanothrix*). *Methanosarcina* typically forms large packets consisting of coccoid (spherical) cell units and uses several methanogenic substrates such as methanol (Reaction (D)), methylamines (Reaction (E)), and sometimes H_2/CO_2 (Reaction (A)). *Methanosarcina* has a typical doubling time of 1–2 days on acetate. *Methanosaeta,* on the other hand, is characterized as a bacillus (rod shaped) and is known to grow only on acetate, with a doubling time of 4–9 days (Zinder 1988). Thus, if a completely mixed anaerobic reactor is operating at a short solids retention time (SRT) (approximately hydraulic retention time, HRT), it is most likely that *Methanosaeta* would wash out from the system and *Methanosarcina* would be the predominant genera. Based on microbial community analysis of 22 full-scale anaerobic digesters, Raskin et al. (1995) reported that *Methanosaeta* spp. were the dominant methanogens. Most of these digesters were operated at an SRT of 20 days or more. Conklin et al. (2006) also reported the predominance of *Methanosaeta* in a majority of anaerobic reactors based on the literature survey. The authors did however find a predominance of *Methanosarcina* in bioreactors with HRTs of 10 days or less or with acetate concentration greater than about 236 mg/L. Noike et al. (1985) observed the predominance of *Methanosarcina* at an SRT of 6.5 days or less and predominant *Methanosaeta* at an SRT of 9.6 days or more in a continuous stirred tank reactor (CSTR).

Table 2.4. Average biokinetic parameters for *Methanosaeta* and *Methanosarcina* under mesophilic conditions.

Biokinetic Parameters	*Methanosaeta*	*Methanosarcina*
Specific substrate uptake rate (k) (mg COD/mg VSS·day)	10.1 (16)[a]	12.2 (5.5)
Half-saturation constant (K_s) (mg COD/L)	49 (19)	280 (77)
Yield coefficient (Y) (mg VSS/mg COD)	0.019 (0.002)	0.048 (0.032)

Source: Conkin et al. (2006). Reprinted with permission.
[a] The number in the parenthesis represents the standard deviation.

Biokinetic parameters (half-velocity constant, yield coefficient, specific growth rate, specific substrate utilization rate, etc.) are helpful in judging the predominance of *Methanosaeta* and/or *Methanosarcina* in the system. The half-velocity constant, yield coefficient, specific growth rate, and specific substrate utilization rate of *Methanosaeta* are lower than that observed with *Methanosarcina.* This evidently suggests that *Methanosaeta* outcompetes *Methanosarcina* at a lower substrate concentration. Low substrate concentration is usually observed in a CSTR, where *Methanosaeta* may predominate. At high substrate concentrations, however, *Methanosarcina* will predominate. The slow growth rates of acetotrophic methanogens relative to other microorganisms in anaerobic treatment systems limit the overall rates of reaction and may eventually lead to accumulation of acetic acid. The growth kinetics of *Methanosarcina* and *Methanosaeta* are presented in Table 2.4.

Methylotrophic pathways catabolize compounds that contain methyl groups, such as methanol (Reaction (D)), mono-, di-, and trimethylamine (Reaction (E)), and dimethyl sulfide (Reaction (F)). The methyl group is transferred to a methyl carrier and reduced to methane. Electrons for this methyl reduction may be obtained by oxidizing a fraction of the methyl groups to CO_2 or by using H_2 as an electron donor (Boone et al. 1993) (see Box 2.3).

Box 2.3

***Methanosarcina*: Engineering Implications**

Methanosarcina has been reported to account for the stability of anaerobic digestion. The common strategies to maintain *Methanosarcina*'s dominance are operation at short SRT or high acetate concentration. Once daily feeding, rapid hydrolysis, and reduced mixing intensity could all lead to an increased acetate concentration and thereby favor the growth of *Methanosarcina.* Two practical approaches of enhancing the growth of *Methanosarcina* during anaerobic digestion are two-phase digestion or pretreatments, e.g., sonication, ozonation, microwave digestion, or heat prior to digestion.

2.3.4.2 Substrates

The most common substrates for the methanogens are H_2/CO_2, formate, acetate, methanol, and methylated amines (Table 2.4). Some species can also use carbon monoxide as a substrate; however, the growth is slime and is not common in nature. Methylmercaptan has also been reported as a methane precursor in an aquatic ecosystem, but shows no methane formation in a pure culture experiment (Oremland 1988). Lately, methylated, reduced sulfur compounds such as dimethylsulfide, methylmercaptan, and dimethyldisulfide have been found to yield methane by the action of methanogens derived from sediments.

2.3.4.3 Stoichiometry

If the chemical composition of the substrate is known, the theoretical CH_4 yield can be accurately predicted using Bushwell equation (Bushwell and Mueller 1952):

$$C_nH_aO_bN_c + \left[n - \frac{a}{4} - \frac{b}{2} + \frac{3c}{4}\right] H_2O \rightarrow \left[\frac{n}{2} + \frac{a}{8} - \frac{b}{4} - \frac{3c}{8}\right] CH_4 + \left[\frac{n}{2} - \frac{a}{8} + \frac{b}{4} + \frac{3c}{8}\right] CO_2 + cNH_3 \tag{2.7}$$

For many laboratory studies where simple substrates are used, it is possible to determine the theoretical methane yield. For municipal and industrial wastewaters, however, it is practically impossible to know the chemical composition of each and every organic fraction. In such circumstances, the maximum theoretical methane yield is estimated based on COD, which is a rough indicator of the total organic matter present in the wastewater. The chemical formula of the organic fraction of primary sludge is $C_{10}H_{19}O_3N$ (Parkin and Owen 1986), while waste-activated sludge (biomass) is represented by $C_5H_7O_2N$ (McCarty 1964). Based on chemical composition, the theoretical methane yields from primary solids and secondary sludge are as follows:

Primary solids

$$C_{10}H_{19}O_3N + 4.5H_2O \rightarrow 6.25CH_4 + 3.75CO_2 + NH_3 = 0.7\ m^3/kg\ VS \tag{2.8}$$

Secondary sludge

$$C_5H_7O_2N + 2H_2O \rightarrow 2.5CH_4 + 2.5CO_2 + NH_3 = 0.5\ m^3/kg\ VS \tag{2.9}$$

Example 2.1

A laboratory-scale anaerobic reactor operating at 35°C is fed synthetic wastewater containing 5 g/L propionic acid at the rate of 150 L/day. What is the theoretical methane yield in L/day?

Solution

The determination of theoretical methane yield is based on the assumption that propionic acid is completely converted to CH_4 gas as represented by Eq. (2.7).

The propionic acid load to the reactor: $5 \times 150 = 750$ g/day

The chemical equation for propionic acid: CH_3CH_2COOH ($C_3H_6O_2$)

From Eq. (2.7), $n = 3$, $a = 6$, $b = 2$, and $c = 0$:

$$C_3H_6O_2 + \left[3 - \frac{6}{4} - \frac{2}{2}\right] H_2O \rightarrow \left[\frac{3}{2} + \frac{6}{8} - \frac{2}{4}\right] CH_4 + \left[\frac{3}{2} - \frac{6}{8} + \frac{2}{4}\right] CO_2$$

$$\Rightarrow C_3H_6O_2 + 0.5H_2O \rightarrow 1.75CH_4 + 1.25CO_2$$

$$74\text{ g} \qquad\qquad 1.75 \times 16\text{ g}$$

i.e., 74 g propionic acid produces 28 g CH_4

=> 750 g propionic acid produces $(28/74) \times 750$

Converting CH_4 mass into an equivalent volume based on the ideal gas law (One mole of any gas at STP (standard temperature and pressure) occupies a volume of 22.4 L.)

$\rightarrow$ 1 mol CH_4 $\sim$ 22.4 L CH_4
$\rightarrow$ 16 g CH_4 $\sim$ 22.4 L CH_4
$\rightarrow$ 1 g CH_4 $\sim$ $22.4/16 = 1.40$ L CH_4
$\rightarrow$ 283.78 g CH_4 $\sim$ 1.40×283.78 L $= 397.29$ L CH_4/day

At temperature 35°C, CH_4 yield $= (308/273) \times 397.29$
$= 448.23$ L/day

The theoretical methane yield $\sim$ 448 L/day

Alternatively, this example can also be solved by converting the propionic acid into equivalent COD as indicated below:

$$C_3H_6O_2 + 3.5O_2 \rightarrow 3CO_2 + 3H_2O$$

$$74\text{ g} \qquad 112\text{ g}$$

i.e., 1 g propionate acid $\sim 112/74 = 1.51351$ g COD

=> 750 g propionate acid $\sim 750 \times 1.51351 = 1{,}135.13$ g COD
$= 283.78$ g CH_4/day

As derived in Example 1.1 (Chapter 1):

1 g COD produces 0.35 L CH_4 at STP

$\therefore$ 1,135.13 g COD produces $0.35 \times 1{,}135.13 = 397.30$ L CH_4/day
At temperature 35°C, CH_4 yield $= (308/273) \times 397.3 = 448.24$ L/day

The theoretical methane yield ~ 448 L/day
Both approaches yield the same amount of methane.

Example 2.2

A pilot-scale anaerobic digester is being proposed to treat 1,000 kg/day of primary sludge at mesophilic temperature (35°C). The volatile solids (VS) content of sludge is 65%. Calculate the maximum methane yield of the sludge in m^3/day.

Solution

The organic (volatile solid) content of the primary sludge is 65%. The total volatile solids that would eventually be converted to methane gas would be:

$\rightarrow 1000 \times 0.65 = 650$ kg VS/day

The chemical formula for primary sludge (organic fraction) is $C_{10}H_{19}O_3N$; so, with $n = 10$, $a = 19$, $b = 3$, and $c = 1$, Eq. (2.7) can be simplified as:

$$C_{10}H_{19}O_3N + 4.5H_2O \rightarrow 6.25CH_4 + 3.75CO_2 + NH_3$$

From above, complete digestion of 1 kg primary sludge organic solids has the potential to produce about 0.5 kg methane (i.e., 0.7 m^3 methane at STP/kg VS).

Total methane yield $= 650 \times 0.5 = 32.5$ kg/day

Again convert CH_4 mass into equivalent volume based on the ideal gas law. (One mole of any gas at STP occupies a volume of 22.4 L.)

$\rightarrow$ 1 mol CH_4 $\sim$ 22.4 L CH_4
$\rightarrow$ 16 g CH_4 $\sim$ 22.4 L CH_4
$\rightarrow$ 1 g CH_4 $\sim 22.4/16 = 1.40$ L CH_4
$\rightarrow$ 32.5 kg/day $\times$ 1,000 g CH_4/1 kg $\sim 1.40 \times 32{,}500$ L $= 45{,}500$ L CH_4/day

At 35°C, CH_4 yield $= (308/273) \times 45{,}500 = 51{,}333$ L/day

The maximum methane yield = 51.3 m^3/day

Note: It is important to point out that in practical applications, only 40–70% of the primary sludge VS is biodegradable (Parkin and Owen 1986). Thus, the actual methane yield would only be 40–70% of the maximum value.

References

Boone, D. R., Whitman, W. B., and Rouvière, P. 1993. Diversity and taxonomy of methanogens. In *Methanogenesis*, edited by J. G. Larry, pp. 35–80. Chapman and Hall, Inc., New York, USA.

Bushwell, A. M., and Mueller, H. F. 1952. Mechanisms of methane fermentation. *Ind. Eng. Chem.* 44:550.

Conklin, A., Stensel, D., and Ferguson, J. 2006. Growth kinetics and competition between *Methanosarcina* and *Methanosaeta* in mesophilic anaerobic digestion. *Water Environ. Res.* 78(5):486–496.

Dolfing, J. 1988. Acetogenesis. In *Biology of Anaerobic Microorganisms*, edited by J. B. Alexander Zehnder, pp. 417–468. John Wiley & Sons, Inc., New York, USA.

Fang, H. P. P. 2000. Microbial distribution in UASB granules and its resulting effects. *Water Sci. Technol.* 42(12):201–208.

Gujer, W., and Zehnder, A. J. B. 1983. Conversion processes in anaerobic digestion. *Water Sci. Technol.* 15:127–167.

Henze, M., and Harremoës, P. 1983. Anaerobic treatment of wastewater in fixed film reactors—a literature review. *Water Sci. Technol.* 15:1–101.

Hungate, R. E. 1967. Hydrogen as an intermediate in a rumen fermentation. *Arch. Microbial.* 59:158–164.

Mackie, R. L., and Bryant, M. P. 1981. Metabolic activity of fatty acid oxidising bacteria and the contribution of acetate, propionate, butyrate and CO_2 to methanogenesis in cattle waste at 40–60°C. *Appl. Environ. Microbiol.* 40:1363–1373.

McCarty, P. L. 1964. Anaerobic waste treatment fundamentals. *Public Works* 95:91–94.

McCarty, P. L., and Smith, D. P. 1986. Anaerobic wastewater treatment: Fourth of a six-part series on wastewater treatment processes. *Environ. Sci. Technol.* 20(12):1200–1206.

Noike, T., Endo G., Chang, J. E., Yaguchi, J. I., and Matsumoto, J. I. 1985. Characteristics of carbohydrate degradation and the rate-limiting step in anaerobic digestion. *Biotechnol. Bioeng.* 27:1482.

Novaes, R. F. V. 1986. Microbiology of anaerobic digestion. *Water Sci. Technol.* 18(12):1–14.

Oremland, R. S. 1988. Biogeochemistry of methanogenic bacteria. In *Biology of Anaerobic Microorganisms*, edited by J. B. Alexander Zehnder, pp. 641–705. John Wiley & Sons, Inc., New York, USA.

Parkin, G. F., and Owen, W. F. 1986. Fundamentals of anaerobic digestion of wastewater sludges. *J. Environ. Eng. ASCE* 112(5):867–920.

Pohland, F. G. 1992. Anaerobic treatment: Fundamental concept, application, and new horizons. In *Design of Anaerobic Processes for the Treatment of Industrial and Municipal Wastes*, edited by J. F. Malina, Jr., and F. G. Pohland, pp. 1–40. Technomic Publishing Co., Inc., Lancaster, USA.

Raskin, L., Zheng, D., Griffin, M. E., Stroot, P. G., and Misra, P. 1995. Characterization of microbial communities in anaerobic bioreactors using molecular probes. *Antonie Leeuwenhoek.* 68:297–308.

Sawyer, C. N., McCarty, P. L., and Parkin, G. F. 2003. *Chemistry for Environmental Engineering and Science*, 5th edn. McGraw-Hill Companies, Inc., New York, USA.

Speece, R. E. 1996. *Anaerobic Biotechnology for Industrial Wastewater Treatments.* Archae Press, Nashvillee, TN, USA.

Speece, R. E., Boonyakitsombut, S., Kim, M., Azbar, N., and Ursillo, P. 2006. Overview of anaerobic treatment: Thermophilic and priopionate implications. *Water Environ. Res.* 78(5):460–473.

van Haandel, A. C., and Lettinga, G. 1994. *Anaerobic Sewage Treatment: A Practical Guide for Regions with a Hot Climate.* John Wiley & Sons, Chichester, England.

Zeikus, J. G. 1981. Microbial intermediary metabolism. In *Anaerobic Digestion*, edited by Hughes et al., 23–25. Proceedings of the 2nd International Symposium on Anaerobic Digestion, Travemunde, Germany.

Zinder, S. H. 1988. Conversion of acetic acid to methane by thermophiles. In *Anaerobic Digestion*, edited by E. R. Hall and P. N. Hobson, pp. 1–12. Proceedings of the 5th International Symposium on Anaerobic Digestion, Bologna, Italy.

CHAPTER

3

Environmental Factors

Samir Kumar Khanal

3.1 Background

Anaerobic microorganisms, especially methanogens, are highly susceptible to changes in environmental conditions. Many researchers evaluate the performance of an anaerobic system based on its methane production rate because methanogenesis is regarded as a rate-limiting step in anaerobic treatment of wastewater. Methanogens' high vulnerability and extremely low growth rate in an anaerobic treatment system require careful maintenance and monitoring of the environmental conditions. Some of these environmental conditions are temperatures, either mesophilic or thermophilic, nutrients and trace mineral concentration, pH (usually in the neutral range), toxicity, and optimum redox conditions. A thorough discussion of these factors follows.

3.2 Temperature

Anaerobic processes, like most other biological systems, are strongly temperature dependent. In an anaerobic system, there exist three optimal temperature ranges for methanogenesis: psychrophilic, mesophilic, and thermophilic. Accordingly, the corresponding methanogens are classified as psychrophiles, mesophiles, and thermophiles. The anaerobic conversion rates generally increase with temperature up to 60°C (Pohland 1992). Anaerobic conversion has its highest efficiency at 5–15°C for psycrophiles, 35–40°C for mesophiles, and about 55°C for thermophiles, with decreased rates between these optima as shown in Fig. 3.1 (Lettinga et al. 2001; van Haandel and Lettinga 1994). Makie and Bryant (1981) suggested that low rates between these optima could be due to a lack of adaptation. Anaerobic processes, however, can still operate in a temperature range of 10–45°C without major changes in the microbial ecosystem (Henze and Harremöes 1983). As a rule of

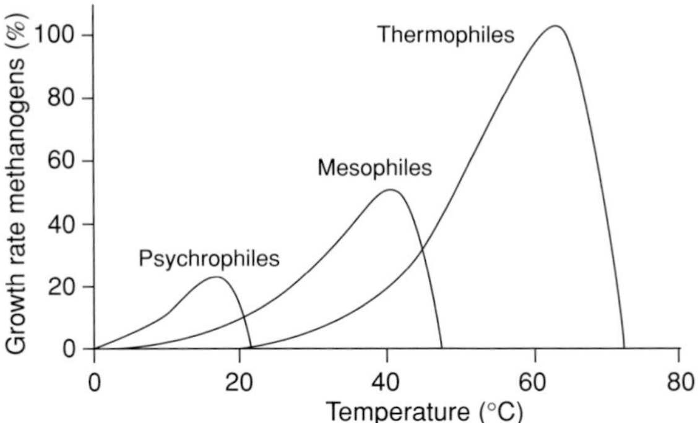

FIG. 3.1. Relative growth rate of psychrophilic, mesophilic, and thermophilic methanogens.
Source: Lettinga et al. (2001). Reprinted with permission.

thumb, the biological activity doubles for every 10°C increase in temperature within the optimal temperature range.

Strong temperature dependence on the maximum substrate utilization rates of microorganisms has been observed by many researchers (Lettinga et al. 2001; van Lier et al. 1997). Most full- and laboratory-scale anaerobic reactors are operated in the mesophilic temperature range. It is therefore logical to elaborate on the activities of mesophilic methanogenic bacteria at different temperatures. Figure 3.2 shows the temperature effects on activities of mesophilic methanogens.

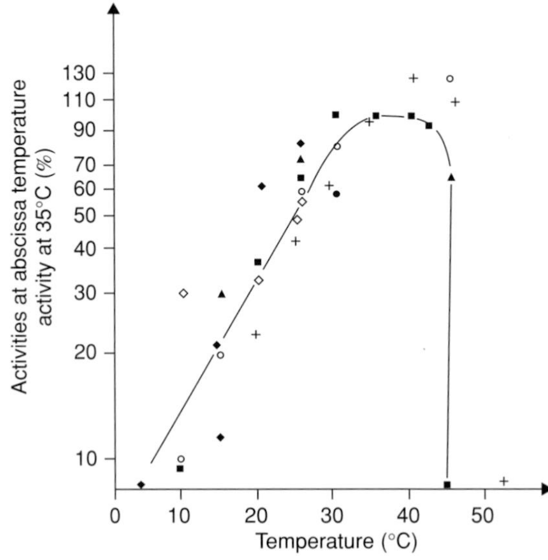

FIG. 3.2. Activities of mesophilic methanogens at different temperatures.
Source: Lettinga et al. (2001). Reprinted with permission.

The digestion rate temperature dependence can be expressed using Arrhenius expression:

$$r_t = r_{30}(1.11)^{(t-30)} \tag{3.1}$$

where t is temperature in °C and r_t, r_{30} are digestion rates at temperature t and 30°C, respectively. Based on Eq. (3.1), the decrease in digestion rate for each 1°C decrease in temperature below the optimum range is 11%. Similarly, the calculated rates at 25 and 5°C are 59 and 7%, respectively, relative to the rate at 30°C.

Temperature has a significant effect on growth kinetic, such as growth yield, decay rate, half-velocity constant, and maximum specific growth rate. Therefore, the effect of temperature on the methane generation rate is the result of the collective effect of temperature on the growth kinetics. Thermophilic processes have a nearly constant methane generation rate, independent of temperature at the range of 50–70°C. The rate is about 25–50% higher at the mesophilic rate, depending on the type of substrates (Henze and Harremöes 1983). Zinder (1988) reported that the hydraulic retention time (HRT) of thermophilic anaerobic digestion can be reduced in comparison to mesophilic processes due to more rapid growth of thermophilic, acid-consuming microorganisms. Thermophilic treatment also leads to the destruction of pathogens. Thermophilic processes, however, possess low net yield (about 50% that of mesophilic ones), thereby resulting in slow start-ups and susceptibility to loading variations, substrate changes, or toxicity. Furthermore, the lysis rate of thermophilic microorganisms is relatively high, and as a result they exist only in an exponential growth phase (Zinder and Mah 1979).

It is recommended that anaerobic treatment should be designed for operation for temperature variations not exceeding 0.6–1.2°C (1–2°F)/day (Manual of Practice—Water Pollution Control Federation (WPCF) 1987). Some research suggests that anaerobic processes are capable of withstanding temperature variation. Activity decreases when the temperature is reduced, but it recovers immediately when the temperature returns to an optimum value as shown in Fig. 3.3. In contrast, a full-scale anaerobic system reported deterioration in treatment performance, with temperature variations as small as 2–3°C. This may be attributed to several factors such as mixing, substrate diffusion limitations, and stratification within the reactor (Grady et al. 1999) (see Box 3.1).

Box 3.1

Heating the reactor to a mesophilic temperature can be achieved through the use of methane gas produced during anaerobic treatment. Dilute wastewater cannot produce sufficient methane gas to heat an anaerobic reactor. A longer contact time or more biomass inventory is needed for efficient treatment of such wastewater.

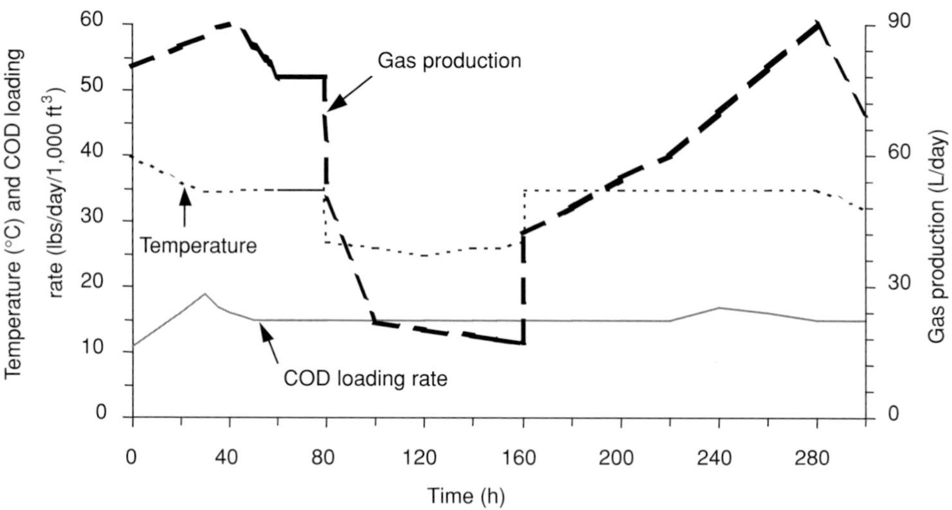

FIG. 3.3. Effect of temperature change on biogas production rate.
Source: Speece (1996). Reprinted with permission.

Anaerobic treatment is not easily applied to dilute wastewater at low temperatures due to insufficient biogas production. This limitation can, however, be overcome by maintaining a long solids retention time (SRT) (Parkin and Speece 1983). The microbial generation time (SRT_{min}) is a function of both growth yield (Y) and maximum substrate utilization rate (k) given by:

$$SRT_{min} = \frac{1}{Yk - b} \quad (3.2)$$

where b is microbial decay rate.

Growth yield of anaerobic microorganisms, and especially methanogens, is extremely low when compared to aerobic microorganisms, and "k" is dependent on temperature. Below 20°C, SRT_{min} may exceed 30 days for methanogens. Thus, efficient treatment of dilute wastewater at a low temperature requires a longer SRT. In a continuous stirred tank reactor, longer SRT (or longer HRT) means bigger reactor size. From an economic standpoint, such reactor configuration is not suitable for treating dilute wastewater. In high-rate reactors, such as upflow anaerobic sludge blanket (UASB), anaerobic filter, and fluidized/expanded-bed reactor, the temperature control becomes less important and dilute wastewater can be treated efficiently due to extremely long SRT irrespective of HRT. Methane fermentation has been reported to proceed effectively at temperatures as low as 10–20°C and chemical oxygen demand (COD) concentrations of 200–600 mg/L in an attached growth system (Switzenbaum and Jewell 1980). Anaerobic membrane bioreactor (AnMBR) can also be employed in treating medium-to low-strength wastewater at low temperatures (Ho et al. 2007). AnMBR maintains a long SRT due to retention of biomass in the bioreactor. A detailed discussion on AnMBR is presented in Chapter 5.

3.3 Operating pH and Alkalinity

Anaerobic treatment performance is adversely affected by slight pH changes away from optimum. Methanogens are more susceptible to pH variation than other microorganisms in the microbial community (Grady et al. 1999). Anaerobes can be grouped into two separate pH groups: acidogens and methanogens. The optimum is 5.5–6.5 for acidogens and 7.8–8.2 for the methanogens. Optimum pH for the combined cultures ranges from 6.8 to 7.4 with neutral pH being the ideal. Since methanogenesis is considered to be the rate-limiting step, it is necessary to maintain the reactor pH close to neutral. Acidogens are significantly less sensitive to low or high pH values, and acid fermentation will prevail over methanogenesis, which may result in souring of the reactor contents (van Haandel and Lettinga 1994) (see Box 3.2).

Box 3.2

Low pH reduces the activity of methanogens, causing accumulation of acetic acid and H_2. At higher partial pressure of H_2, propionic acid–degrading bacteria will be severely inhibited, thereby causing excessive accumulation of higher-molecular-weight volatile fatty acids (VFAs), particularly propionic and butyric acids, which then slow down the production of acetic acid dropping the pH further. If the situation is left uncorrected, the process may eventually fail. This condition is known as a "sour" or "stuck."

Methanogenic activity (acetate utilization rate) versus pH is shown in Fig. 3.4. The drastic drop in methanogenic activity at pH 8.0 and above could be due to a shift of NH_4^+ to more toxic unionized form NH_3 (Seagren et al. 1991). A detailed discussion on ammonia toxicity is presented later.

Speece (1996) reported reduced but stable methane generation when the pH was lowered from the optimum value. Even at low pH of 5.0, methanogenesis was found sustainable and maintained at about 25% of that observed at neutral pH. Considerable time was needed to restore the methane generation to the original level once the pH was restored to neutral value, as shown in Fig. 3.5.

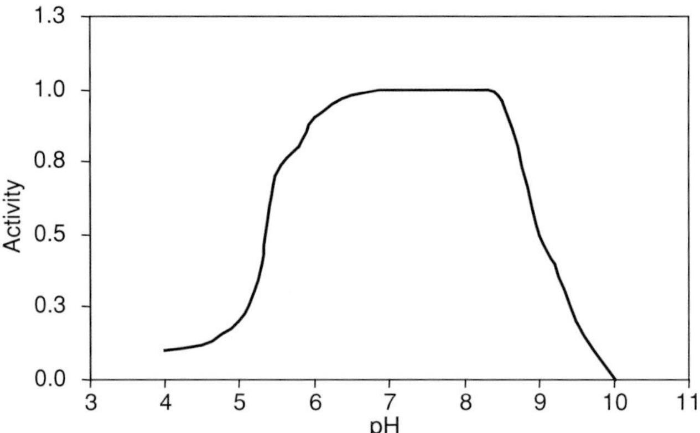

FIG. 3.4. pH dependence of methanogenic activity.
Source: Speece (1996). Reprinted with permission.

In anaerobic treatment process, the drop in pH is often caused by the accumulation of VFAs and/or excessive generation of carbon dioxide. One of the first options to resolve the problem is to reduce the volumetric organic loading rate (VOLR) to the point where the accumulated VFAs are allowed to be consumed faster than produced. Once the excess VFAs are exhausted, the pH of the reactor will return to a normal operating range and the methanogens begin to rejuvenate. The VOLR can then be increased gradually as the process recovers to full loading capacity. Under extreme circumstances, the decrease in VOLR should be coupled

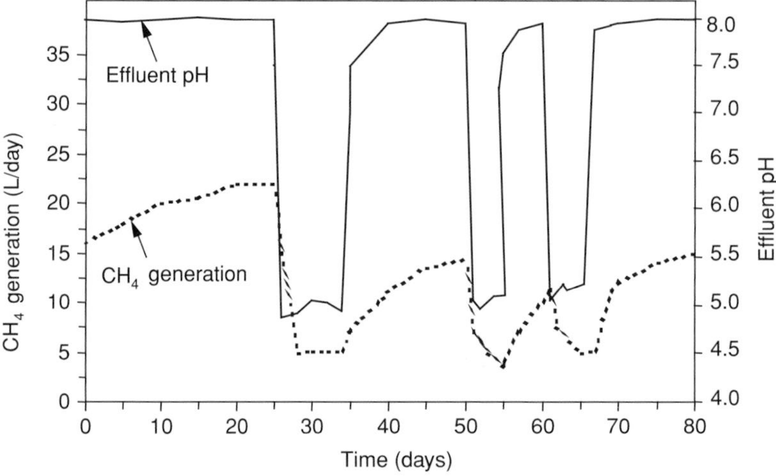

FIG. 3.5. Effect of pH changes on methane generation rate.
Source: Speece (1996). Reprinted with permission.

with the supplementation of chemicals for pH adjustment. (A detailed discussion is presented later.)

Another recently explored option is the periodic dosing of oxygen in anaerobic treatment system (Khanal and Huang 2006). Limited oxygenation helps eliminate the excess VFAs drastically by the facultative microbes. These microbes are less susceptible to changes in pH. Since methanogens are vulnerable to abrupt pH changes away from the optimum value, the anaerobic treatment system needs sufficient buffering capacity (alkalinity) to mitigate the pH change. The pH of an anaerobic system, operating within an acceptable range, is primarily controlled by self-produced alkalinity or natural alkalinity. The destruction of organic matter (COHNS), primarily the proteins, releases ammonia–N. Each mole of organic nitrogen theoretically generates one equivalent of alkalinity (Moosburger et al. 1990). Ammonia–N reacts with carbon dioxide produced during the biochemical reaction to produce ammonium bicarbonate, which contributes to alkalinity as indicated in the following:

$$RCHNH_2COOH + 2H_2O \rightarrow RCOOH + NH_3 + CO_2 + 2H_2 \tag{3.3}$$

$$NH_3 + H_2O + CO_2 \rightarrow \underset{\text{(Alkalinity)}}{NH_4^+ + HCO_3^-} \tag{3.4}$$

Only wastes containing high organic nitrogen (e.g., protein) can adequately contribute alkalinity. Many carbohydrate-rich wastes (e.g., brewery, molasses, potato, and starch) do not contribute alkalinity because they lack organic nitrogen. The successful anaerobic treatment of such wastewaters requires supplementation of alkalinity.

Anaerobic treatment of high-sulfate/sulfite wastewater also generates alkalinity because of sulfate/sulfite reduction (Eqs (3.5) and (3.6)). Theoretically, reduction of 1 g SO_4 generates 1.04 g of alkalinity as $CaCO_3$ (Greben et al. 2000).

$$H_2 + SO_4^{2-} + CO_2 \rightarrow HS^- + HCO_3^- + 3H_2O \tag{3.5}$$

$$CH_3COO^- + SO_4^{2-} \rightarrow HS^- + 2HCO_3^- \tag{3.6}$$

Khanal and Huang (2006) found that anaerobic treatment of high-sulfate wastewater resulted in an increase in effluent alkalinity from 4,464 to 6,000 mg/L as $CaCO_3$ when the influent sulfate was increased from 1,000 to 3,000 mg/L. Increase in pH also followed the same trend (Fig. 3.6). At 3,000 mg/L influent sulfate, alkalinity supplementation was reduced by one half. Thus, anaerobic treatment of such wastewater reduces costs related to alkalinity supplementation (see Box 3.3).

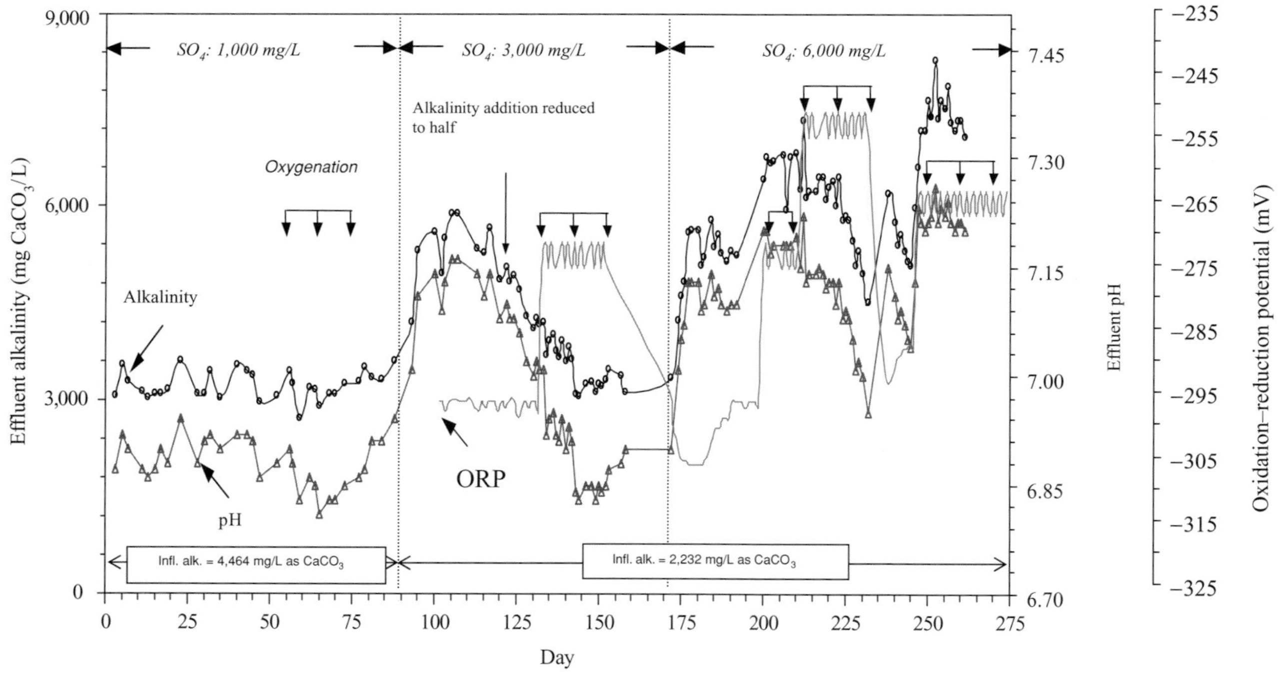

FIG. 3.6. Effluent alkalinities and pHs at different influent sulfate levels.

Box 3.3

Treatment of Thin Stillage: Alkalinity Contribution from Proteinous Matter
Once the beer mash is distilled, the whole stillage is centrifuged to separate the solids. The centrate known as thin stillage is considered as a waste stream. Thin stillage has unique characteristics of very high total COD of 100 g/L and volatile solids (VS) of 60 g/L, with a very low pH of 4.0–4.5 and zero alkalinity. Such characteristics make anaerobic treatment of thin stillage extremely challenging. Laboratory-scale tests conducted at Iowa State University showed that such a waste stream can be effectively treated anaerobically. During thermophilic anaerobic treatment, significant VS reduction of 90% at an SRT of 20 days was achieved. Methane yield was also high with a typical yield of 0.40 L CH_4/g $COD_{removed}$ during steady-state operation. Effluent VFAs were low for a thermophilic anaerobic digester with less than 200 mg/L as acetic acid. The final alkalinity in the bioreactor reached about 4 g/L as $CaCO_3$ and the effluent pH was around 7.24. Following the successful start-up, the bioreactors were operated without supplementation of alkalinity. High protein concentration (organic nitrogen = 1,700 mg N/L) was the primary source of alkalinity during anaerobic digestion.

When VFAs start to accumulate in the anaerobic reactor, they are neutralized by the alkalinity present in the reactor and maintain a stable pH as shown by the equation as follows:

$$HCO_3^- + HAc \Leftrightarrow H_2O + CO_2\uparrow + Ac^- \qquad (3.7)$$

Under unstable conditions, there is an excessive loss of alkalinity coupled with the generation of carbon dioxide as shown by Eq. (3.7). Thus, loss of alkalinity and generation of CO_2 lead to pH decreases. The relationship between alkalinity, pH, and gas-phase carbon dioxide is given by (Grady et al. 1999):

$$\text{Alkalinity, as } CaCO_3 = 6.3 \times 10^{-4}\left[\frac{pCO_2}{10^{-pH}}\right] \qquad (3.8)$$

where pCO_2 is the partial pressure of carbon dioxide in the gas phase in atmospheres.

In the neutral pH range, which is of great interest for anaerobic treatment, alkalinity mainly exists in the bicarbonate form. The relationship between pH, bicarbonate alkalinity concentration, and gas-phase carbon dioxide is shown in Fig. 3.7.

Figure 3.7 shows that as long as gas-phase carbon dioxide is below 10%, bicarbonate alkalinity of 250 mg/L as $CaCO_3$ is sufficient to maintain the desirable

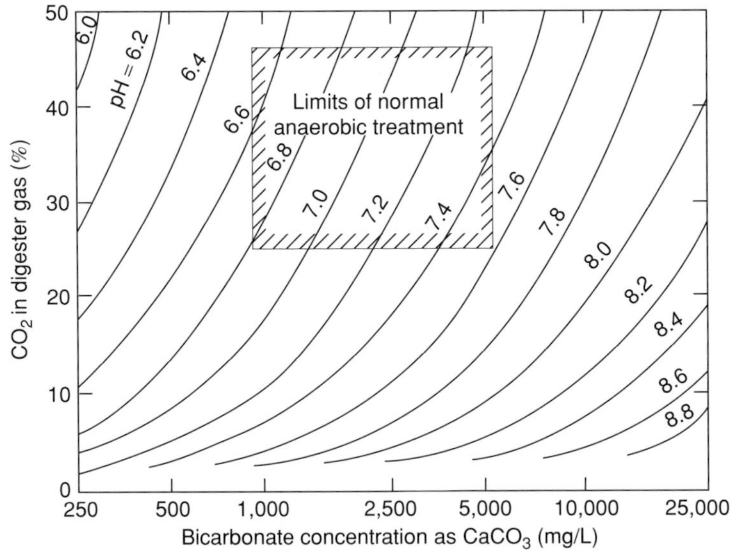

Fig. 3.7. Relationship between pH, bicarbonate, and carbon dioxide at 35°C and 1 atm pressure.
Source: Sawyer et al. (2003). Reprinted with permission.

pH of 6.8 or above for anaerobic treatment. Gas-phase carbon dioxide may exceed 30% depending on the waste characteristics. Thus, bicarbonate alkalinity exceeding 1,000 mg/L as $CaCO_3$ is needed to maintain pH above 6.8. Normally, the alkalinity varies from 1,000 to 5,000 mg/L as $CaCO_3$ in anaerobic treatment processes (Metcalf and Eddy 2003). The stability of an anaerobic treatment system can also be judged by VFA/ALK (alkalinity) ratio. For anaerobic treatment, the VFA/ALK ratio of 0.1–0.25 is usually considered favorable without the risk of acidification. Increase in the ratio beyond 0.3–0.4 indicates digester upset, and corrective measures are required. At a ratio of 0.8 or above, significant pH reductions and inhibition of methanogenesis occur, thereby resulting in digester failure (Manual of Practice—WPCF 1987). The anaerobic bioreactor operating at the lower limit of safety is primarily governed by the capability of operator to spot and rectify performance deterioration quickly (see Box 3.4).

Box 3.4

Determination of Alkalinity When Gas-Phase CO_2 and pH are Known
In an anaerobic process, gas-phase carbon dioxide, pH, and alkalinity are closely related as discussed earlier. By regular measurement of reactor pH and carbon dioxide content in the biogas, one can easily calculate the alkalinity. The following discussion details a procedure for alkalinity determination.

The alkalinity can be estimated based on the carbonate equilibrium equations (Snoeyink and Jenkins 1980). The solubility of CO_2 in the liquid phase is given by Henry's law:

$$P_g = Hx_g \tag{3.9}$$

where P_g is partial pressure of CO_2 (atm), H is Henry's law constant (atm/mol fraction), and x_g is equilibrium mole fraction of dissolved CO_2.

$$x_g = \frac{\text{mol gas}(n_g)}{\text{mol gas}(n_g) + \text{mol water}(n_w)} \tag{3.10}$$

The aqueous or dissolved $(CO_2)_{aq}$ reacts reversibly with water and forms carbonic acid:

$$(CO_2)_{aq} + H_2O \Leftrightarrow H_2CO_3 \tag{3.11}$$

In Eq. (3.11), H_2CO_3 and $(CO_2)_{aq}$ are difficult to differentiate. So these two terms are lumped up together and represented by $H_2CO_3{}^*$:

$$H_2CO_3{}^* = (CO_2)_{aq} + H_2CO_3 \tag{3.12}$$

Since the concentration of H_2CO_3 is relatively small, $H_2CO_3{}^*$ is equal to $(CO_2)_{aq}$. $H_2CO_3{}^*$ is a diprotic acid and dissociates in two steps—first to bicarbonate and then to carbonate. The first dissociation is given by:

$$H_2CO_3{}^* \Leftrightarrow H^+ + HCO_3{}^- \tag{3.13}$$

The corresponding equilibrium equation is as follows:

$$\frac{[H^+][HCO_3{}^-]}{[H_2CO_3{}^*]} = K_1 \tag{3.14}$$

The second dissociation is from bicarbonate to carbonate and is defined by:

$$HCO_3{}^- \Leftrightarrow H^+ + CO_3{}^{2-} \tag{3.15}$$

$$\frac{[H^+][CO_3{}^{2-}]}{[HCO_3{}^-]} = K_2 \tag{3.16}$$

The Henry's law constant and equilibrium constants for CO_2 and water can be found in *Handbook of Chemistry and Physics* (2001), 82nd edn. CRC Press, Boca Raton, FL, USA.

pH and partial pressure of CO_2 (concentration in the biogas) can be measured regularly. Based on regular measurement of partial pressure of CO_2, carbonic acid or aqueous CO_2 can be estimated using Eq. (3.9). Using Eqs (3.14) and (3.16) and pH, the concentration of bicarbonate and carbonate species can be deduced.

The alkalinity is given by:

$$\text{Alkalinity} = [HCO_3^-] + 2[CO_3^{2-}] + [OH^-] - [H^-] \tag{3.17}$$

In Eq. (3.17), the unit for all species is mol/L and $[OH^-] = Kw/[H^+]$. An example of calculating the alkalinity based on pH and CO_2 data is illustrated later.

Alkalinity supplementation using chemicals such as sodium bicarbonate, sodium carbonate, ammonium hydroxide, gaseous ammonia, lime, sodium, and potassium hydroxide is required in many cases to maintain an optimum pH in the bioreactor. Sodium bicarbonate is preferred because of its high solubility, long-lasting impact, and low toxicity. In addition, direct addition of bicarbonate ions results in a direct pH increase contributing gas-phase carbon dioxide. The downside of sodium bicarbonate is cost. Lime is the cheapest option and does not increase the Ca^{2+} toxicity through precipitation of Ca^{2+} as a $CaCO_3$. It can, however, cause serious scaling problems in the bioreactor. Hydroxide ions from slaked lime react with carbon dioxide to form alkalinity as shown in the following:

$$Ca(OH)_2 + 2CO_2 \Leftrightarrow Ca^{2+} + 2HCO_3^- \tag{3.18}$$

Similarly, the addition of sodium carbonate contributes the alkalinity as indicated in the following:

$$Na_2CO_3 + H_2O + CO_2 \Leftrightarrow 2Na^+ + 2HCO_3^- \tag{3.19}$$

In Eqs (3.18) and (3.19), it is apparent that hydroxide-based chemicals consume 2 mol of carbon dioxide to produce 2 mol of bicarbonate, whereas carbonate-based chemicals consume just 1 mol of carbon dioxide to produce the same alkalinity. Thus, if the hydroxide demand for pH control is too high, it may lead to development of negative pressure within the reactor. The excessive negative pressure could be disastrous for two reasons: firstly, air drawn into the reactor can create an explosive mixture of oxygen and methane; and secondly, a sudden collapse of the reactor itself due to excessive negative pressure. Another advantage of carbonate over hydroxide addition is less dramatic pH fluctuations.

It is important to note that quite often effluent pH is taken to be a measure of anaerobic operating pH. As the effluent enters the ambient environment, changes in the partial pressure of dissolved acid gases, especially, carbon dioxide, result in pH changes.

The desired operating pH in the reactor can be achieved either by adjusting the pH of the influent feed or by controlling the pH in the reactor *per se*. Prior knowledge of required chemical additions to the influent is needed to achieve the desired pH effect, whereas in the latter case such prior knowledge is not necessary. The reactor is usually monitored with an online pH-measuring device connected to a relay logic controller. The desired pH set point is programmed and the chemical addition (acid or base) takes place automatically. The pH probe senses the reactor pH and signals the controller to either start or stop the chemical addition pump. Although online pH control is highly desirable, it is expensive and requires meticulous attention.

3.4 Nutrients

Like all biochemical operations, both macronutrients (nitrogen and phosphorus) and micronutrients (trace minerals) are required for the anaerobic processes to support the synthesis of new biomass. The amount of nitrogen and phosphorus required for biomass synthesis can be calculated by assuming the empirical formula for an anaerobic bacterial cell as $C_5H_7O_2N$ (Speece and McCarty 1964). The cell mass consists of about 12% nitrogen, which means that about 12 g nitrogen is needed for every 100 g anaerobic biomass produced. Phosphorus demand accounts for 1/7–1/5 of nitrogen demand. As a rule of thumb, it is assumed that about 10% of the organic matter (COD) removed (i.e., 0.10 kg volatile suspended solids (VSS)/kg COD removed) during an anaerobic process is utilized for biomass synthesis; this can be used to calculate the nitrogen and phosphorus needs. Another approach for calculating the macronutrient requirements is based on wastewater strength (COD). The theoretical minimum COD/N/P ratios of 350:7:1 for highly loaded (0.8–1.2 kg COD/kg VSS·day) and 1,000:7:1 for lightly loaded (<0.5 kg COD/kg VSS·day) anaerobic systems can be used to calculate the nitrogen and phosphorus needs (Henze and Harremöes 1983). Nitrogen is most commonly supplemented as urea, aqueous ammonia, or ammonium chloride, and phosphorus as phosphoric acid or a phosphate salt.

In addition to nitrogen and phosphorus, several other trace nutrients are identified as essential for anaerobic microorganisms. Trace metals such as iron, cobalt, molybdenum, selenium, calcium, magnesium, sulfide zinc, copper, manganese, tungsten, and boron in the mg/L level and vitamin B_{12} in μg/L level have been found to enhance methane production (Speece 1988). Some of the trace metals and their roles in anaerobic process are discussed as follows:

Nickel: The importance of Ni on the growth of methanogens was first reported by Schönheit et al. (1979). Ni is particularly important because it is a structural constituent of factor F430, which is found only in methanogenic bacteria (Whitman and Wolfe 1980).

Cobalt: Co is also important because it is also the structural constituent of vitamin B_{12}, which catalyzes the methanogenesis.

Nickel, cobalt, and other trace minerals are essential for methanol degradation in a UASB reactor under mesophilic conditions. A cobalt concentration of 0.05 mg/L was reported optimal for methylotrophic methanogens and homoacetogens (Weijma and Stams 2001). The higher cobalt needs for these trophic groups is likely due to the involvement of cobalt-containing corrinoids methylotransferases in the initial step of methanol conversion.

3.5 Toxic Materials

The toxicity of anaerobic processes is mediated by the substances present in the influent waste stream or through byproducts of the metabolic activities of the microorganisms. Ammonia, heavy metals, halogenated compounds, cyanide, and phenol are the examples of the former, while ammonia, sulfide, and long-chain fatty acids (LCFAs) belong to the latter group. It is interesting to report that many of the anaerobic bacteria are also capable of degrading refractory organics (Stronach et al. 1986). In some cases, toleration is manifested by acclimation to toxicants. These observations provide a considerable cause for optimism about the feasibility of anaerobic treatment in difficult situations, particularly for industrial wastewaters that contain significant concentrations of toxic compounds (Parkin and Speece 1983; Pohland 1992).

3.5.1 Ammonia

Ammonia may be present in the influent or produced during the anaerobic degradation of organic nitrogenous compounds such as proteins or amino acids. Protein usually contains 16% nitrogen. Many agri-based industries, especially those that are related to raising and processing livestock, also generate waste streams with high ammonia levels. The anaerobic treatment of such wastes is often unsuccessful due to high ammonia levels (Farina et al. 1988). An ammonia level exceeding 4,000 mg N/L is reported inhibitory during anaerobic digestion of cattle manure (Angelidaki and Ahring 1994). With adaptation, acetate-utilizing bacteria (acetoclastic methanogens) tolerate free ammonia levels up to 700 mg N/L (Angelidaki and Ahring 1994). On the other hand, free ammonia levels as low as 100–150 mg N/L inhibited unadapted cultures (De Baere et al. 1984). Hansen et al. (1997) reported that a free ammonia concentration exceeding 1,100 mg N/L caused inhibition in batch cultures at a pH of 8.0 and a temperature of 55°C, and higher free ammonia levels decreased growth. McCarty (1964) reported that at ammonia nitrogen exceeding 3,000 mg/L, the ammonium ion itself became toxic independent of pH (Table 3.1).

Table 3.1. Ammonia nitrogen concentration and its effect on anaerobic treatment.

Ammonia–N (mg/L)	Effects
50–100	Beneficial
200–1,000	No adverse effect
1,500–3,000	Inhibitory effect at higher pH values
Above 3,000	Toxic

At low concentrations, ammonia buffers pH decreases. Ammonia nitrogen exists in two forms depending on the pH, as given by the following equilibrium equation:

$$NH_4^+ \leftrightarrow NH_3 + H^+ \tag{3.20}$$

$$\text{Total ammonia nitrogen} = NH_4^+ + NH_3 (\text{p}K_a = 9.2455 \text{ at } 25^\circ\text{C}) \tag{3.21}$$

From Eqs (3.20) and (3.21), the % distribution of NH_3 as a function of pH is given by:

$$\%NH_3 = \frac{100}{1 + ([H^+]/K_a)} \tag{3.22}$$

At neutral pH, free NH_3–N represents about 0.5% of the total ammonia nitrogen ($NH_3 - N + NH_4^+ - N$). Free ammonia is the most toxic form of ammonia nitrogen to methanogens, as the free ammonia or unionized molecule can penetrate the cell membrane. Concentrations as low as 100 mg/L as N are sufficient to inhibit methanogenesis. On the other hand, ammonium ion (NH_4^+) concentrations as high as 7,000–9,000 mg/L as N have been treated successfully with an acclimated culture, without a toxic response (Grady et al. 1999). By maintaining a neutral reactor pH, ammonia toxicity can be avoided.

It is important to note that an anaerobic reactor operating at higher temperatures is more sensitive to ammonia toxicity than one operating in the mesophilic range. Poggi-Varaldo et al. (1997) reported that thermophilic cultures were more susceptible to ammonia nitrogen than the mesophilic ones. Temperature reduction from 55 to 46°C results in an increased biogas yield in high-ammonia-loaded reactors (Angelidaki and Ahring 1994). The fraction of unionized ammonia increases with temperature, increasing toxicity (Gallert and Winter 1997). Acclimation is another factor that can influence the degree of ammonia inhibition. Adaptation of methanogens to high concentrations of ammonia increases the ammonia tolerance of these microbes. The mesophilic system studied by Parkin and Miller (1982) performed well at 9.0 g/L of total ammonia nitrogen after acclimation.

3.5.2 Sulfide Toxicity

Sulfide is produced during the anaerobic treatment of sulfur-rich waste streams by sulfate-reducing bacteria. Industrial waste streams from tanneries, petrochemical refineries, coal gasification, etc., contain a significant sulfide load. Sulfide is also produced during the degradation of sulfur-containing organic matter (proteins) found in waste such as swine manure. The unionized sulfide (H_2S) is considered more toxic to methanogens than the ionized form (HS^-). A detailed discussion of sulfide control is covered in Chapter 7.

3.5.3 Heavy Metals

Heavy metals find their way to wastewater and sludge from industrial activities, for example, electroplating, tanneries, and other metal-processing industries. Soluble heavy metals are regarded to be more critical to anaerobic process failure than insoluble forms (Stronach et al. 1986). The generation of sulfide benefits anaerobic treatment by reducing metal toxicity through formation of insoluble metal sulfides, with the exception of chromium. Approximately 0.5 mg of sulfide is needed to precipitate 1.0 mg of heavy metal. Heavy metal toxicity follows the following order: Ni > Cu > Pb > Cr > Zn (Hayes and Theis 1978), with iron considered more beneficial than detrimental because it mediates sulfide toxicity.

3.5.4 Short-Chain Fatty Acids

The level of VFAs is an indicator of the health of an anaerobic treatment system. The term "volatile" indicates that they can be recovered by distillation at atmospheric pressure. During anaerobic degradation, complex organic matter is hydrolyzed and fermented into low-molecular-weight compounds, including short-chain fatty acids (SCFAs) (C-2–C-6). This includes primarily acetic, propionic, and butyric acids and lesser amounts of isobutyric, valeric, isovaleric, and caproic acids. In a healthy anaerobic system, the VFA concentration in the effluent is relatively low and usually in the range of 50–250 mg HAc/L (Sawyer et al. 2003). When the symbiotic relationship between acidogens and methanogens breaks down, VFAs accumulate. The inhibition of methanogens due to toxicity (sulfide, ammonia, heavy metals, synthetic organics, etc.), changes in environmental factors (pH, temperature, and oxidation–reduction potential (ORP)), or nutrient-limiting conditions causes an accumulation of acetate and hydrogen. Excessive hydrogen partial pressure severely inhibits propionic acid–degrading bacteria, resulting in the accumulation of propionic acid. Studies suggest that VFA concentrations exceeding 2,000 mg HAc/L inhibit methanogens, but neither acetic or butyric acid at concentration exceeding 10,000 mg/L inhibits methane formation at neutral pH. Propionic acid is inhibitory at a concentration of 6,000 mg/L at neutral pH. This extremely high concentration is unlikely to be found in any anaerobic process operated at a neutral pH (Grady et al. 1999).

It now appears that VFAs will be of little concern as long as the pH remains within the optimal range for the growth of methanogens (6.8–7.4). Like sulfide and ammonia, unionized form of VFAs will inhibit methanogens when present at concentrations of 30–60 mg/L.

3.5.5 Long-Chain Fatty Acids

Wastewater and sludge from edible oil refinery, slaughterhouses, wool scouring, meat packing, restaurants, and dairy processing contain high concentrations of lipids. Municipal sludge also contains lipids. Lipids are an important organic component of waste in the anaerobic process. They generate the highest theoretical amount of methane when compared to other components (Pereira et al. 2003). LCFAs are produced by the hydrolysis of lipids such as fats, oils, and greases during anaerobic treatment. Some typical examples of LCFAs are palmitic, stearic, capric, linoleic, and oleic acids. An inhibitory effect of LCFAs on anaerobes was reported by McCarty in early 1960s (McCarty 1964). Hanaki et al. (1981) reported that LCFAs were toxic to hydrogen-producing acetogenic bacteria as well as acetotrophic and hydrogenotrophic methanogens with the former being more susceptible. LCFAs were also found to be toxic to some, but not nearly all carbohydrate-fermentative bacteria. Rinzema et al. (1994) also studied toxic effect of capric acid on methanogenic granules. They found that capric acid concentrations of (ca) 750–1,008 mg/L were lethal to acetotrophic methanogens.

3.5.6 Synthetic Organic Compounds

Synthetic organic compounds are recalcitrant in both aerobic and anaerobic processes. The recalcitrance is often linked with toxicity or inhibition of methanogens (Pohland 1992). The common structural elements that exhibit toxicity are halogens, aldehydes, double bonds, and aromatic compounds. Biodegradation of these compounds occurs with sufficient acclimation. Fully acclimated cultures can tolerate these organics at concentrations 50 times higher than inhibiting concentrations at initial exposure.

3.6 Redox Potential or Oxidation–Reduction Potential

Morris (1975) reported that to obtain the growth of an obligate anaerobe in any medium, the culture ORP value should be maintained around –200 to –350 mV at pH 7.0. Many reported ORP values focus on the optimum growth requirement for a pure culture of methanogens. It is well established that methanogens require a highly reducing environment with redox potential as low as –400 mV (Archer and Harris 1986; Hungate 1967). When culturing methanogens, strong reducing agents such as sulfide, cysteine, or titanium III are added to poise the medium at a proper ORP. A limited number of studies have been conducted that evaluated the effect of ORP on methanogenesis using a mixed culture in an anaerobic treatment of wastewater (Gupta et al. 1994).

Example 3.1

Calculate alkalinity when the reactor pH is 6.87 and gas-phase CO_2 is 64.2%.

Solution

Step 1: Determination of $(CO_2)_{aq}$ *or* $H_2CO_3{}^*$

From Eq. (3.9), the equilibrium mole fraction of dissolved CO_2 is given by:

$$x_g = \frac{0.642}{0.2095 \times 10^4} = 3.064 \times 10^{-4}$$

Since 1 L water contains 1,000/18 = 55.6 g/mol, the concentration of CO_2 in mol/L is given by Eq. (3.10):

$$55.6 \times 3.064x^{-4} = 0.01704 \text{ mol/L}$$

$$\therefore H_2CO_3{}^* = 0.01704 \text{ mol/L}$$

Step 2: Determination of $HCO_3{}^-$

For pH 6.87, $[H^+] = 1.349 \times 10^{-7}$ mol/L and $K_1 = 4.87 \times 10^{-7}$ mol/L, using Eq. (3.14), we have:

$$[HCO_3{}^-] = K_1 \times \frac{[H_2CO_3{}^*]}{[H^+]} = 4.87 \times 10^{-7} \times \frac{0.01704}{1.349 \times 10^{-7}} = 0.0615 \text{ mol/L}$$

Step 3: Determination of $CO_3{}^{2-}$

For pH 6.87, $[H^+] = 1.349 \times 10^{-7}$ mol/L and $K_2 = 5.58 \times 10^{-11}$ mol/L, using Eq. (3.16), we have:

$$[CO_3{}^{2-}] = K_2 \times \frac{[HCO_3{}^-]}{[H^+]} = 5.58 \times 10^{-11} \times \frac{0.0615}{1.349 \times 10^{-7}} = 2.51 \times 10^{-5} \text{ mol/L}$$

Step 4: Determination of alkalinity

Substituting $CO_3{}^{2-}$, $HCO_3{}^-$, H^+, and OH^- in Eq. (3.17), we have:
Alkalinity, mol/L = $(0.0615 + 2 \times 2.51 \times 10^{-11} + 1.53 \times 10^{-7} - 1.35 \times 10^{-7})$

Alkalinity = 0.06155 mol/L = 3,077.6 mg/L as $CaCO_3$

Example 3.2

An anaerobic reactor treating carbohydrate-rich wastewater encountered a severe souring problem with the VFA/ALK ratio of 0.6. The plant operator decided to use lime to rectify the problem. Comment on the various aspects of lime addition.

Comments

The addition of hydrated lime produces alkalinity (calcium bicarbonate). When the point of maximum solubility of calcium bicarbonate is reached, the addition of lime will generate insoluble calcium carbonate precipitate. This will produce no additional alkalinity, but continue to consume carbon dioxide. If the gas-phase carbon dioxide is below 10%, pH will increase uncontrollably. The chemical reactions are shown as follows:
Lime (500–1,000 mg/L)

$$Ca(OH)_2 + 2CO_2 \rightarrow Ca(HCO_3)_2$$

Lime (>1,000 mg/L)

$$Ca(OH)_2 + CO_2 \rightarrow CaCO_3 \downarrow + H_2O$$

The other effect of lime addition is creation of vacuum or negative pressure in the bioreactor due to dissolution of gas-phase carbon dioxide. This may result in air intrusion and, in severe cases, the collapse of the bioreactor.

References

Angelidaki, I., and Ahring, B. K. 1994. Anaerobic thermophilic digestion of manure at different ammonia loads: Effect of temperature. *Water Res.* 28(3):727–731.

Archer, D. B., and Harris, J. E. 1986. Methanogenic bacteria and methane production in various habitats. In *Anaerobic Bacteria in Habitats Other Than Man*, edited by E. M. Barnes and G. C. Mead, pp. 185–223. Blackwell Scientific Publications, Oxford, UK

De Baere, L. A., Devocht, M., van Assche, P., and Verstraete, W. 1984. Influence of high NaCl and NH_4Cl salt levels on methanogenic associations. *Water Res.* 18(5):543–548.

Farina, R., Boopathy, R., Hartmann, A., and Tilche, A. 1988. *Ammonia Stress During Thermophilic Digestion of Raw Laying Hen Wastes*, pp. 111–117. Proceedings of the Fifth International Symposium on Anaerobic Digestion.

Gallert, C., and Winter, J. 1997. Mesophilic and thermophilic anaerobic digestion of source-sorted organic waste: Effect of ammonia on glucose degradation and methane production. *Appl. Microbiol. Biotechnol.* 48:405–410.

Grady, C. P. L., Jr, Daigger, G. T., and Lim, H. C. 1999. *Biological Wastewater Treatment*, 2nd edn. Marcel Dekker, Inc., New York, USA.

Greben, H. A., Maree, J. P., and Mnqanqeni, S. 2000. Comparison between sucrose, ethanol and methanol as carbon and energy sources for biological sulphate reduction. *Water Sci. Technol.* 41:247–253.

Gupta, A., Flora, J. R. V., Gupta, M., Sayles, G. D., and Suidan, M. K. (1994). Methanogenesis and sulfate reduction in chemostats. I. Kinetic studies and experiments. *Water Res.* 28(4):781–793.

Hanaki, K., Matsuo, T., and Nagase, M. 1981. Mechanisms of inhibition caused by long-chain fatty acids in anaerobic digestion process. *Biotechnol. Bioeng.* XXIII:1591–1610.

Hansen, K. H., Angelidaki, I., Ahring, B. K. 1997. Anaerobic digestion of swine manure: Inhibition by ammonia. *Water Res.* 32:5–12.

Hayes, T. D., and Theis, T. L. 1978. The distribution of heavy metals in anaerobic digestion. *J. Water Pollut. Control Fed.* 50:61–72.

Henze, M., and Harremoës, P. 1983. Anaerobic treatment of wastewater in fixed film reactors—a literature review. *Water Sci. Technol.* 15:1–101.

Ho, J., Khanal, S. K., and Sung, S. 2007. Anaerobic membrane bioreactor for treatment of synthetic municipal wastewater at ambient temperature. *Water Sci. Technol.* 55(7):79–86.

Hungate, R. E. 1967. A roll tube method for cultivation of strict anaerobes. In *Methods in Microbiology*, edited by J. R. Norris and D. W. Ribbons, pp. 117–132. Academic Press, New York, USA.

Khanal, S. K., and Huang, J.-C. 2006. Online oxygen control for sulfide oxidation in anaerobic treatment of high sulfate wastewater. *Water Environ. Res.* 78(4):397–408.

Lettinga, G., Rebac, S., and Zeeman, G. 2001. Challenges of psychrophilic anaerobic wastewater treatment. *Trends Biotechnol.* 19(9):363–370.

Makie, R. L., and Bryant, M. P. 1981. Metabolic activity of fatty acid oxidising bacteria and the contribution of acetate, propionate, butyrate and CO_2 to methanogenesis in cattle waste at 40–60°C. *Appl. Environ. Microbiol.* 40:1363–1373.

Manual of Practice (MOP)—Water Pollution Control Federation (WPCF)# 16 1987. *Anaerobic Sludge Digestion*, 2nd edn. Water Pollution Control Federation, Alexandria, VA, USA.

McCarty, P. L. November 1964. Anaerobic waste treatment fundamentals. Part III: Toxic materials and their control. *Public Works* 91–94.

Metcalf and Eddy (2003). *Wastewater Engineering: Treatment and Reuse*, 4th edn, McGraw-Hill, Inc., New York, USA.

Moosburger, R. E., Lowenthal, R. E., and Marais, G. R. 1990. Pelletisation in a UASB system with protein (casein) substrate. *Water S.A.* 16:171–178.

Morris, J. G. 1975. The physiology of obligate anaerobiosis. *Adv. Microb. Physiol.* 12:169–246.

Parkin, G. F., and Miller, S. W. 1982. *Response of Methane Fermentation to Continuous Addition of Selected Industrial Toxicants.* Proceedings of 37th Purdue Industrial Waste Conference, West Lafayette, Indiana.

Parkin, G. F., and Speece, R. E. 1983. Attached versus suspended growth anaerobic reactors: Response to toxic substances. *Water Sci. Technol.* 15:261–289.

Pereira, M. A., Cavaleiro, A. J., Mota, M., and Alves, M. M. 2003. Accumulation of long-chain fatty acids onto anaerobic sludge under steady state and shock loading conditions: Effect on acetogenic and methonogenic activity. *Water Sci. Technol.* 48:33–40.

Poggi-Varaldo, H. M, Medina, E. A., Fernandez-Villagomez, G., and Caffarel-Mendez, S. 1997. *Inhibition of Mesophilic Solid Substrate Anaerobic Digestion (Dass) by Ammonia-Rich Wastes.* Proceedings of 52nd Purdue Industrial Waste Conference, West Lafayette, Indiana.

Pohland, F. G. 1992. Anaerobic treatment: Fundamental concept, application, and new horizons. In *Design of Anaerobic Processes for the Treatment of Industrial and Municipal Wastes*, edited by J. F. Malina, Jr, and F. G. Pohland, pp. 1–40. Technomic Publishing Co., Inc., Lancaster, USA.

Rinzema, A., Boone, M., van Knippenberg, K., and Lettinga, G. 1994. Bactericidal effect of long chain fatty acids in anaerobic digestion. *Water Environ. Res.* 66(1):40–49.

Sawyer, C. N., McCarty, P. L., and Parkin, G. F. 2003. *Chemistry for Environmental Engineering and Science*, 5th edn. McGraw-Hill, Inc., New York, USA.

Schönheit, P., Moll, J., and Thauer, R. K. 1979. Nickel, cobalt and molybdenum requirement for growth of *Methanobacterium thermoautotrophicum. Arch. Microbiol.* 123:105–107.

Seagren, E. A., Levine, A. D., and Dague, R. R. 1991. High pH effects. In *Anaerobic Treatment of Liquid Industrial Byproducts*, pp. 377–386. 45th Purdue Industrial Waste Conference Proceedings. Lewis Publishers, Inc., Chelsea, MI, USA.

Snoeyink, V. L., and Jenkins, D. 1980. *Water Chemistry.* John Wiley & Sons, New York, USA.

Speece, R. E. 1988. Advances in anaerobic biotechnology for industrial wastewater treatment. In *Anaerobic Treatment of Industrial Wastewaters*, edited by M. F. Torpy, pp. 1–6. Noyes Data Corporation, Park Ridge, NJ, USA.

Speece, R. E. 1996. Anaerobic biotechnology for industrial wastewater treatments. Archae Press, Nashvillee, TN, USA.

Speece, R. E., and McCarty, P. L. 1964. Nutrient requirements and biological solids accumulation in anaerobic digestion. *Adv. Water Pollut. Control Res.* 2:305–322.

Stronach, S. M., Rudd, T., and Lester, J. N. 1986. *Anaerobic Digestion Processes in Industrial Wastewater Treatment*, pp. 59–92. Springer-Verlag, Berlin, Germany.

Switzenbaum, M. S., and Jewell, W. J. 1980. Anaerobic attached-film expanded-bed reactor treatment. *J. Water Pollut. Control Fed.* 52:1953–1965.

van Haandel, A. C., and Lettinga, G. 1994. *Anaerobic Sewage Treatment: A Practical Guide for Regions with a Hot Climate.* John Wiley & Sons, Chichester, England.

van Lier, J. B., Rebac, S., Lens, P., and Lettinga, G. 1997. Anaerobic treatment of partly acidified wastewater in a two-stage expanded granular sludge bed (EGSB) system at 8°C. *Water Sci. Technol.* 36(6):317–324.

Weijma, J., and Stams, A. J. M. 2001. Methanol conversion in high-rate anaerobic reactors. *Water Sci. Technol.* 44(8):7–14.

Whitman, W. B., and Wolfe, R. S. 1980. Presence of nickel in factor F430 from *Methanobacterium bryantii. Biochem. Biophys. Res. Commun.* 92:1196–1201.

Zinder, S. H. 1988. Conversion of acetic acid to methane by thermophiles. In *Anaerobic Digestion*, edited by E. R. Hall and P. N. Hobson, pp. 1–12. Proceedings of the 5th International Symposium on Anaerobic Digestion, Bologna, Italy.

Zinder, S. H., and Mah, R. A. 1979. Isolation and characterization of a thermophilic strain of *Methanosarcina* unable to use H_2-CO_2 for methanogenesis. *Appl. Environ. Microbiol.* 38:996–1008.

CHAPTER

4

Kinetics and Modeling in Anaerobic Processes

Keshab Raj Sharma

4.1 Background

Anaerobic process is a microbial-mediated process in which the organic matter is metabolized to methane and carbon dioxide through several biochemical steps. The anaerobic technology has been applied for decades, primarily to sludge digestion and stabilization. In recent years, the focus has been shifted to the low-cost treatment of high/low-strength wastewaters and also to energy generation through the production of methane and hydrogen.

Despite the widespread application, the design, operation, and control of anaerobic processes is normally based on empirical guidelines. With the development of several anaerobic process models in recent years, this scenario is rapidly changing. The mathematical models have found their application as a valuable tool in design, operation, control, and optimization of various anaerobic processes. A model can be used to describe the process quantitatively, and hence make accurate predictions of long-term system performance.

Several anaerobic process models ranging from simple kinetic models (Pavlostathis and Gossett 1986) to more complicated structured models such as Anaerobic Digestion Model No. 1 (ADM1) (Batstone et al. 2002) are available. Many of these models have been developed for specific applications in terms of reactor types and substrate characteristics. The ADM1 model published by International Water Association (IWA) is a standardized and inclusive model, which can be used in conjunction with other wastewater treatment models such as activated sludge models (ASMs) published by IWA (Henze et al. 1987) as a common approach in terms of units and nomenclature has been adopted. The purpose of this chapter is to familiarize the reader with basic concepts of modeling, introduce ADM1, and illustrate formulation and application of the model as applied to bioenergy production.

4.2 Basic Elements

4.2.1 Material Balance

Material balance is the most important element and forms the basis of any modeling work. It is used to describe the uptake or generation of a particular material (state variable in modeling term) within a defined system boundary (Fig. 4.1) in relation to the flow entering and leaving the system. This can be represented mathematically as follows:

$$\frac{dM_x}{dt} = \dot{m}_{x,\text{in}} - \dot{m}_{x,\text{out}} + \dot{r} \qquad (4.1)$$

where M_x is the mass (M) of the material (x) in the system, $\dot{m}_{x,\text{in}}$ is the mass flow rate (MT^{-1}) entering the system, $\dot{m}_{x,\text{out}}$ is the mass flow rate (MT^{-1}) leaving the system, and $\dot{r}$ is the net mass generation rate (MT^{-1}). The term $\dot{r}$ will be negative if the material is consumed.

For a completely mixed reactor, the concentration of the material in output stream is same as that in the reactor. Assuming that volume of the reactor does not change with time, Eq. (4.1) above can be written as follows:

$$V\frac{dC_x}{dt} = Q_{\text{in}} \cdot C_{x,\text{in}} - Q_{\text{out}} \cdot C_x + V \cdot r_x \qquad (4.2)$$

where V is the volume of reactor (L^3), $C_{x,\text{in}}$ is the concentration of material x in the input stream (ML^{-3}), C_x is the concentration in the reactor (ML^{-3}), and Q_{in} and

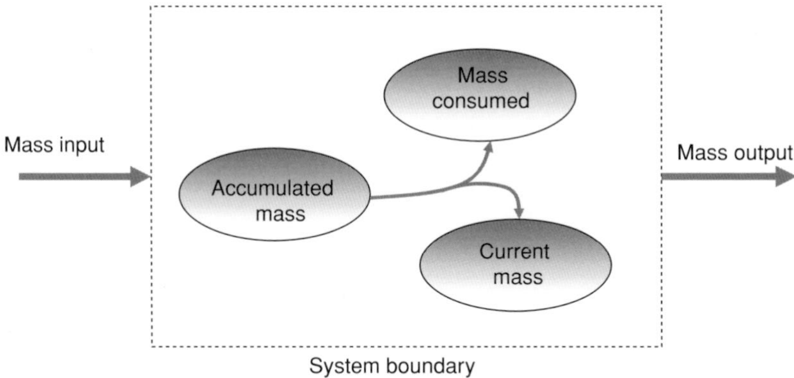

FIG. 4.1. System boundaries and mass balance.

Q_{out} are the flow rates (L^3T^{-1}) of the input and output streams, respectively. All the terms here except C_x and r_x are known beforehand. The purpose of any modeling work is thus to formulate the generation rate (r_x) and estimate the concentration of the material (C_x). Complication arises from the fact that the generation rate is normally the function of the concentration.

The rate of change of concentration in the system (dC_x/dt) becomes zero under steady-state conditions. The Eq. (4.2) then becomes an algebraic equation, which can be solved either analytically or numerically depending on the complexity of the generation term.

$$Q_{in} \cdot C_{x,in} - Q_{out} \cdot C_x + V \cdot r_x = 0 \tag{4.3}$$

4.2.2 Kinetics

If we look carefully at the mass balance equations above, all the terms except the generation rate are affected by reactor dimensions and flow rate. The rate of generation (or consumption) is governed by the kinetics of the process involving the material (or state variable) of interest. As such, the process kinetics plays a vital role in mass balance. A number of relationships to describe the biological process rates are available, and these range from simple zero-order expression to highly complicated expressions. Earlier models developed for single-culture systems used Monod kinetics of various forms. Recent models for mixed-culture system developed to describe dynamics of anaerobic process use Monod kinetics in which the growth rate is assumed to be the function of the limiting substrate concentration. Furthermore, Monod kinetics with multiple substrates is used in cases where more than one substrate affects the reaction rate. Some of the processes are also modeled using first-order kinetics. Depending on the nature of biological reaction, inhibition terms can also be included as required.

4.2.2.1 First-Order Kinetics

The first-order kinetics is typically used for hydrolytic processes in which the rate of hydrolysis is assumed to be proportional to the concentration of hydrolyzable substrate.

$$\frac{dC_x}{dt} = -k_h \cdot C_x \tag{4.4}$$

It is assumed that no diffusion limitation exists for the transport of solubilized matter out of the cell, and hence no distinction between the intracellular and extracellular hydrolysis is made. In such a case, the hydrolysis constant (k_h) represents the sum of the intracellular and extracellular hydrolysis.

4.2.2.2 Monod Kinetics

The rate of microbial growth can be described by a relatively simple empirical model proposed by Monod, which accounts for the effect of growth-limiting substrate on bacterial growth. The microbial growth rate for growth-related Monod kinetics is presented as follows:

$$\mu = \mu_{\max} \cdot \frac{S}{K_s + S} \cdot X \tag{4.5}$$

where $\mu_{\max}$ is the maximum specific growth rate and can be calculated from substrate consumption rate as $\mu_{\max} = k_m \cdot Y$.

The Monod equation relates the growth rate to the concentration of a single growth-controlling substrate via two parameters: the maximum specific growth rate ($\mu_{\max}$) and the substrate affinity constant (K_s). The affinity constant represents the substrate concentration at which the growth rate (μ) becomes one half of the maximum growth rate ($\mu_{\max}$).

In substrate-related Monod kinetics, the specific substrate consumption rate (k_m) is used instead of specific growth rate, and growth rate is presented as follows:

$$\rho = k_m \cdot Y \cdot X \cdot \frac{S}{K_s + S} \tag{4.6}$$

The substrate consumption rate is affected by inhibition caused by other substrates or products as well as by pH. The substrate or product inhibition could be either competitive or noncompetitive depending on the substrate of interest. The inhibitory effect can be taken into account by adding substrate inhibition functions to Eq. (4.6) as follows:

$$\rho = k_m \cdot Y \cdot X \cdot \frac{S}{K_s + S} \cdot I_1 \cdot I_2 \cdots I_n \tag{4.7}$$

where $I \cdots I_n$ are the substrate inhibition functions exclusive to the growth process being considered.

4.2.3 Physical–Chemical Processes

Two major physical–chemical factors that affect the anaerobic process are gas flow rate and the pH. The gas flow rate affects the transfer of gaseous components (such as CH_4 and CO_2) from liquid phase to gas phase, thereby affecting overall equilibrium. On the other hand, the effects of pH are many folds. The pH affects biological activity, mineral precipitation, and also the transfer of gases from liquid to gas phase. The anaerobic process model should therefore include the effects of both gas flow and the pH.

4.2.4 Biochemical Conversion Processes

The processes involved in anaerobic digestion are catalyzed by intra- or extracellular enzymes and use available organic matter (both soluble and particulate). These processes are well documented and a general agreement exists (Batstone 2006; Batstone et al. 2002; Gujer and Zehnder 1983). The major steps involved in the anaerobic processes are:

1. *Disintegration:* Extracellular process converting composite particulate materials to carbohydrates, proteins, and lipids
2. *Hydrolysis:* Extracellular process converting particulates into soluble monomers
3. *Acidogenesis:* Conversion of monomers to bicarbonates, alcohols, hydrogen, and organic acids through fermentation
4. *Acetogenesis:* Oxidation of alcohols and organic acids to hydrogen and acetate
5. *Methanogenesis:* Formation of methane from hydrogen and acetate

The processes 3–5 are intracellular processes mediated by microorganisms and these result in biomass growth.

4.3 Stepwise Approach to Modeling

In this section a step-by-step illustration of how an anaerobic process model can be developed using the concepts explained in earlier sections is presented. The explanation presented in this section follows the modeling approach used in IWA models for wastewater treatment processes.

4.3.1 Process Matrix

Very first step of building a model is the development of a process matrix. The process matrix provides complete information on the interactions of the system components. As such, all the processes and components are presented in a matrix form as shown in Table 4.1. This is an example for the conversion of acetate into methane gas. The first step in setting up the process matrix is to identify the relevant components. In this case, the processes of interest are uptake of acetate, which results in the growth of acetate consumers, and the production of methane gas and the decay of acetate consumers resulting into the formation of composites. The system components are therefore (1) composite particulate matter, (2) acetate,

Table 4.1. Process matrix for acetate uptake.

	Component i				Process Rate (ρ_j)
	1	2	3	4	
j Process	X_c	S_{ac}	S_{ch4}	X_{ac}	$ML^{-3}\ T^{-1}$
1. Uptake of acetate		-1	$(1-Y_{ac})$	Y_{ac}	$k_{m,ac} \times \frac{S_{ac}}{K_S+S_{ac}} \times X_{ac}$
2. Decay of X_{ac}	1			-1	$k_{dec,Xac} \times X_{ac}$
	Composites (kg COD m^{-3})	Total acetate (kg COD m^{-3})	Methane gas (kg COD m^{-3})	Acetate consumers (kg COD m^{-3})	

(3) methane, and (4) acetate consumers. It is a common practice to denote particulate components by notation X and the soluble components by S. For simplicity, additional components such as carbonate and nitrogen are excluded in this example.

The components are broadly divided into two categories: soluble components (S_i) and particulate components (X_i). The components are assigned the index i, while the processes are assigned the index j. In the example above, i ranges from 1 to 4, while j ranges from 1 to 2. The kinetic expressions for each process (ρ_j) are recorded in the last column. The kinetic parameters used to describe the process kinetics in this case are as follows:

1. Maximum acetate uptake rate ($k_{m,ac}$)
2. Half-saturation coefficient for acetate uptake (K_s)
3. First-order decay rate for acetate consumers ($k_{dec,xac}$)

The mass relationship between the system components in a process is given by stoichiometric coefficients (v_{ij}). For example, uptake of acetate (–1) results in the growth of acetate consumers (Y_{ac}) and production of methane ($1 - Y_{ac}$). The only stoichiometric coefficient in the example above is, therefore, the acetate yield coefficient (Y_{ac}). Since all the components are expressed in a common unit (chemical oxygen demand or COD in case of organic compounds and biomass), the sum of the stoichiometric coefficients for a process must be equal to zero.

4.3.2 Reaction Rates

The concentration of a component within a system is affected by a number of processes. In the above example, the concentration of acetate consumers is affected by both acetate consumption and their decay. All these processes are listed in the column representing the component of interest.

The mass balance for any closed system consists of input, output, and reaction terms. The input and output terms are the transport terms and are defined by the flow rate and physical properties of the system, while the reaction term can be obtained from the process matrix. The system reaction term r_i is simply the summation of the products of stoichiometric coefficients (v_{ij}) and the kinetic rate expression (ρ_j).

$$r_i = \sum_j v_{ij}\rho_j \tag{4.8}$$

The rate of reaction for the acetate concentration in the system:

$$r_{S_{ac}} = -k_{m,ac} \cdot \frac{S_{ac}}{K_s + S_{ac}} \cdot X_{ac} \tag{4.9}$$

The rate for the concentration of acetate consumers:

$$r_{X_{ac}} = Y_{ac} \cdot k_{m,ac} \cdot \frac{S_{ac}}{K_s + S_{ac}} \cdot X_{ac} - k_{dec,xac} \cdot X_{ac} \tag{4.10}$$

The rate for the concentration of methane:

$$r_{S_{CH4}} = (1 - Y_{ac}) \cdot k_{m,ac} \cdot \frac{S_{ac}}{K_s + S_{ac}} \cdot X_{ac} \tag{4.11}$$

The rate for the concentration of composites:

$$r_{X_c} = k_{dec,xac} \cdot X_{ac} \tag{4.12}$$

4.3.3 Model Formulation

The mass balance equations for each of the components of the model should be developed as explained in Section 4.2.1. The process rates explained in the previous section should be used for this purpose. The following mass balance equations can be developed for a completely mixed reactor.

$$\frac{dS_{ac}}{dt} = \frac{Q_{in} \cdot S_{ac,in} - Q_{out} \cdot S_{ac}}{V} - k_{m,ac} \cdot \frac{S_{ac}}{K_s + S_{ac}} \cdot X_{ac} \tag{4.13}$$

$$\frac{dX_{ac}}{dt} = \frac{Q_{in} \cdot X_{ac,in} - Q_{out} \cdot X_{ac}}{V} + Y_{ac} \cdot k_{m,ac} \cdot \frac{S_{ac}}{K_s + S_{ac}} \cdot X_{ac} - k_{dec,xac} \cdot X_{ac} \tag{4.14}$$

$$\frac{dS_{CH4}}{dt} = \frac{Q_{in} \cdot S_{CH4,in} - Q_{out} \cdot S_{CH4}}{V} + (1 - Y_{ac}) \cdot k_{m,ac} \cdot \frac{S_{ac}}{K_s + S_{ac}} \cdot X_{ac} \tag{4.15}$$

$$\frac{dX_c}{dt} = \frac{Q_{in} \cdot X_{c,in} - Q_{out} \cdot X_c}{V} + k_{dec,xac} \cdot X_{ac} \tag{4.16}$$

The process model consists of a system of first-order differential equations (ODEs) consisting of mass balance equations for all components of interest. The time series of the concentration of components in the model can be obtained by solving the above system of ODEs using numerical methods. The values of kinetic parameters and initial concentrations for the components (concentration at time $t = 0$) must be provided for the solution of the system of ODEs.

4.3.4 Units

The models dealing with processes involved in wastewater treatment use COD as the unit of components. This is mainly because the COD is used in wastewater characterization and also in the balance of carbon oxidation state. Concentration of components with no COD (e.g., CO_2, HCO_3^-, NH_4^+, NH_3) is expressed in molar unit (mol/L).

4.3.5 Modeling of Biochemical Conversion Processes

The AMD1, which is a structured model consisting of various steps involved in an anaerobic process, forms the basis for the following discussion. The AMD1 uses the same nomenclature and the units as the ASMs developed by IWA. This facilitates the integration of the models in a biological system consisting of aerobic, anoxic, and anaerobic components.

4.3.6 Variables and Parameters

Four main types of variables and parameters are used in the model.

4.3.6.1 Stoichiometric Coefficients

These are the coefficients describing the relationships between the components (variables) of the model. Some examples of the stoichiometric coefficients are the substrate and biomass yield coefficients, and nitrogen and carbon contents of the particulate components.

4.3.6.2 Kinetics Rates and Parameters

These are the parameters used to describe the rate of a process in the model. Examples include maximum specific substrate uptake rates, half-saturation constants, and biomass decay rates.

4.3.6.3 Dynamic State Variables

These are the model components requiring the solution of differential equations. The values of the components vary with time, and the variation is defined by the first derivative of concentration (dC/dt).

4.3.6.4 Discrete State Variables

The values of some of the components of the model are obtained by solving algebraic equations. The concentration can be based on the concentrations of dynamic state variables, but the model must be solved outside the dynamic model. Hydrogen ion concentration is a typical example of discrete state variable.

4.3.6.5 Equilibrium Coefficients and Constants

These are coefficients and constants required for solving the equilibrium reactions used for pH calculations and also for liquid–gas mass transfer reactions. pK_a values, gas law constant (H_{gas}), and Henry's law coefficient (K_H) are some examples.

4.3.7 Biochemical Processes

The biochemical processes form the core of the anaerobic process model. The steps of biochemical conversion included in the ADM1 model are presented in Fig. 4.2. Some key considerations of the model are as follows:

- Waste contains complex particular matter and assumed to be homogeneous, which disintegrates into carbohydrates, proteins, and lipids.
- All extracellular steps are assumed to be of first order. An empirical function representing cumulative effect of multistep process has been employed.
- All the intracellular processes involve substrate uptake, biomass growth, and decay.

The biochemical processes included in the ADM1 model are the following:

1. Disintegration of particulates
2. Hydrolysis of carbohydrates, proteins, and lipids
3. Substrate uptake (resulting in biomass growth)
4. Biomass decay

Disintegration and hydrolysis are the two extracellular processes that mediate the conversion of complex organic matter into soluble substrate. Composite particulate material is converted into particulate carbohydrates, proteins, and lipids, and

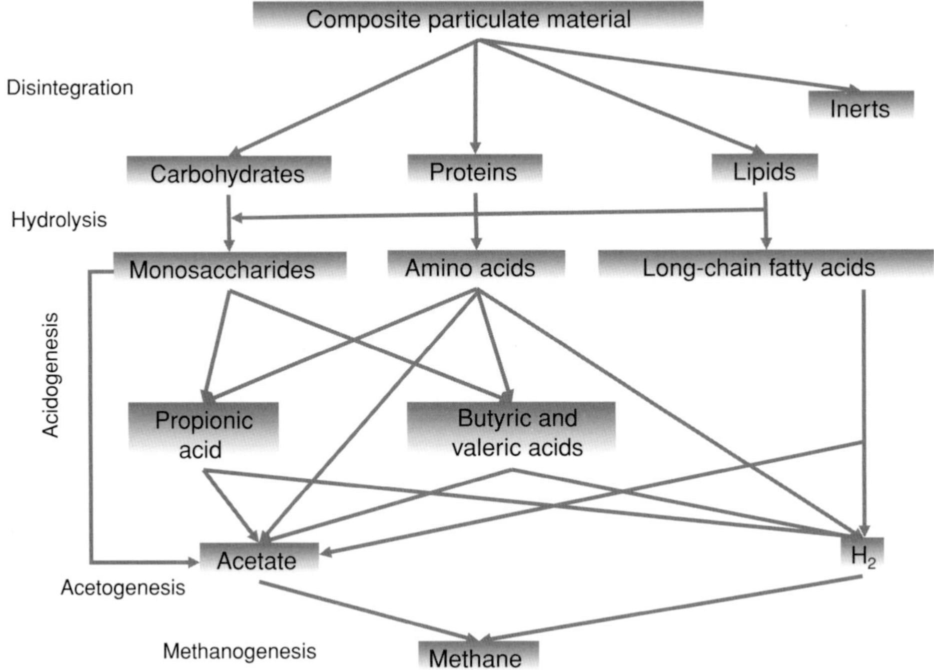

FIG. 4.2. Organic degradation pathway in an anaerobic process.

inert particulate and inert soluble material during disintegration. The first three products of disintegration, namely carbohydrates, proteins, and lipids, serve as the substrates for hydrolysis, resulting in the production of monosaccharides, amino acids, and long-chain fatty acids (LCFAs), respectively. These are consumed by microorganisms in the subsequent steps resulting in biomass growth. Decay of biomass occurs simultaneously with its growth.

4.3.8 Biomass Population

ADM1 uses substrate uptake-related kinetics rather than growth-related kinetics. This is done to decouple growth from uptake and allow variable yields. Following eight groups of microorganisms are considered for substrate uptake. The decay of biomass is also considered for each of these microbial groups:

1. Sugar degraders (monosacccharides → butyrate/propionate/acetate/H_2)
2. Amino acid degraders (amino acids → valerate/butyrate/propionate/acetate/H_2)
3. Long-chain fatty acid degraders (LCFA → acetate/H_2)
4. Valerate and butyrate degraders (valerate/butyrate → propionate/acetate/H_2)
5. Propionate degraders (propionate/H_2 → acetate/H_2)

6. Acetate degraders (acetate → methane)
7. Hydrogen degraders (hydrogen → methane)
8. Sulfate reducers (sulfate → sulfide)

The sulfate reducers are not included in the original ADMI model, but have been added later as an extension.

Example 4.1

Under steady-state conditions, the composition of headspace gas in an anaerobic reactor is as follows: $CH_4 = 60\%$, $H_2 = 1\%$, $CO_2 = 30\%$, and others = 9%. Calculate the rate of transfer of CH_4 from liquid to gas phase when the liquid-phase CH_4 concentration is 30 mg/L and the temperature is 35°C. Assume k_{La} value of 2/h.

Solution

Partial pressure of CH_4 in headspace $p_{CH_4} = 0.60$ atm

$$
\begin{aligned}
K_{H,CH_4} \text{ at } 35^\circ\text{C} &= K_{H,CH4} \text{ at } 25^\circ\text{C} \times e^{-C(1/308-1/298)} \\
&= 1.4 \times 10^{-3} \times e^{-1{,}700(1/308-1/298)} \\
&= 1.685 \times 10^{-3} \text{ mol/L} \cdot \text{atm}
\end{aligned}
$$

Saturation concentration of methane:

$$
\begin{aligned}
S_{sat} &= K_{H,CH_4} \times p_{CH_4} \\
&= 1.685 \times 10^{-3} \text{ mol/L} \cdot \text{atm} \times 0.60 \text{ atm} \\
&= 1.011 \times 10^{-3} \text{ mol/L} \\
&= 1.011 \text{ mmol/L} \times 16 \text{ mg/mmol} \\
&= 16.20 \text{ mg/L}
\end{aligned}
$$

$$
\begin{aligned}
\text{Rate of transfer of } CH_4 &= k_{La} \times (S_L - S_{sat}) \\
&= (2)1/\text{h} \times (30.0 - 16.2) \text{ mg/L} \\
&= 27.6 \text{ mg/L}
\end{aligned}
$$

4.3.9 Kinetics of Biochemical Processes

The matrix of the processes and components in ADM1 model (Batstone et al. 2002) are presented in Tables 4.2 and 4.3. The kinetic expressions used here are taken directly from the ADM1 model. An additional process of sulfate reduction and related components, sulfate (SO_4), inorganic sulfide (IS), and the concentration of sulfate reducers (XSO_4), has been added as proposed by Batstone (2006).

Since the first-order reaction kinetics is reported to be the appropriate way to describe enzymatic hydrolysis (Pavlostathis and Gossett 1986), the same has been used for disintegration and hydrolysis. For all the processes involving microorganisms,

Table 4.2. Biochemical rate coefficients (υ_{ij}) and kinetic rate equations (ρ_{ij}) for soluble components ($i = 1 - 14$; $j = 1 - 21$).

	Component i														
	1	2	3	4	5	6	7	8	9	10	11	12	13	14	Process Rate (ρ_j) (ρ_j, kg COD m^{-3} day^{-1})
j Process	S_{su}	S_{aa}	S_{fa}	S_{va}	S_{bu}	S_{pro}	S_{ac}	S_{h_2}	S_{IS}	S_{so_4}	S_{ch4}	S_{IC}	S_{IN}	S_I	
1 Disintegration														$f_{SI,Xc}$	$k_{dis} X_c$
2 Hydrolysis of carbohydrates	1														$k_{hyd,ch} X_{ch}$
3 Hydrolysis of proteins		1													$k_{hyd,pr} X_{pr}$
4 Hydrolysis of lipids	$1 - f_{fa,li}$		$f_{fa,li}$												$k_{hyd,li} X_{li}$
5 Uptake of sugars	-1				$(1 - Y_{su}) f_{bu,su}$	$(1 - Y_{su}) f_{pro,su}$	$(1 - Y_{su}) f_{ac,su}$	$(1 - Y_{su}) f_{h_2,su}$				$- \sum_{i=9,13-21} C_i v_{i,5}$	$-Y_{su} N_{bac}$		ρ_5
6 Uptake of amino acids		-1		$(1 - Y_{aa}) f_{va,aa}$	$(1 - Y_{aa}) f_{bu,aa}$	$(1 - Y_{aa}) f_{pro,aa}$	$(1 - Y_{aa}) f_{ac,aa}$	$(1 - Y_{aa}) f_{h_2,aa}$				$- \sum_{i=9,13-21} C_i v_{i,6}$	$N_{aa} - Y_{aa} N_{bac}$		ρ_6
7 Uptake of LCFA			-1				$(1 - Y_{fa})0.70$	$(1 - Y_{fa})0.30$					$-Y_{fa} N_{bac}$		ρ_7
8 Uptake of valerate				-1		$(1 - Y_{c_4})0.54$	$(1 - Y_{c_4})0.31$	$(1 - Y_{c_4})0.15$					$-Y_{c_4} N_{bac}$		ρ_8
9 Uptake of butyrate					-1		$(1 - Y_{c_4})0.80$	$(1 - Y_{c_4})0.20$					$-Y_{c_4} N_{bac}$		ρ_9
10 Uptake of propionate						-1	$(1 - Y_{pro})0.57$	$(1 - Y_{pro})0.43$				$- \sum_{i=9,13-21} C_i v_{i,10}$	$-Y_{pro} N_{bac}$		ρ_{10}
11 Uptake of acetate							-1				$(1 - Y_{ac})$	$- \sum_{i=9,13-21} C_i v_{i,11}$	$-Y_{ac} N_{bac}$		ρ_{11}
12 Uptake of hydrogen								-1			$(1 - Y_{h_2})$	$- \sum_{i=9,13-21} C_i v_{i,12}$	$-Y_{h_2} N_{bac}$		ρ_{12}
13 Sulfate reduction								-1	$(1 - Y_{so_4})$	$-(1 - Y_{so_4})/64$		$-Y_{so_4} C_{bac}$	$-Y_{so_4} N_{bac}$		ρ_{13}
14 Decay of X_{su}															$k_{dec,Xsu} \cdot X_{su}$
15 Decay of X_{aa}															$k_{dec,Xaa} \cdot X_{aa}$
16 Decay of X_{fa}															$k_{dec,Xfa} \cdot X_{fa}$
17 Decay of X_{c_4}															$k_{dec,Xc_4} \cdot X_{c_4}$
18 Decay of X_{pro}															$k_{dec,Xpro} \cdot X_{pro}$
19 Decay of X_{ac}															$k_{dec,Xac} \cdot X_{ac}$
20 Decay of X_{h_2}															$k_{dec,Xh_2} \cdot X_{h_2}$
21 Decay of X_{so_4}															$k_{dec,Xso_4} \cdot X_{so_4}$
	Monosaccharides (kg COD m^{-3})	Amino acids (kg COD m^{-3})	LCFAs (kg COD m^{-3})	Total valerate (kg COD m^{-3})	Total butyrate (kg COD m^{-3})	Total propionate (kg COD m^{-3})	Total acetate (kg COD m^{-3})	Hydrogen gas (kg COD m^{-3})	Sulfide (kg COD m^{-3})	Sulfate (M)	Methane gas (kg COD m^{-3})	Inorganic carbon (M)	Inorganic nitrogen (M)	Soluble inerts (kg COD m^{-3})	

Note: C_i is the carbon content (kmol kg/ COD) of component i.

Table 4.3. Biochemical rate coefficients (υ_{ij}) and kinetic rate equations (ρ_{ij}) for particulate components (i = 15–27; j = 1–21).

	Component i													
j Process	15 X_c	16 X_{ch}	17 X_{pr}	18 X_{li}	19 X_{su}	20 X_{aa}	21 X_{fa}	22 X_{c_4}	23 X_{pro}	24 X_{ac}	25 X_{h_2}	26 X_{so_4}	27 X_I	Process Rate (ρ_j) (ρ_j, kg COD $m^{-3}day^{-1}$)
1 Disintegration	-1	$f_{ch,Xc}$	$f_{pr,Xc}$	$f_{li,Xc}$									$f_{SI,Xc}$	$k_{dis} X_c$
2 Hydrolysis of carbohydrates		-1												$k_{hyd,ch} X_{ch}$
3 Hydrolysis of proteins			-1											$k_{hyd,pr} X_{pr}$
4 Hydrolysis of lipids				-1										$k_{hyd,li} X_{li}$
5 Uptake of sugars					Y_{su}									ρ_5
6 Uptake of amino acids						Y_{aa}							ρ_6	
7 Uptake of LCFA							Y_{fa}							ρ_7
8 Uptake of valerate								Y_{c_4}						ρ_8
9 Uptake of butyrate								Y_{c_4}						ρ_9
10 Uptake of propionate									Y_{pro}					ρ_{10}
11 Uptake of acetate										Y_{ac}				ρ_{11}
12 Uptake of hydrogen											Y_{h_2}			ρ_{12}
13 Sulfate reduction												Y_{so_4}		ρ_{13}
14 Decay of X_{su}	1				-1									$k_{dec,Xsu} \cdot X_{su}$
15 Decay of X_{aa}	1					-1								$k_{dec,Xaa} \cdot X_{aa}$
16 Decay of X_{fa}	1						-1							$k_{dec,Xfa} \cdot X_{fa}$
17 Decay of X_{c_4}	1							-1						$k_{dec,Xc_4} \cdot X_{c_4}$
18 Decay of X_{pro}	1								-1					$k_{dec,Xpro} \cdot X_{pro}$
19 Decay of X_{ac}	1									-1				$k_{dec,Xac} \cdot X_{ac}$
20 Decay of X_{h_2}	1										-1			$k_{dec,Xh_2} \cdot X_{h_2}$
21 Decay of X_{so_4}	1											-1		$k_{dec,Xso_4} \cdot X_{so_4}$
	Composites (kg COD m^{-3})	Carbohydrates (kg COD m^{-3})	Proteins (kg COD m^{-3})	Lipids (kg COD m^{-3})	Sugar degraders (kg COD m^{-3})	Amino acid degraders (kg COD m^{-3})	LCFA degraders (kg COD m^{-3})	Valerate and butyrate degraders (kg COD m^{-3})	Propionate degraders (kg COD m^{-3})	Acetate degraders (kg COD m^{-3})	Hydrogen degraders (kg COD m^{-3})	Sulfate reducers (kg COD m^{-3})	Particulate inerts (kg COD m^{-3})	

Monod kinetics with respect to their primary substrate is used. The decay of cell mass is modeled using first-order kinetics.

The process rates in Tables 4.2 and 4.3 are as follows:

$$\rho_5 = k_{m,aa} \cdot \frac{S_{aa}}{K_{s,aa} + S_{aa}} \cdot X_{aa} \cdot I_1 \tag{4.17}$$

$$\rho_6 = k_{m,su} \cdot \frac{S_{su}}{K_{s,su} + S_{su}} \cdot X_{su} \cdot I_1 \tag{4.18}$$

$$\rho_7 = k_{m,fa} \cdot \frac{S_{fa}}{K_{s,fa} + S_{fa}} \cdot X_{fa} \cdot I_2 \tag{4.19}$$

$$\rho_8 = k_{m,C_4} \cdot \frac{S_{va}}{K_{s,C_4} + S_{va}} \cdot X_{C_4} \cdot \frac{1}{1 + \frac{S_{bu}}{S_{va}}} \cdot I_2 \tag{4.20}$$

$$\rho_9 = k_{m,C_4} \cdot \frac{S_{bu}}{K_{s,C_4} + S_{bu}} \cdot X_{C_4} \cdot \frac{1}{1 + \frac{S_{va}}{S_{bu}}} \cdot I_2 \tag{4.21}$$

$$\rho_{10} = k_{m,pro} \cdot \frac{S_{pro}}{K_{s,pro} + S_{bu}} \cdot X_{pro} \cdot I_2 \tag{4.22}$$

$$\rho_{11} = k_{m,ac} \cdot \frac{S_{ac}}{K_{s,ac} + S_{ac}} \cdot X_{ac} \cdot I_3 \tag{4.23}$$

$$\rho_{12} = k_{m,h_2} \cdot \frac{S_{h_2}}{K_{s,h_2} + S_{h_2}} \cdot X_{h_2} \cdot I_1 \tag{4.24}$$

$$\rho_{13} = k_{m,SO_4} \cdot \frac{S_{SO_4}}{K_{s,SO_4} + S_{SO_4}} \cdot \frac{S_{h_2}}{K_{s,h_2} + S_{h_2}} \cdot X_{SO_4} \cdot I_1 \tag{4.25}$$

The inhibition factors in the above expressions are as follows:

$$I_1 = I_{pH} I_{IN,lim} \tag{4.26}$$

$$I_2 = I_{pH} I_{IN,lim} I_{h_2} \tag{4.27}$$

$$I_3 = I_{pH} I_{IN,lim} I_{NH_3,Xac} \tag{4.28}$$

Assuming that pH inhibits the process both at high and low pH, the following empirical relationship can be used for pH inhibition factor.

$$I_{pH} = \frac{1 + 2 \times 10^{0.5(pH_{LL} - pH_{UL})}}{1 + 10^{(pH - pH_{UL})} + 10^{(pH_{LL} - pH)}} \tag{4.29}$$

where pH_{LL} and pH_{UL} are the lower and upper pH limits at which 50% inhibition occurs.

Free ammonia and hydrogen inhibition can be modeled as follows:

$$I_{h2}, I_{NH3,Xac} = \frac{1}{1 + \frac{S_I}{K_I}} \qquad (4.30)$$

where S_I is the concentration of the inhibitor and K_I is the inhibition parameter.

The factor for the inhibition due to the limitation of inorganic nitrogen can be calculated using the following expression:

$$I_{IN,lim} = \frac{1}{1 + \frac{S_I}{K_I}} \qquad (4.31)$$

As in any structured model, a large number of parameters are involved in ADM1 model. Furthermore, the values of these parameters vary with the process conditions. The parameter values at mesophilic conditions would be different from those at thermophilic conditions. Similarly, a process treating dilute wastewaters would have different kinetic values as compared to the one treating high-solids waste streams. Since this is not possible to present a comprehensive description of the parameters here, the reader is advised to refer to ADM1 model for details (Batstone et al. 2002).

4.4 Modeling of pH Change

Many biochemical reactions in an anaerobic process involve production or consumption of proton, thereby affecting the pH of the system. As a consequent, the concentration of components involved in chemical equilibria will change, and this would affect not only the physical–chemical processes such as liquid–gas mass transfer of gaseous components, but also the rate of biochemical reactions. In an anaerobic system, significant changes in the pH are expected and hence the anaerobic model should be able to predict the change in pH accurately.

The chemical dissociation reactions occur very fast as compared to the biochemical reactions. It takes from few hours to days to complete a biochemical reaction, while it is the matter of only few seconds for chemical dissociation reactions. Therefore, a model of a biological system with varying pH will have different time scales, thus making the model stiff. The solution of stiff model generally encounters computational problems (the major one being a very long simulation time). To avoid the problem, the fast dissociation reactions can be assumed to occur instantaneously (steady-state approach). This can be done by solving the system of algebraic equations representing the equilibria (Volcke 2006).

Table 4.4. Chemical equilibrium systems in pH calculation.

S. No.	Acid Type	Dissociation	Systems	Equilibrium Reactions
1.	Monoprotic acid with monovalent positive charge	Dissociates into neutral base form	Ammonia	$NH_4^+ \leftrightarrow H^+ + NH_3$
2.	Monoprotic neutral acid	Dissociates into base with monovalent negative charge	Acetate	$HAc \leftrightarrow H^+ + Ac^-$
3.	Diprotic neutral acid	Dissociates in two steps into base with divalent negative charge	Carbonate	$H_2CO_3 \leftrightarrow H^+ + HCO_3^-$ $HCO_3^- \leftrightarrow H^+ + CO_3^{2-}$
			Sulfide	$H_2S \leftrightarrow H^+ + HS^-$ $HS^- \leftrightarrow H^+ + S^{2-}$
4.	Triprotic neutral acid	Dissociates in three steps into base with trivalent negative charge	Phosphate	$H_3PO_4 \leftrightarrow H^+ + H_2PO_4^-$ $H_2PO_4^- \leftrightarrow H^+ + HPO_4^{2-}$ $HPO_4^{2-} \leftrightarrow H^+ + PO_4^{3-}$
5.	Water	Dissociates into H^+ and OH^-	Water	$H_2O \leftrightarrow H^+ + OH^-$

The first step in this approach is to identify the chemical equilibrium reactions. The equilibrium reactions involved in an anaerobic process are listed in Table 4.4.

The chemical equilibrium reactions listed in Table 4.4 are assumed to be in steady state and the following equations apply:

$$\frac{[NH_3] \cdot [H^+]}{[NH_4{}^+]} = K_{1,NH_3} \quad (4.32)$$

$$\frac{[A_c{}^-] \cdot [H^+]}{[HA_c]} = K_{1,HAc} \quad (4.33)$$

$$\frac{[HS^-] \cdot [H^+]}{[H_2S]} = K_{1,H_2S} \quad (4.34)$$

$$\frac{[S^{2-}] \cdot [H^+]}{[HS^-]} = K_{2,H_2S} \quad (4.35)$$

$$\frac{[HCO_3{}^-] \cdot [H^+]}{[H_2CO_3]} = K_{1,CO_3} \quad (4.36)$$

$$\frac{[CO_3{}^{2-}] \cdot [H^+]}{[HCO_3{}^-]} = K_{2,CO_3} \quad (4.37)$$

$$\frac{[H_2PO_4{}^-] \cdot [H^+]}{[H_3PO_4]} = K_{1,PO_4} \quad (4.38)$$

$$\frac{[HPO_4{}^{2-}]\cdot[H^+]}{[H_2PO_4{}^-]} = K_{2,PO_4} \quad (4.39)$$

$$\frac{[PO_4{}^{3-}]\cdot[H^+]}{[HPO_4{}^{2-}]} = K_{3,PO_4} \quad (4.40)$$

$$[H^+]\cdot[OH^-] = K_w \quad (4.41)$$

In the above equations, the K term on the right-hand side of the equation is the equilibrium constant of the corresponding reaction. The equilibrium constants for the reactions listed above can be found in any standard physical chemistry textbook. The equilibrium constant varies with temperature, and this variation can be modeled using the Arrhenius equation. In a system where the temperature varies with time, the value of K needs to be calculated at each time step.

The equilibrium calculations are based on the total concentrations of all equilibrium forms of the component of interest. The lumped components are calculated as follows:

Total ammonia (TNH_3)

$$[TNH_3] = [NH_3] + [NH_4{}^+] \quad (4.42)$$

Total acetate (TAc)

$$[TA_c] = [HA_c] + [A_c{}^-] \quad (4.43)$$

Total sulfide (TH_2S)

$$[TH_2S] = [H_2S] + [HS^-] + [S^{2-}] \quad (4.44)$$

Total carbonate (TCO_3)

$$[TCO_3] = [H_2CO_3] + [HCO_3{}^-] + [CO_3{}^{2-}] \quad (4.45)$$

Total phosphate (TPO_4)

$$[TPO_4] = [H_3PO_4] + [H_2PO_4{}^-] + [HPO_4{}^{2-}] + [PO_4{}^{3-}] \quad (4.46)$$

The concentrations of each of the equilibrium forms of the components can be expressed in terms of the lumped total concentration and the hydrogen ion concentration ($[H^+]$). Following relationship can be obtained by rearranging the expressions in Table 4.4 and Eqs (4.42–4.46).

$$[A_c^-] = \frac{K_{1,HAc}}{K_{1,HAc} + [H^+]} \cdot [TA_c] \tag{4.47}$$

$$[NH_4^+] = \frac{[H^+]}{K_{1,NH_3} + [H^+]} \cdot [TNH_3] \tag{4.48}$$

$$[HS^-] = \frac{[H^+]K_{1,H_2S}}{[H^+]^2 + [H^+]K_{1,H_2S} + K_{1,H_2S}K_{2,H_2S}} \cdot [TH_2S] \tag{4.49}$$

$$[S^{2-}] = \frac{K_{1,H_2S}K_{2,H_2S}}{[H^+]^2 + [H^+]K_{1,H_2S} + K_{1,H_2S}K_{2,H_2S}} \cdot [TH_2S] \tag{4.50}$$

$$[HCO_3^-] = \frac{[H^+]K_{1,CO_3}}{[H^+]^2 + [H^+]K_{1,CO_3} + K_{1,CO_3}K_{2,CO_3}} \cdot [TCO_3] \tag{4.51}$$

$$[CO_3^{2-}] = \frac{K_{1,CO_3}K_{2,CO_3}}{[H^+]^2 + [H^+]K_{1,CO_3} + K_{1,CO_3}K_{2,CO_3}} \cdot [TCO_3] \tag{4.52}$$

$$[H_2PO_4^-] = \frac{[H^+]^2 K_{1,PO_4}}{[H^+]^3 + [H^+]^2 K_{1,PO_4} + [H^+]K_{1,PO_4}K_{2,PO_4} + K_{1,PO_4}K_{2,PO_4}K_{3,PO_4}} \cdot [TPO_4] \tag{4.53}$$

$$[HPO_4^{2-}] = \frac{[H^+]K_{1,PO_4}K_{2,PO_4}}{[H^+]^3 + [H^+]^2 K_{1,PO_4} + [H^+]K_{1,PO_4}K_{2,PO_4} + K_{1,PO_4}K_{2,PO_4}K_{3,PO_4}} \cdot [TPO_4] \tag{4.54}$$

$$[PO_4^{3-}] = \frac{K_{1,PO_4}K_{2,PO_4}K_{3,PO_4}}{[H^+]^3 + [H^+]^2 K_{1,PO_4} + [H^+]K_{1,PO_4}K_{2,PO_4} + K_{1,PO_4}K_{2,PO_4}K_{3,PO_4}} \cdot [TPO_4] \tag{4.55}$$

In addition to the species in Eqs (4.47–4.55), $[OH^-]$ should be known and this can be calculated from Eq. (4.41). Expressions for propionate, butyrate, and valerate can also be developed using a similar approach if needed.

A charged balanced equation can be set up as follows:

$$\Delta_{charge} = [H^+] - [OH^-] - [A_c^-] + [NH_4^+] - [HS^-] - 2 \times [S^{2-}] - [HCO_3^-] - 2 \times [CO_3^{2-}] - [H_2PO_4^-] - 2 \times [HPO_4^{2-}] - 3 \times [PO_4^{3-}] + [T_{cat0}] + [T_{cat}] \tag{4.56}$$

where T_{cat0} and T_{cat} are the lumped components representing the concentration of net positive charges of the system that are not affected by pH change. The difference between these two is that T_{cat0} consists of charges not involved in biological reactions (e.g., Na^+), whereas T_{cat0} consists of charges involved in biological reactions (e.g., SO_4^{2-}). For electroneutrality, the sum of charges on right-hand side of Eq. (4.56) must be equal to zero.

The concentration of T_{cat} will vary with time as the concentration of relevant charged species vary. Hence, this needs to be included in the model as a state variable. The concentration of T_{cat0} should be taken as a constant and can be calculated from known pH of the feed as follows:

1. Calculate the concentration of the charged species in the feed using the expression given in Eqs (4.47–4.55) using known pH and the total concentrations of the species involved.
2. Calculate the concentration of T_{cat} from known concentration of relevant charged species in the feed.
3. Rearrange Eq. (4.56) making net charge equal to zero. The concentration of T_{cat0} can then be calculated as follows:

$$
\begin{aligned}
[T_{cat0}] = {} & -[H^+] + [OH^-] + [A_c^-] - [NH_4^+] + [HS^-] + 2x[S^{2-}] + [HCO_3^-] \\
& + 2 \times [CO_3^{2-}] + [H_2PO_4^-] + 2 \times [HPO_4^{2-}] + 3 \times [PO_4^{3-}] - [T_{cat}]
\end{aligned}
\tag{4.57}
$$

The charge balance equation can be set up using the concentrations of charged species and making net charge equal to zero as follows:

$$
\begin{aligned}
[H^+] - [OH^-] & - \frac{K_{1,HAc}}{K_{1,HAc} + [H^+]} \cdot [TA_c] + \frac{[H+]}{K_{1,NH_3} + [H^+]} \cdot [TNH_3] \\
& - \frac{[H^+]K_{1,H_2S}}{[H^+]^2 + [H^+]K_{1,H_2S} + K_{1,H_2S}K_{2,H_2S}} \cdot [TH_2S] \\
& - 2 \times \frac{K_{1,H_2S}K_{2,H_2S}}{[H^+]^2 + [H^+]K_{1,H_2S} + K_{1,H_2S}K_{2,H_2S}} \cdot [TH_2S] \\
& - \frac{[H^+]K_{1,CO_3}}{[H^+]^2 + [H^+]K_{1,CO_3} + K_{1,CO_3}K_{2,CO_3}} \cdot [TCO_3] \\
& - 2 \times \frac{K_{1,CO_3}K_{2,CO_3}}{[H^+]^2 + [H^+]K_{1,CO_3} + K_{1,CO_3}K_{2,CO_3}} \cdot [TCO_3]
\end{aligned}
$$

$$- \frac{[H^+]^2 K_{1,PO_4}}{[H^+]^3 + [H^+]^2 K_{1,PO_4} + [H^+] K_{1,PO_4} K_{2,PO_4} + K_{1,PO_4} K_{2,PO_4} K_{3,PO_4}} \cdot [TPO_4]$$
$$- 2 \times \frac{[H^+] K_{1,PO_4} K_{2,PO_4}}{[H^+]^3 + [H^+]^2 K_{1,PO_4} + [H^+] K_{1,PO_4} K_{2,PO_4} + K_{1,PO_4} K_{2,PO_4} K_{3,PO_4}} \cdot [TPO_4]$$
$$- 3 \times \frac{K_{1,PO_4} K_{2,PO_4} K_{3,PO_4}}{[H^+]^3 + [H^+]^2 K_{1,PO_4} + [H^+] K_{1,PO_4} K_{2,PO_4} + K_{1,PO_4} K_{2,PO_4} K_{3,PO_4}} \cdot [TPO_4]$$
$$+ [T_{cat0}] + [T_{cat}] = 0 \quad (4.58)$$

Since the hydrogen ion concentration is the only unknown in the above equation, the equation needs to be solved numerically for hydrogen ion concentration for which sum of all charges (Δ_{charge}) is zero. Newton–Raphson method is generally used for this purpose. The method starts from an initial guess of $[H^+]_0$ that deviates from the actual value of $[H^+]$, which satisfies the condition of electroneutrality.

$$[H^+] = [H^+]_0 + \delta[H^+] \quad (4.59)$$

A new estimate for $[H^+]$ can be made using the following expression:

$$[H^+]_1 = [H^+]_0 - \frac{\Delta_{charge,0}}{\frac{d\Delta_{charge,0}}{d[H^+]_0}} \quad (4.60)$$

In the next step, another estimate for $[H^+]$ is made as follows:

$$[H^+]_2 = [H^+]_1 - \frac{\Delta_{charge,1}}{\frac{d\Delta_{charge,1}}{d[H^+]_1}} \quad (4.61)$$

This interactive procedure continues until the estimated $[H^+]$ satisfies the charge balance. The derivative term in the above equations can be calculated either numerically or analytically. Numerical estimate of the derivative can be made by making a small increment in hydrogen ion concentration and calculating net charge concentration at this new hydrogen ion concentration. The derivative can then be calculated by dividing the change in the charge concentration by the increment in $[H^+]$. As an alternative approach, the derivatives of the left-hand side terms in Eq. (4.58) with respect to $[H^+]$ can be obtained analytically.

Example 4.2

Calculate pH of a solution when 0.01 M acetic acid is added to pure water. Compare the pH with a case when sodium acetate is added instead of acetic acid.

Solution

When acetic acid is added:

The dissociation reactions are the following:

$$HAc \leftrightarrow H^+ + Ac^-$$

$$H_2O \leftrightarrow H^+ + OH^-$$

Total acetate concentration, TAc = 0.01 M

$$[Ac^-] = \frac{K}{K + [H^+]} \cdot TAc$$

Charge balance:

$$\begin{aligned}
\Delta\text{charge} &= [H^+] - [OH^-] - [Ac^-] \\
&= [H^+] - \frac{Kw}{[H^+]} - \frac{K}{K + [H^+]} \cdot TAc \\
\frac{d\Delta\text{charge}}{d[H^+]} &= 1 + \frac{Kw}{[H^+]^2} + \frac{K}{(K + [H^+])^2} \cdot TAc \\
Kw &= 10^{-14} \\
K &= 10^{-4.75} (pK_a = 4.75)
\end{aligned}$$

Let us start with pH of 7 as an initial guess.

$[H^+] = 10^{-7}$
Δcharge $= -0.009944$
$d\Delta$charge$/d[H^+] = 558.069$
$[H^+]_{new} = [H^+] - \Delta\text{charge}/(d\Delta\text{charge}/d[H^+])$
$[H^+]$ for next iteration $= 10^{-7} + 0.009944/558.069 = 1.792 \times 10^{-5}$
Δcharge $= -0.00496$
$d\Delta$charge$/d[H^+] = 140.52$
$[H^+]$ for next iteration $= 1.792 \times 10^{-5} + 0.00496/140.52 = 5.324 \times 10^{-5}$
Δcharge $= -0.00245$
$d\Delta$charge$/d[H^+] = 36.25$

$[H^+]$ for next iteration $= 5.324 \times 10^{-5} + 0.00245/36.25 = 1.208 \times 10^{-4}$
Δcharge $= -0.00116$
$d\Delta \text{charge}/d[H^+] = 10.26$
$[H^+]$ for next iteration $= 1.208 \times 10^{-4} + 0.00116/10.26 = 2.341 \times 10^{-4}$
After several iteration,
$[H^+] = 4.129 \times 10^{-4}$
Δcharge $= 0$
The $[H^+]$ in the solution that satisfies the charge balance is therefore $-\log 10(4.129 \times 10^{-4}) = 3.38$.

When sodium acetate is added:
The dissociation reactions are the following:

$NaAc \leftrightarrow Na^+ + Ac^-$
$H^+ + Ac^- \leftrightarrow HAc$
$H_2O \leftrightarrow H^+ + OH^-$

Total acetate concentration, TAc $= 0.01$ M
$[Na^+] = 0.01$ M

$$[Ac^-] = \frac{K}{K + [H^+]} \cdot \text{TAc}$$

Charge balance:

$$\begin{aligned}\Delta\text{charge} &= [H^+] + [Na^+] - [OH^-] - [Ac^-] \\ &= [H^+] + [Na^+] - \frac{Kw}{[H^+]} - \frac{K}{K + [H^+]} \cdot \text{TAc}\end{aligned}$$

$$\frac{d\Delta\text{charge}}{d[H^+]} = 1 + \frac{Kw}{[H^+]^2} + \frac{K}{(K + [H^+])^2} \cdot \text{TAc}$$

The difference between this and the previous case is that the addition of sodium needs to be considered in the charge balance.

Let us start with pH of 8.0 as initial guess.
$[H^+] = 10^{-8}$
Δcharge $= 4.63 \times 10^{-6}$
$d\Delta \text{charge}/d[H^+] = 662.71$
$[H^+]$ for next iteration $= 10^{-8} - 4.63 \times 10^{-6}/662.71 = 3.013 \times 10^{-9}$
Δcharge $= -1.622 \times 10^{-6}$
$d\Delta \text{charge}/d[H^+] = 1{,}664.59$
$[H^+]$ for next iteration $= 3.013 \times 10^{-9} + 1.622 \times 10^{-6}/1{,}664.59 = 3.987 \times 10^{-9}$
Δcharge $= -2.622 \times 10^{-7}$
$d\Delta \text{charge}/d[H^+] = 1{,}192.06$

$[H^+]$ for next iteration $= 3.987 \times 10^{-9} + 2.622 \times 10^{-7}/1{,}192.06 = 4.207 \times 10^{-9}$.
After several iteration,
$[H^+] = 4.214 \times 10^{-9}$
$\Delta\text{charge} = 0$

The $[H^+]$ in the solution that satisfies the charge balance is therefore $-\log 10(4.214 \times 10^{-9}) = 8.38$.

4.5 Modeling of Energy Generation

One of the key advantages of anaerobic process is the positive net energy production. Energy is mainly produced as biogas and to a lesser extent as microbial heat. The rate of the production of biogas and its composition can be modeled as illustrated in the following section. The ADM1 model can be used to simulate the biogas production and thus the energy generation in an anaerobic process (Lubken et al. 2007). A portion of the energy that is generated is consumed by the process itself. Energy is needed to heat the reactor contents and compensate for energy loss due to irradiation. In addition, the operation of pumps and stirrers requires energy. The net energy production would be the energy generated minus the total energy loss.

The model of energy generation involves an additional state variable, net energy production (P_{net}). The dynamic change in net energy production (kWh/day) can be modeled as follows (Lubken et al. 2007):

$$\frac{dP_{net}}{dt} = \left(p_{elect}^{prod} - p_{pump}^{loss} - p_{stir}^{loss}\right) + \left(p_{therm}^{prod} - p_{rad}^{loss} - p_{sub\_heat}^{loss}\right) + p_{micro\_heat}^{prod} \quad (4.62)$$

where P_{net} is the net energy production rate (kWh/day), p_{elect}^{prod} is the electric energy production rate (kWh/day), p_{therm}^{prod} is the thermal energy production rate (kWh/day), p_{pump}^{loss} is the mechanical power of the pump (kWh/day), p_{stir}^{loss} is the mechanical power of the stirrer (kWh/day), p_{rad}^{loss} is the energy loss due to irradiation (kWh/day), $p_{sub\_heat}^{loss}$ is the energy required for heating (kWh/day), and $p_{micro\_heat}^{prod}$ is microbial heat production (kWh/day).

Each of the terms in Eq. (4.62) can be expressed as follows (Lubken et al. 2007):

$$p_{elect}^{prod} = q_G\, p_{CH_4} H_c \eta_{elect} \quad (4.63)$$

$$p_{therm}^{prod} = q_G\, p_{CH_4} H_c \eta_{therm} \quad (4.64)$$

where q_G is the biogas production rate (m^3/day), p_{CH4} is the methane content of the biogas (% by volume), H_c is the calorific value of methane, η_{elect} is the electric efficiency factor (about 35%), and η_{therm} is the thermal efficiency factor (about 50%).

$$p_{pump}^{loss} = Q_{in} H \rho g t_p \frac{1}{\eta} \tag{4.65}$$

where Q_{in} is the flow rate, H is the pumping head, ρ is the density of media being pumped, g is the gravitational acceleration, t_p is the time of pumping (h/day), and η is the efficiency factor depending on the type of pump used.

$$p_{stir}^{loss} = V_{liq} S t_s \tag{4.66}$$

where V_{liq} is the liquid volume (m^3), S is the specific power of stirrer (kWm^{-3}), and t_s is the time of stirring (h/day).

$$p_{rad}^{loss} = K_{heat\_trans} \left[\left(T_{liq} - T_{ambient} \right) V_{liq} + \left(T_{gas} - T_{ambient} \right) \right] \cdot \left(V_{tot} - V_{liq} \right) \cdot \frac{A}{V_{tot}} \cdot \frac{24}{100} \tag{4.67}$$

where K_{heat_trans} is the heat transfer coefficient (Wh $m^{-2}h^{-1}K^{-1}$), T_{liq} is the temperature of the liquid in the reactor (K), $T_{ambient}$ is the ambient temperature (K), T_{gas} is the gas temperature (K), V_{tot} is the total digester volume (m^3), and A is the surface area of the reactor (m^2).

$$p_{sub\_heat}^{loss} = Q_{in} c (T_{reactor} - T_{substrate}) \frac{1}{3.6} \tag{4.68}$$

where c is the heat capacity of the substrate (kJ $kg^{-1}K^{-1}$), $T_{reactor}$ is the temperature of the reactor (K), and $T_{substrate}$ is the substrate temperature (K).

$$p_{mic\_heat}^{prod} = \sum_{j=5-12} \Delta E_j f_j \rho_j V_{liq} \frac{1}{3.6} \tag{4.69}$$

where ΔE_j is the amount of energy released to the environment due to microbial activity of process j (kJ/mol), f_j is the molar mass/g COD of the product of process j (mol g/COD), and ρ_j is the process kinetic rate (Tables 4.2 and 4.3). ΔE_j values for the processes with a detailed description on its derivation can be found in Lubken et al. (2007).

4.5.1 Model Application to a Continuously Mixed Reactor

The underlining concepts discussed earlier are applicable to any reactor configuration. However, if the process involves biofilm, this would involve many complications as the transport limitations due to diffusion will have to be considered in the model. Also, the difference in hydraulics will make the models different in these cases. For simplicity, let us assume that the reactor is a continuous-flow stirred tank reactor (CSTR), and continuous flow occurs. A schematic of a typical single-tank reactor with all system parameters is presented in Fig. 4.3.

For each state variable in the model, mass balance equation for liquid phase can be written as follows:

$$\frac{d(VS_{L,i})}{dt} = Q_{in}S_{in,i} - Q_{out}S_{L,i} + V\sum_{j=1-21}\rho_j v_{i,j} \tag{4.70}$$

$$\frac{d(VX_{L,i})}{dt} = Q_{in}X_{in,i} - Q_{out}X_{L,i} + V\sum_{j=1-21}\rho_j v_{i,j} \tag{4.71}$$

In the above expressions, the term on left-hand side is the rate of change of mass of the component (MT^{-1}), the first term on right is the rate at which the mass is entering the tank, the second term is the rate at which the mass is leaving the tank, and the last term is the rate of generation calculated as explained in Section 4.3.2.

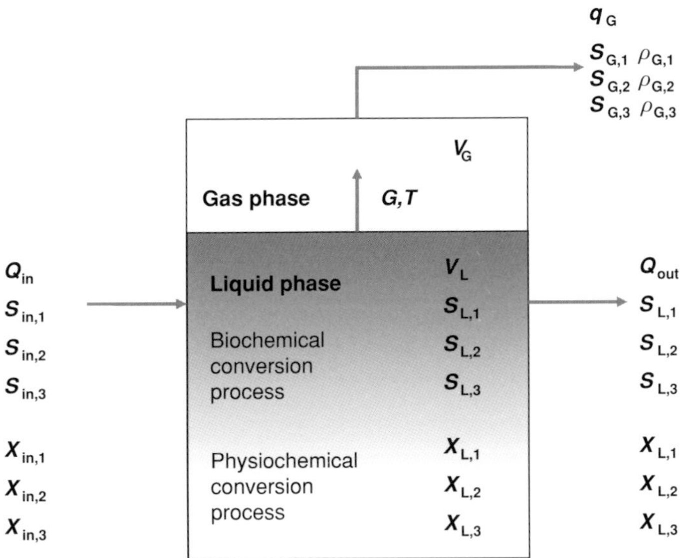

FIG. 4.3. Schematic of a typical CSTR reactor.

If the volume is constant, $Q_{in} = Q_{out} = Q$, and the expression becomes:

$$\frac{dS_{L,i}}{dt} = \frac{Q}{V}\left(S_{in,i} - S_{L,i}\right) + \sum_{j=1-21} \rho_j v_{i,j} \tag{4.72}$$

$$\frac{dX_{L,i}}{dt} = \frac{Q}{V}\left(X_{in,i} - X_{L,i}\right) + \sum_{j=1-21} \rho_j v_{i,j} \tag{4.73}$$

The solid components in biofilm or high-rate reactors with very high biomass concentration spend a lot more time in the system than the liquid components. The residence time of the solid components would be higher than the hydraulic retention time, and this can be taken care of by introducing a term $t_{res,X}$, which is the residence time of solids components above hydraulic retention time (Q/V).

$$\frac{dX_{L,i}}{dt} = \frac{QX_{in,i}}{V} + \frac{X_{L,i}}{t_{res,X} + \frac{V}{Q}} + \sum_{j=1-21} \rho_j v_{i,j} \tag{4.74}$$

The rate of transfer of gases from liquid phase to gas phase will depend on factors such as reactor configuration and design, partial pressure of the gas in the headspace, and temperature. The rate can be expressed as follows:

$$\rho_g = k_{La} \cdot \left(S_{L,g} - S_{sat,g}\right) \tag{4.75}$$

where k_{La} is the liquid–gas mass transfer coefficient, $S_{L,g}$ is the concentration of gas in the liquid phase, and $S_{sat,g}$ is the gas saturation concentration that can be estimated using Henry's law as follows:

$$S_{sat,g} = K_{H,g} \cdot p_g \tag{4.76}$$

where $K_{H,g}$ is the Henry's coefficient (mol/L · atm) and p_g is the partial pressure of gas (atm of absolute pressure). Four gaseous components involved in an anaerobic process are methane, hydrogen, hydrogen sulfide, and carbon dioxide modeled as inorganic carbon. The Henry's coefficient for these gases at 298°K is given in Table 4.5.

Table 4.5. Henry's coefficients for gases in water at 298°K.

Gas	$K_{H,g}$ (mol/L · atm)	C (°K)
CH_4	1.4×10^{-3}	1,700
H_2	7.8×10^{-4}	500
CO_2	3.4×10^{-2}	2,400
H_2S	1.0×10^{-1}	2,100

When the temperature of a system changes, the Henry's constant will also change. This dependency of Henry's constant on temperature is given by Van't Hoff equation:

$$K_H(T) = K_H(T\theta) \cdot e^{(-C\cdot(1/T-1/T\theta))} \tag{4.77}$$

where $K_H(T)$ is the Henry's constant at temperature T, $K_H(T\theta)$ is the Henry's constant at standard temperature $T\theta$, and C is a constant, the value of which is presented in Table 4.5.

Based on the discussion above, the kinetic rates for liquid–gas transfer for the four gaseous components can be calculated as follows:

$$\rho_{T,H_2} = k_{L,a}\left(S_{L,H_2} - 16K_{H,H_2}\,\rho_{gas,H_2}\right) \tag{4.78}$$

$$\rho_{T,H_2S} = k_{L,a}\left(S_{L,H_2S} - 64K_{H,H_2S}\,\rho_{gas,H_2S}\right) \tag{4.79}$$

$$\rho_{T,CH_4} = k_{L,a}\left(S_{L,CH_4} - 64K_{H,CH_4}\,\rho_{gas,CH_4}\right) \tag{4.80}$$

$$\rho_{T,CO_2} = k_{L,a}\left(S_{L,CO_2} - K_{H,CO_2}\,\rho_{gas,CO_2}\right) \tag{4.81}$$

These rates are based on liquid volume and need to be multiplied by a factor of V_L/V_G if these are to be used for the gas-phase calculations. The gas-phase mass balance for the gaseous components for a constant gas volume can be written as follows:

$$\frac{dS_{G,i}}{dt} = -\frac{S_{G,i}q_G}{V_G} + \rho_{T,i}\frac{V_L}{V_G} \tag{4.82}$$

The gas flow rate can be calculated for restricted flow through an orifice using the following expression:

$$q_G = k_p\,(P_G - P_{atm}) \tag{4.83}$$

where P_{atm} is the atmospheric pressure in bar and k_p is the pipe resistance coefficient in $m^3\ day^{-1} bar^{-1}$.

The pressure of each gas can be calculated using the ideal gas law as follows:

$$\begin{aligned} \rho_{G,H_2} &= S_{G,H_2}RT/16 \\ \rho_{G,H_2S} &= S_{G,H_2S}RT/64 \\ \rho_{G,CH_4} &= S_{G,CH_4}RT/64 \\ \rho_{G,CO_2} &= S_{G,CO_2}RT \end{aligned} \tag{4.84}$$

The gas-phase pressure can be calculated from partial pressures in Eq. (4.84) as follows:

$$P_G = \rho_{G,H_2} + \rho_{G,H_2S} + \rho_{G,CH_4} + \rho_{G,CO_2} \tag{4.85}$$

References

Batstone, D. J. 2006. Mathematical modelling of anaerobic reactors treating domestic wastewater: Rational criteria for model use. *Rev. Environ. Sci. Biotechnol.* 5:57–71.

Batstone, D. J., Keller, J., Angelidaki, I., Kalyuzhnyi, S. V., Pavlostathis, S. G., Rozzi, A., Sanders, W. T. M., Siegrist, H., and Vavilin, V. A. 2002. *Anaerobic Digestion Model No. 1 (ADM1), IWA Task Group for Mathematical Modelling of Anaerobic Digestion Processes.* IWA Publishing, London.

Gujer, W., and Zehnder, A. J. B. 1983. Conversion processes in anaerobic digestion. *Water Sci. Technol.* 15(8–9):127–167.

Henze, M., Grady, C. P. L., Gujer, W., Marais, G. v. R., and Matsuo, T. 1987. *Activated Sludge Model No. 1 (ASM1), IAWPRC Task Group on Mathematical Modelling on Design and Operation of Biological Wastewater Treatment.* IWA Publishing, London.

Lubken, M., Wichern, M., Schlattmann, M., Gronauer, A., and Horn, H. 2007. Modeling the energy balance of an anaerobic digester fed with cattle manure and energy crops. *Water Res.* 41:4085–4096.

Pavlostathis, S. G., and Gossett, J. M. 1986. A kinetic model for anaerobic digestion of biological sludges. *Biotechnol. Bioeng.* 28:1519–1530.

Volcke, E. 2006. *Modelling, Analysis and Control of Partial Nitritation in a Sharon Reactor.* PhD thesis, Faculty of Agriculture and Applied Biological Sciences, Ghent University, Belgium.

CHAPTER

5

Anaerobic Reactor Configurations for Bioenergy Production

Samir Kumar Khanal

5.1 Background

Selection of the appropriate bioreactor type and configuration is particularly critical to maximize metabolic, nonoxidative bioenergy production. A reactor appropriate for bioenergy production may not be appropriate for waste treatment.

Biomass retention capacity is an important consideration when selecting a suitable bioreactor because anaerobes grow slowly during metabolic generation of methane, hydrogen, ethanol, and butanol. It is frequently essential to select a bioreactor configuration that decouples the hydraulic retention time (HRT) from the solids retention time (SRT). Such decoupling can maintain a significantly high SRT/HRT ratio and prevents washout of slow-growing anaerobes. Other considerations include feedstock types (solid, liquid, or gaseous), product inhibition, bioenergy recovery, and mass transfer limitations.

This chapter covers various approaches for decoupling HRT and SRT, classification of different anaerobic bioreactor configurations with particular emphasis on bioenergy generation, and design considerations for anaerobic bioreactors.

5.2 Strategies for Decoupling HRT and SRT

Decoupling SRT and HRT enhances the organic loading rate and enables reactor size reductions. There are four approaches for decoupling SRT from HRT as outlined in Table 5.1.

Decoupling is extremely difficult for high-solids feed streams. Such feed stream is often digested in a completely mixed reactor in which HRT = SRT. To maximize bioenergy production, therefore, a long detention time is needed. Pretreatment of

Table 5.1. Different approaches to decouple HRT from SRT.

Approaches	Biomass Retention Mechanisms	Anaerobic Reactor Types
Biomass immobilization in attached growth systems	Anaerobes attach to the support media (e.g., plastic, gravel, sand, and activated carbon) to form biofilms	Anaerobic filter; rotating anaerobic contactor; expanded/fluidized bed reactor
Granulation and floc formation	Anaerobic microorganisms agglomerate to form granules and flocs that settle well in the bioreactor	Upflow anaerobic sludge blanket reactor; static granular bed reactor; anaerobic-sequencing batch reactor; anaerobic baffled reactor
Biomass recycling	Feed with high-suspended solids (e.g., meat-packing waste and wood fiber) enables microorganisms to attach to solids, thus forming settleable flocs, which are then recycled back to the reactor	Anaerobic contact reactor/Clarigester
Biomass retention	Membrane integration into an anaerobic reactor retains biomass	Anaerobic membrane bioreactor

feed streams may reduce the detention time and enhance the bioenergy production potential as discussed in Chapter 11.

5.3 Classification of Anaerobic Bioreactors

Anaerobic reactors can be classified as low rate or high rate as shown in Fig. 5.1.

Low-rate anaerobic reactors are unmixed. Temperature, SRT, and other environmental conditions are not regulated. The organic loading rate is low in the range of 1–2 kg COD/m^3·day. These reactor configurations are not suitable for bioenergy production. Some anaerobic ponds and lagoons, however, are covered and are mixed to enhance biogas production and recovery.

High-rate anaerobic systems maintain a very high biomass level in the bioreactor. Environmental conditions are well maintained to optimize bioreactor performance. The organic loading rates vary from 5 to 30 kg COD/m^3·day or even higher. High-rate anaerobic reactors are more appropriate for bioenergy production and are discussed in greater detail in the following section.

5.3.1 High-Rate Anaerobic Digester

5.3.1.1 General Description

High-rate anaerobic digesters are essentially a continuous stirred tank reactor (CSTR), operated under mesophilic or thermophilic conditions. Temperature- and acid-phase anaerobic digesters are other examples of high-rate anaerobic digesters, commonly employed for methane fermentation of high-solids (TS 1–6%)

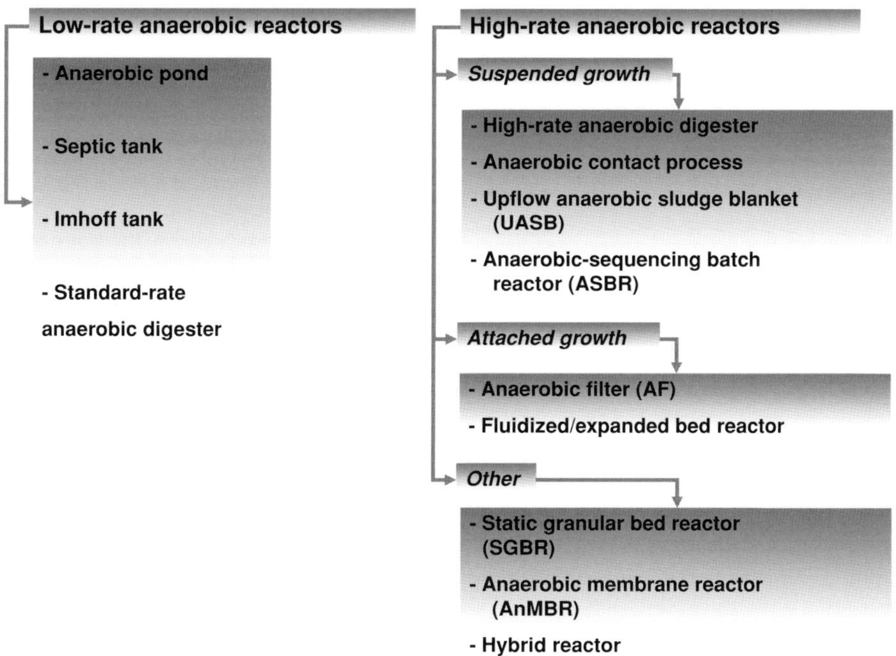

FIG. 5.1. Classification of anaerobic reactors.

wastes and residues, for example, municipal sludge, animal manure, and other biowastes. CSTR is also a common reactor configuration for hydrogen, ethanol, and butanol fermentation. Important considerations for anaerobic digestion of municipal wastewater residuals are outlined in Table 5.2.

Table 5.2. Important considerations for anaerobic digester operation.

Component	Remarks
Seed inocula	Anaerobically digested sludge or biomass from a digester treating similar waste streams
Start-up	Feed rate at 20% of the design volatile solids loading capacity for the first 20 days; gradual increase in the loading rate between 30 and 40 days; start-up time 30–40 days (seed available), or 60–90 days (without seed available)
Important monitoring parameters	pH, alkalinity, volatile organic acids, and biogas production rate
HRT (or SRT)	15–30 days (high-rate digester); 30–60 days (low-rate digester)
Alkalinity	1,500–3,000 mg/L as $CaCO_3$
VFA/ALK (alkalinity) ratio	0.1–0.2 for a healthy digester
Volatile solids loading rate	1.6–4.8 $kg/m^3 \cdot day$ (100–300 lb/1,000 $ft^3 \cdot day$)
Biogas production rate	0.75–1.12 $m^3/kg\ VS_{removed}$ (12–18 $ft^3/lb\ VS_{removed}$)
Biogas composition	55–70% CH_4 and 25–35% CO_2

Source: Adapted from Manual of Practice (MOP) # 16 1987.

5.3.1.2 Design Considerations

Although anaerobic digester design is typically based on an empirical approach, fundamental principles can be used to size a digester.

Fundamental Principles

For a suspended growth process (e.g., CSTR), the net growth of methanogens (r'_M) can be written as follows:

$$r'_M = \mu_{max,M} \frac{S}{S + K_{s,M}} X_m - b_M X_M \tag{5.1}$$

where r'_M is net growth rate of methanogens ($ML^{-3}T^{-1}$), S is concentration of growth-limiting substrate (acetic acid) (ML^{-3}), $\mu_{max,M}$ is maximum specific growth rate of methanogens (T^{-1}), X_M is concentration of methanogens (ML^{-3}), $K_{s,M}$ is half-velocity constant (ML^{-3}), and b_M is decay coefficient for methanogens (T^{-1}).

The net specific growth rate (μ'_M) can be obtained by dividing Eq. (5.1) by X_M as follows:

$$\mu'_M = \mu_{max,M} \frac{S}{S + K_{s,M}} - b_M \tag{5.2}$$

The mass balance of digester methanogens as shown in Fig. 5.2 can be written as follows:

$$\frac{dX_M}{dt} V_r = QX_0 - QX_M + r'_M V_r \tag{5.3}$$

For the steady-state condition, $\frac{dX_M}{dt} = 0$, when the methanogenic biomass (X_M) entering the digester = 0, Eq. (5.3) can be simplified to:

$$QX_M = r'_M V_r \tag{5.4}$$

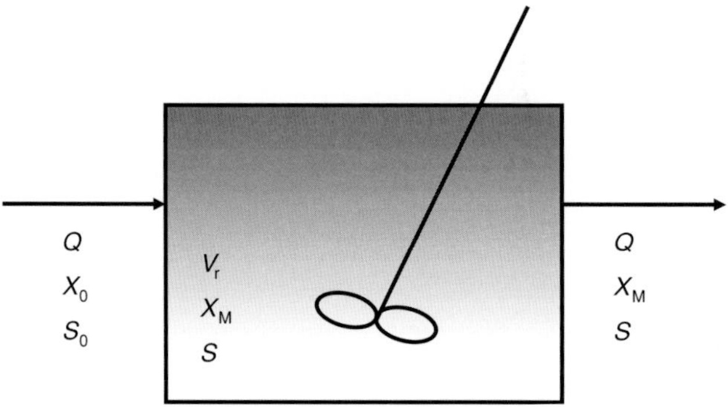

Fig. 5.2. Schematics of a completely mixed bioreactor.

Combining Eqs (5.1) and (5.4), we have:

$$\frac{Q}{V_r} = \frac{1}{\theta_c} = \mu_{max,M}\frac{S}{S + K_{s,M}} - b_M \tag{5.5}$$

From Eqs (5.5) and (5.2), the solids retention time (θ_c) can be calculated as a function of net specific growth rate. Knowing θ_c, the digester volume can be easily calculated:

$$\theta_c = \frac{1}{\mu'_M} \tag{5.6}$$

Empirical Approach

Solids retention time (SRT): The optimum retention time needed for effective digestion can be assessed from laboratory and pilot-scale studies or by evaluation of existing operating plants based on the maximum bioenergy production as a function of SRT. The retention time may vary from 15 to 30 days for mesophilic digestion and from 5 to 15 days for thermophilic digestion. Digester size can be estimated by knowing the volume of waste and residues produced. It is important to note that this approach does not take into consideration the waste characteristics.

Volatile solids (VS) loading rate: VS loading rate is the most commonly adopted approach for sizing an anaerobic digester. This approach does account for feedstock characteristics. The typical VS loading rate for mesophilic digestion is presented in Table 5.2. For a thermophilic digester, the VS loading rate could be twice that of mesophilic conditions.

Volatile solids reduction: VS destruction can be estimated using the following empirical equation (Metcalf and Eddy 2003):

$$V_d = 13.7 \ln(\text{SRT}) + 18.9 \tag{5.7}$$

where V_d is volatile solids destruction (%) and SRT is solids retention time (day).

In Eq. (5.7), the VS reduction correlates to SRT, which can then be used to calculate the digester volume.

Example 5.1

A high-rate anaerobic digester is employed to stabilize primary and secondary sludge under mesophilic conditions. The primary sludge flow rate is 500 m^3/day, with a total solids (TS) content of 5%, 68% of which is volatile. The secondary sludge flow rate is 1,250 m^3/day at a TS content of 1%, 75% of which is volatile. Assume that the specific gravity of the primary and secondary sludge is 1.02 and 1.01, respectively. The minimum SRT of the digester should be 12 days, and the allowable VS loading rate is 2.5 kg VS/m^3· day. How much total VS load (in kg/day) is fed to the digester? Determine the digester volume in m^3.

Solution

(i) Find the mass of sludge produced:
Dry solids (kg/day), $W = V \times \rho \times S \times P$
V = volume of sludge produced (m^3/day)
ρ = density of water (kg/m^3)
S = specific gravity of sludge
P = % of solids expressed in fraction
Primary sludge dry solid mass (kg/day) = 500 × 1,000 × 05 × 1.02 = 25,500 kg/day
Volatile solids in primary sludge (kg VS/day) = 25,500 × 0.68 = 17,340 kg VS/day
Secondary sludge dry solid mass (kg/day) = 1,250 × 1,000 × .01 × 1.01 = 12,625 kg/day
Volatile solids in secondary sludge (kg VS/day) = 12, 635 × 0.75 = 9,469 kg VS/day
Total volatile solids = 17,340 + 9,469 = 26,809 kg VS/day

(ii) Size of digester
Based on SRT:
Total volume of sludge = 500 + 1, 250 = 1,750 m^3/day
Given SRT = 12 days
Digester volume = 1, 750 × 12 = 21,000 m^3
Based on volatile solids loading:
Total volatile solids = 26,809 kg VS/day
Given VS loading = 2.5 kg VS/m^3· day
Digester volume = 26,809/2.5 = 10,724 m^3

Note: Since the volume based on SRT is much higher than that based on VS loading rate, SRT governs the design of digester. Thus by choosing the larger volume, the possibility of biomass washout is minimized.

5.3.2 Anaerobic Contact Process

5.3.2.1 General Description

The anaerobic contact process (ACP) is essentially a CSTR with an external settling tank to settle biomass, as illustrated in Fig. 5.3. The settled biomass is recycled back to maintain long SRT. The degassifier enables removal of biogas bubbles (CO_2 and CH_4) attached to the sludge particles; otherwise the sludge may float to the surface. ACP is particularly useful for high-suspended solids waste streams (e.g.,

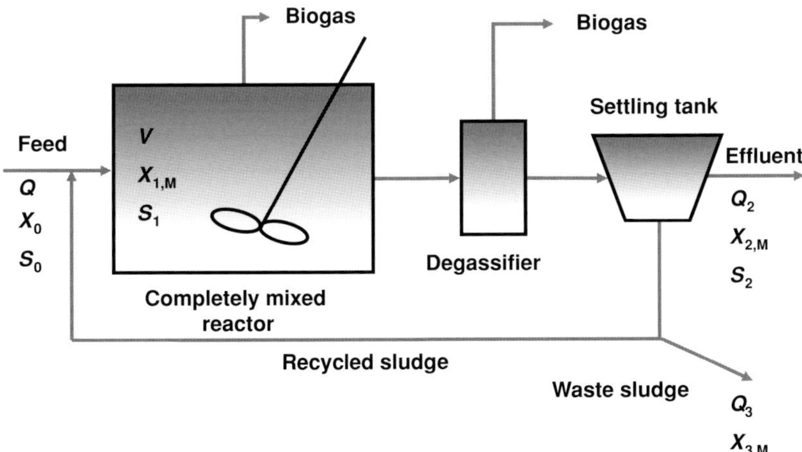

FIG. 5.3. Schematics of anaerobic contact process.

meat-packing waste, wood fiber). Microorganisms are able to attach themselves to particulates, forming settleable solids in the process.

Typical reactor biomass concentrations are 4–6 g/L, with maximum concentrations as high as 25–30 g/L, depending on the settleability of sludge. The loading rate ranges from 0.5 to 10 kg COD/m^3·day. The required SRT can be maintained by controlling the recycle rate similarly to an activated sludge process.

5.3.2.2 Design Considerations

Since ACP is a suspended growth process, the design approach is very similar to that described in Section 5.3.1.2, except that biomass is allowed to settle and is recycled in the process. Therefore, the mass balance of ACP methanogens, as shown in Fig. 5.3, includes both settled effluent and waste sludge in outflow streams.

The mass balance for ACP methane producers is given by (Henze et al. 2002):

$$\mu'_M X_M V_r = Q_2 X_{2,M} + Q_3 X_{3,M} \tag{5.8}$$

From Eqs (5.6) and (5.8), the anaerobic reactor volume is given by:

$$V_r = \frac{Q_2 X_{2,M} + Q_3 X_{3,M}}{X_M} \theta_c \tag{5.9}$$

In Eq. (5.9), θ_c can be obtained from Fig. 5.4, depending on operating temperature; X_M is a parameter that can be selected based on the process; and the term

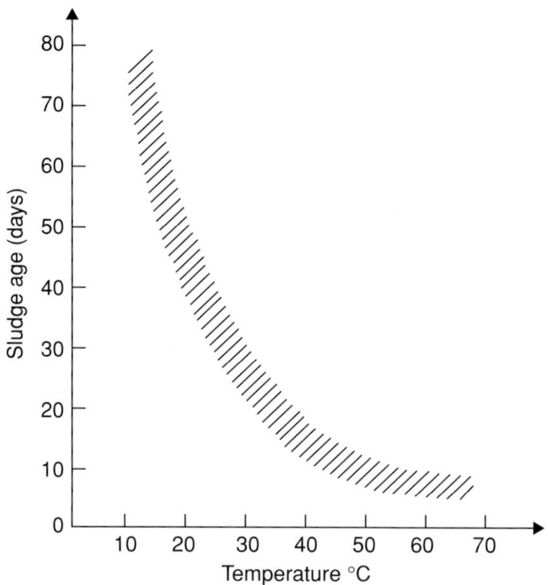

Fig. 5.4. Needed solids retention time at different temperature for anaerobic processes. *Source:* Henze et al. (2002). Reprinted with permission.

$(Q_2X_{2,M}+ Q_3X_{3,M})$ is the biomass (sludge) production rate (P_x), which can be calculated using the following expression:

$$P_x = \frac{Y_{obs}\,Q\,(S_0 - S_2)}{(1 + b_M\theta_c)} \tag{5.10}$$

where Y_{obs} is observed yield in kg biomass-COD/kg COD (0.05–0.10 biomass-COD/kg $COD_{removed}$).

5.3.3 Anaerobic Filter

The first published report describing the application of an anaerobic filter (biofilm) for wastewater treatment was reported by Young and McCarty (1969). The authors evaluated the suitability of an anaerobic filter to treat soluble organic feed. Depending on feeding mode, an anaerobic filter can be classified as an upflow anaerobic filter (UAF), a downflow anaerobic filter (DAF), or a multifed anaerobic filter (MFAF) (Fig. 5.5). Recirculation is generally not recommended for maximum bioenergy recovery. A detailed discussion of each filter type is presented in the following sections.

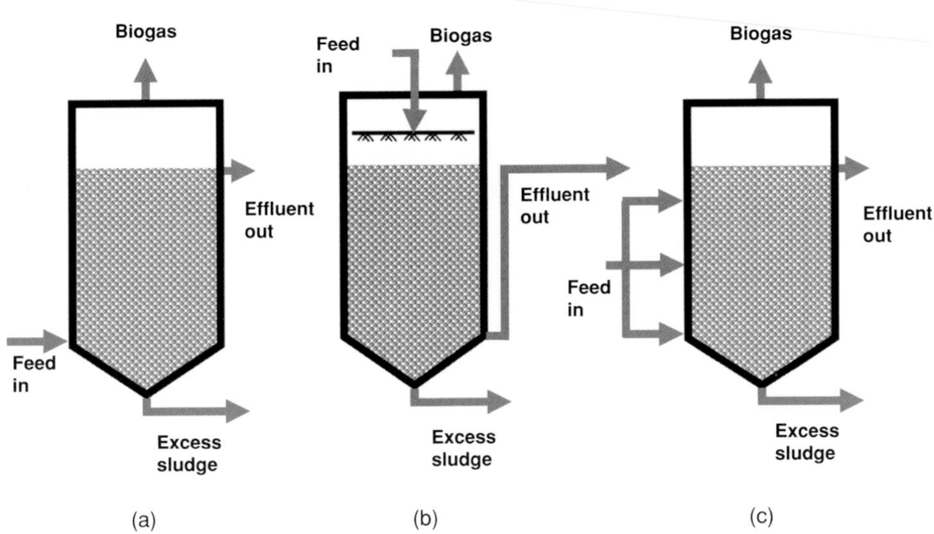

FIG. 5.5. Schematic diagram of anaerobic filter: (a) upflow anaerobic filter, (b) downflow anaerobic filter, and (c) multifed anaerobic filter.

5.3.3.1 Upflow Anaerobic Filter

In a UAF, wastewater is distributed across the bottom and the flow proceeds upward through a bed of rocks or plastic media. The entire filter bed is submerged.

Although UAF is a fixed-film reactor, a significant portion of the biomass remains entrapped within the interstices between the media. The nonattached biomass forms bigger floc and eventually takes a granular shape due to the rolling action of rising gas bubbles. Thus, nonattached biomass contributes significantly to biological activity. Biofilm growth on support media in a UAF is shown in Fig. 5.6.

Originally, rocks were employed as packing media in anaerobic filters. But due to a very low void volume (40–50%), serious clogging problem occurred. In the present day, medium is often synthetic plastic or ceramic tiles of different configurations. The void volume of plastic media ranges from 80 to 95% and provides a high specific surface area, typically 100 m^2/m^3 or higher which enhances biofilm growth. Figure 5.7 shows plastic media of different configurations.

Since an anaerobic filter retains a large amount of biomass, a long SRT can be maintained regardless of HRT. Typically, HRT varies from 0.5 to 4 days and the loading rate varies from 5 to 15 kg $COD/m^3 \cdot$ day. Wasting of biomass may be needed periodically to minimize clogging and short-circuiting. Hydrodynamic conditions play an important role in biomass retention within the void space. The flow regime is often quasi-plug flow.

FIG. 5.6. Biofilm growth on support media.

5.3.3.2 Downflow Anaerobic Filter

A DAF is similar to a UAF except that biomass is truly attached to the media. Loosely held biomass gets washed of the reactor. The specific surface area of media plays a more important role in a DAF than in a UAF. Clogging is less of a problem with a DAF, and it can accommodate feed streams with some suspended solids.

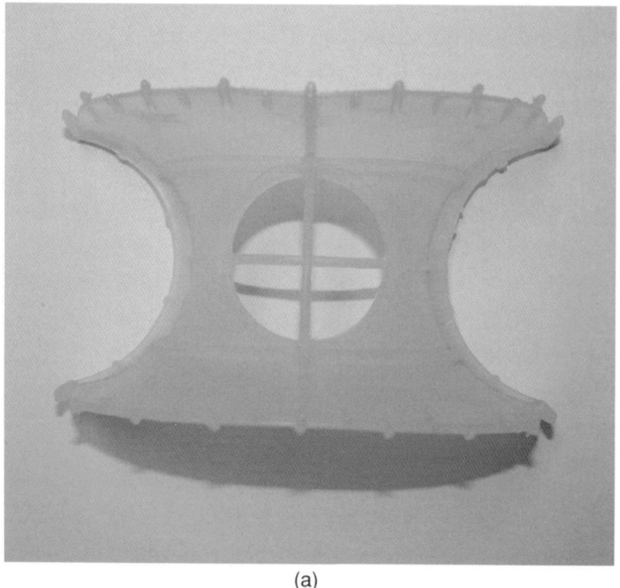

(a)

FIG. 5.7.

(b)

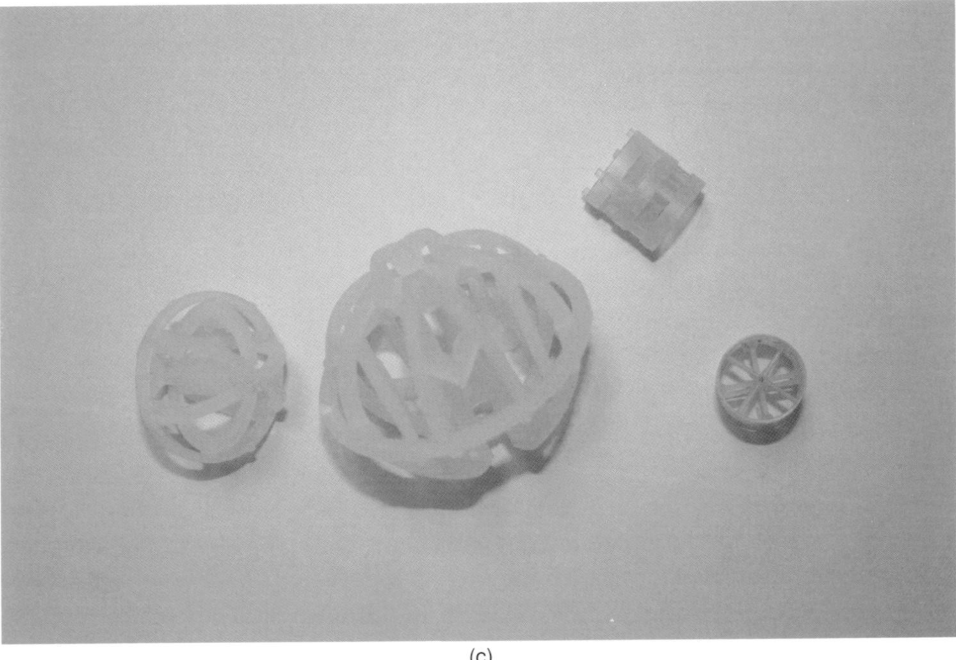

(c)

FIG. 5.7. Plastic media of different configurations.

Although a DAF has a low biomass inventory, the specific activity of its biomass is relatively high. A DAF is more suitable for treating sulfate-rich wastewater, in which sulfate reduction occurs in the upper zone and methanogenesis in the lower zone. This physical separation of two microbial communities reduces the inhibitory effect of sulfide on methanogens.

5.3.3.3 Multifed Anaerobic Filter

In an MFAF, the feed enters the bioreactor through several points along filter depth. The merits of this strategy include (Puñal et al. 1998) the following:

- Homogenous biomass distribution throughout the bed, unlike the stratification of hydrolytic, acidogenic, and methanogenic groups in a single-fed system.
- Maintenance of a completely mixed regime throughout the reactor, thus preventing short-circuiting and accumulation of volatile fatty acids (VFAs).
- Uniform substrate concentration throughout the reactor, which prevents heavy growth of biomass in the bottom of the reactor, thereby minimizing clogging of the filter bed.
- Effective utilization of the whole filter bed, with a working volume of 87%, compared to only 65% for a filter with a single feed point.

5.3.3.4 Design Considerations

Deriving a mathematical expression for the design of an anaerobic filter involves many variables. Importantly, a significant portion of the biomass remains unattached within the void spaces. An empirical approach based on the volumetric organic loading rate is the most common method of designing an anaerobic filter and is given by:

$$\mathrm{VOLR} = \frac{C_0 \cdot Q}{V_r} \tag{5.11}$$

where VOLR is volumetric organic loading rate (kg COD/m^3·day), C_0 is feed COD concentration (kg/m^3 or mg/L), Q is feed flow rate (m^3/day), and V_r is bioreactor volume (m^3).

C_0 and Q are known parameters for a given feed stream. VOLR is obtained based on laboratory and pilot-scale studies that examine maximum bioenergy production (instead of removal efficiency) as a function of loading rate. The reactor volume can be easily determined using Eq. (5.11). If the bioreactor volume does not account for void ratio (ε), the total working volume (V_T) should be determined as follows:

$$V_T = \frac{V_r}{\varepsilon} \tag{5.12}$$

Example 5.2

A UAF has been employed to produce biomethane from distillery slops at 35°C. The influent flow rate is 1,000 m^3/day and soluble chemical oxygen demand (SCOD) concentration is 25 g/L. Based on pilot testing, maximum biogas production was obtained at VOLR of 15 kg/m^3·day.

(i) Design an anaerobic filter process using spherical polypropylene support media. The porosity of the bioreactor with the packing media is 85%.
(ii) Calculate the maximum CH_4 generation rate in m^3/day. What would be the biogas generation rate at 80% COD removal efficiency where 12% of the removed COD is utilized for biomass synthesis and the biogas contains 70% CH_4 by volume?
(iii) What is the total energy that could be generated from CH_4 in kWh?

(i) Anaerobic filter design
From Eq. (5.11), the working volume of anaerobic filter is:

$$V_r = \frac{25\ \text{kg/m}^3 \times 1{,}000\ \text{m}^3/\text{day}}{15\ \text{kg/m}^3/\text{day}} = 1{,}667\ \text{m}^3$$

(*Note:* Assume that the volume accounts for space occupied by packing media.)
Provide four anaerobic filters of 6-m diameter and 15-m height.

(ii) Maximum CH_4 generation and biogas production
Maximum COD removed $= 25$ kg COD/$m^3 \times 1{,}000$ m^3/day $= 25{,}000$ kg/day
At STP, 1 kg COD produces 0.35 m^3 of CH_4
Total CH_4 production at STP $= 0.35 \times 25{,}000 = 8{,}750$ m^3/day
Maximum CH_4 production at 35°C $= 8{,}750 \times ((273 + 35)/273) = 9{,}872$ m^3/day
COD removed at 80% efficiency $= 0.80 \times 25{,}000$ kg/day $= 20{,}000$ kg/day
Since, 12% of removed COD is utilized in biomass synthesis, the COD available for CH_4 production $= 0.88 \times 20{,}000 = 17{,}600$ kg/day
At STP, CH_4 production from the remaining COD $= 0.35 \times 17{,}600 = 6{,}160$ m^3/day
At 35°C, CH_4 production $= 6{,}160 \times (273 + 35)/273 = 6{,}950$ m^3/day
Biogas production with 70% methane $= 6{,}950/0.7 = 9{,}929$ m^3/day

(iii) Total energy generated from CH_4
1 kg COD generates = 1.16 kWh (see Chapter 1)
Available COD for methane generation = 17,600 kg/day
Total energy generation $= 17{,}600 \times 1.16 = 20{,}416$ Kwh/day

5.3.4 Upflow Anaerobic Sludge Blanket

5.3.4.1 General Description

Upflow anaerobic sludge blanket (UASB) was developed in the 1970s by Lettinga and coworkers in the Netherlands to treat sugar-rich soluble wastewater (Lettinga et al. 1980). UASB is essentially a suspended growth system in which proper hydraulic and organic loading conditions are maintained in order to facilitate the formation of dense biomass aggregates known as granules. The diameter of the granules varies from 1 to 3 mm. Granules have superior settling characteristics

(~60 m/h). A UASB reactor is operated at a superficial upflow velocity less than 2 m/h. This way SRT and HRT can be easily decoupled. Thus, an extremely long SRT of 200 days can be achieved at HRT as low as 6 h (Hulshoff Pol et al. 2004). Due to its ability to absorb extremely high VOLR (up to 50 kg COD/m^3·day), UASB is ideally suited for biomethane production from a high-strength soluble feed stream. Detailed discussion on the theory of granulation is presented elsewhere (Hulshoff Pol et al. 2004).

5.3.4.2 Working Principle of UASB Reactor

Feed is uniformly distributed through a specially designed distributor at the bottom of the reactor. The active anaerobic granules in the reaction zone metabolize the organic matter to biogas. Large and dense granules remain suspended within the sludge bed due to superficial upflow velocity and rising biogas bubbles. Granules with entrapped gas enter into the gas–solid separator (GSS), where the gas bubbles get detached as they hit the inclined wall (Fig. 5.8). The granules then slide back into the reactor. The biogas is collected through a gas collection system. The liquid and smaller-size granules enter the settling zone, which is designed in such a way that the superficial upflow velocity decreases significantly as the liquid moves upward (due to a gradual increase in surface area). This facilitates settling of small and light granules back into the reactor. The effluent is collected in a series of weirs placed at the top of the reactor.

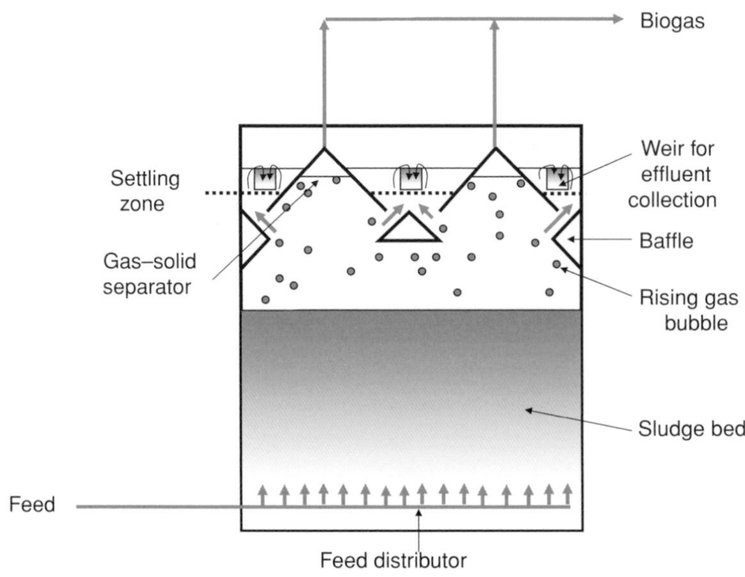

FIG. 5.8. Schematic diagram of upflow anaerobic sludge blanket reactor.

5.3.4.3 Design Considerations

To maximize bioenergy production from high-strength feed streams, the UASB reactor design should be based on the maximum allowable VOLR. Either an empirical or theoretical approach could be used to determine VOLR. The empirical approach employs pilot-scale testing to obtain the optimum VOLR corresponding to maximum methane production, as illustrated in earlier section. The theoretical approach based on specific sludge activity (or specific substrate utilization rate) (U) is discussed here (Lettinga and Hulshoff Pol 1992).

$$\mathrm{VOLR} = X_o\, f_p\, f_o\, S_F\, U \tag{5.13}$$

where X_o is biomass concentration in the reactor (kg/m^3 or mg/L), f_p is contact factor between sludge particles and feed (unitless), f_o is contact factor between substrate and active biomass (unitless), and S_F is safety factor (unitless).

Specific sludge activity (U) is given by:

$$U = \frac{k_{max} S}{K_s + S} \tag{5.14}$$

where k_{max} is maximum specific substrate utilization rate (kg COD/kg biomass·day), S is concentration of growth-limiting substrate (kg/m^3 or mg/L), and K_s is half-velocity constant (kg/m^3 or mg/L).

k_{max} can also be expressed as μ_{max}/Y_{obs}, where Y_{obs} is observed yield coefficient. From Eqs (5.13) and (5.14), we have the following:

$$\mathrm{VOLR} = X_o\, f_p\, f_o\, S_F \frac{k_{max} S}{K_s + S} \tag{5.15}$$

In the above equation, the desired X_o can be maintained by controlling the sludge wastage rate; S is the effluent SCOD concentration that can be set; and k_{max} and K_s are the biokinetic parameters of sludge (granules) that can be obtained from literature or determined experimentally. S_F can be chosen based on the design engineer's experience. The "f" factor is governed by the effectiveness of the feed distribution factor. For VOLR exceeding 5–10 kg COD/m^3·day, the f factor approaches unity (Lettinga and Hulshoff Pol 1992). Furthermore, mixing associated with biogas production at higher VOLR also facilitates better contact between feed and the biomass.

It is important that the superficial velocity (v_a) should be maintained below the washout point of granules. The superficial velocity can be calculated by:

$$v_a = \frac{H}{\theta} \tag{5.16}$$

where H is reactor height (m) and θ is hydraulic retention time (h).

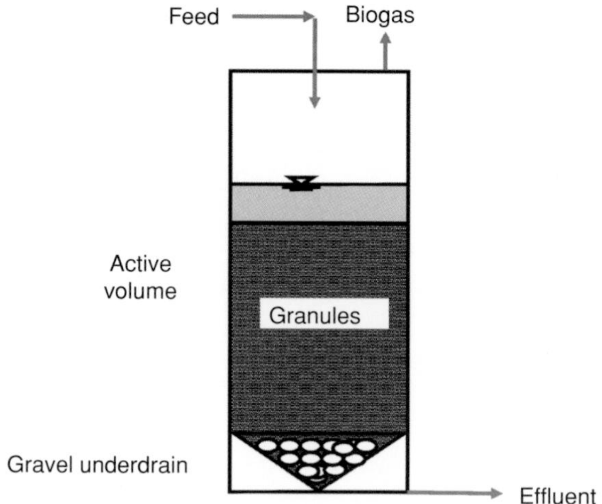

Fig. 5.9. Static granular bed reactor.

A velocity of 5 m/h is recommended for successful operation of a UASB reactor. A higher velocity may also be possible, especially if the granules are large and dense. In a UASB reactor, where a superficial velocity >6 m/h is applied is known as an expanded granular sludge bed (EGSB). The higher velocity helps expand or partially fluidize the sludge bed. Since EGSB was developed to treat low-strength wastewater and wastewater with some particulate matter (Kato et al. 1994), its potential for bioenergy production is limited and is not discussed further.

5.3.5 Static Granular Bed Reactor

Static granular bed reactor (SGBR) is a patented process developed at Iowa State University by Ellis and Mach (Mach 2000). It also employs granules similar to a UASB reactor. SGBR, however, is operated in downflow mode without recirculation as shown in Fig. 5.9. It does not require an elaborate feed distribution system and solid–gas separator. One major concern is head loss due to solids buildup for feed with suspended solids (>2,000 mg/L). SGBR, therefore, is mainly suitable for low-particulate feed streams. Several pilot-scale studies are currently being conducted to examine its applicability for bioenergy production from high-strength feed streams. For scale up from pilot-scale studies, design parameters such as HRT, VOLR, and head loss of the SGBR system need to be examined.

5.3.6 Anaerobic Sequential Batch Reactor

5.3.6.1 General Description

The anaerobic sequential batch reactor (ASBR) was developed by Dague and coworkers at Iowa State University in the early 1990s (Dague et al. 1992). The

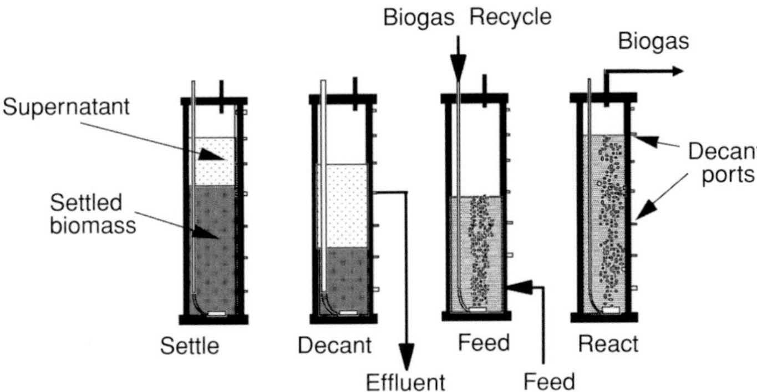

FIG. 5.10. Operating steps for the anaerobic-sequencing batch reactor. *Source:* Dague et al. (1992). Reprinted with permission.

ASBR system was developed as a high-rate anaerobic reactor to treat high-strength and medium solids content feeds (TS: 1–4%). Because of sequential operation, a single reactor can be used as a reaction vessel and as a settling tank, achieving high biomass levels in the reactor regardless of HRT. The ASBR process retains biomass due to bioflocculation followed by biogranulation similar to a UASB reactor. The first full-scale ASBR plant was built in 1997 at Excel Corporation in Ottumwa, Iowa, USA, to treat meat-packing plant wastewater. The ASBR process is highly suitable for bioenergy production from animal manure and other biowastes with medium TS contents.

5.3.6.2 Working Principle of ASBR

As illustrated in Fig. 5.10, the ASBR operation includes four steps: settle, decant, feed, and react. The sequencing frequency and the feed volume processed with each sequence determine the hydraulic loading (HRT) and the strength of the waste establishes the VOLR. The reactor biomass concentration is an important variable affecting biomass settleability. Initial food-to-microorganism (F/M) ratio affects bioflocculation and biochemical reactions.

In a continuously fed, completely mixed reactor operating at steady state, the substrate concentration surrounding the microorganisms is constant. In a batch-fed reactor, the substrate concentration is high immediately after feeding and continuously declines until the reactor is fed again, as illustrated in Fig. 5.11. The substrate concentration just prior to feeding in the batch-fed system is lower than that at any time in the continuously fed system. Thus, the batch-fed system achieves a more efficient biomass flocculation and solids separation than the continuously fed system.

The phenomenon described is one of the key characteristics of the ASBR process. The other key characteristic is that, at any given biomass concentration in the reactor, the F/M ratio is at its highest level immediately after the feed cycle is

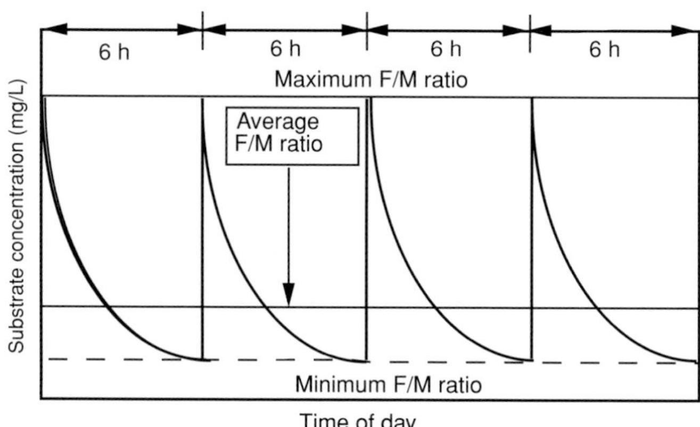

FIG. 5.11. Effect of batch feeding on substrate concentration.
Source: Sung and Dague (1995). Reprinted with permission.

completed. This provides a high driving force for metabolic activity and high overall rates of waste conversion to methane, in accordance with Monod kinetics (Sung and Dague 1995). At the beginning of the settling cycle, the substrate concentration is at its lowest level, resulting in a low rate of internal gassing, an ideal condition for solids separation. In addition, a low F/M ratio in this cycle enhances bioflocculation, which increases the settleability of biomass in the reactor.

5.3.7 Expanded- and Fluidized-Bed Reactor

The expanded-bed reactor (EBR) is an attached growth system with some suspended biomass. The microbes attached on biocarriers such as sand, granular-activated carbon, shredded tire, or other synthetic plastic media are responsible for degradation of organic matter. The biocarriers are expanded by the upflow velocity of feed and through effluent recirculation (Fig. 5.12). In an EBR, sufficient upflow velocity is maintained to expand the bed by 15–30% (Hall 1992). The biocarriers are partly supported by fluid flow and partly by contact with adjacent biocarriers, and they tend to maintain the same relative position within the bed.

The fluidized-bed reactor (FBR) is configured similarly to the EBR. The FBR, however, is truly a fixed-film reactor, as suspended biomass tends to wash out from the reactor due to an extremely high upflow velocity. The bed expansion is 25–300% of the settled bed volume in the FBR. This requires a much higher upflow velocity of 10–25 m/h. The biocarriers are supported entirely by the upflow liquid velocity and therefore are able to move freely in the bed (Fig. 5.12). FBR is free from clogging and short-circuiting problems and provides better substrate diffusion within the biofilm.

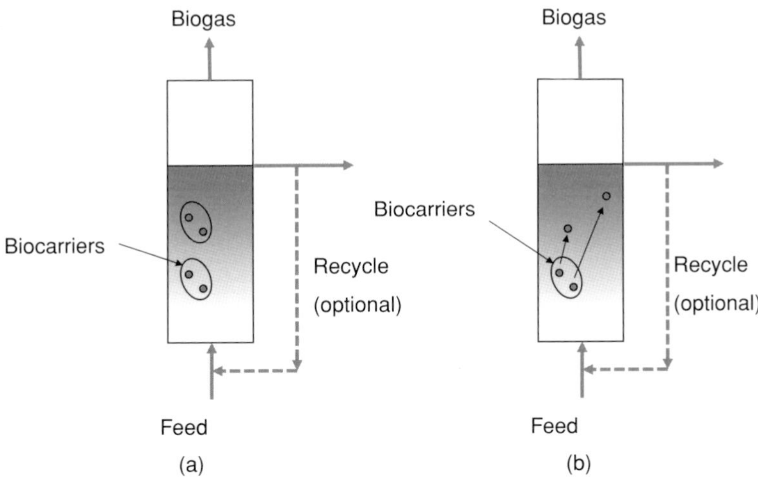

FIG. 5.12. Schematics of (a) expanded bed reactor and (b) fluidized bed reactor.

Although FBR is common in methane fermentation, it has shown superior butanol productivity in comparison to other reactor configurations (batch-fed and CSTR). The butanol production rate of FBR was 4.5–15.8 g/L · h as compared to 0.10–0.38 g/L·h in batch reactors (Qureshi et al. 2005).

5.3.8 Anaerobic Membrane Bioreactor

The anaerobic membrane bioreactor (AnMBR) integrates a membrane unit within the reactor or in an external loop to aid in solid–liquid separation. An AnMBR is capable of retaining biomass and thus can be operated at an extremely long SRT regardless of HRT, which is a prerequisite for successful operation of anaerobic processes. With significant advancements in membrane technology and subsequent reductions in membrane costs in recent years (Visvanathan et al. 2000), membranes have great potential in anaerobic biotechnology for renewable energy generation. This is particularly important for feed streams with high particulate matter (such as stillage from biofuel plants, sludge, and food waste), which traditionally have been digested in a CSTR. This configuration, however, does not decouple SRT and HRT. Thus, integration of a membrane may overcome the drawbacks of CSTR.

In AnMBR, a membrane can be placed in an external loop (Fig. 5.13a) or immersed (submerged) within the reactor (Fig. 5.13b). The former is highly recommended for anaerobic processes as it eliminates the possibility of air intrusion to the reactor during membrane disassembly for cleaning. By controlling cross-flow velocity, cake formation on the membrane surface is minimized. The biomass-rich retentrate is recycled back to the reactor to maintain the needed SRT.

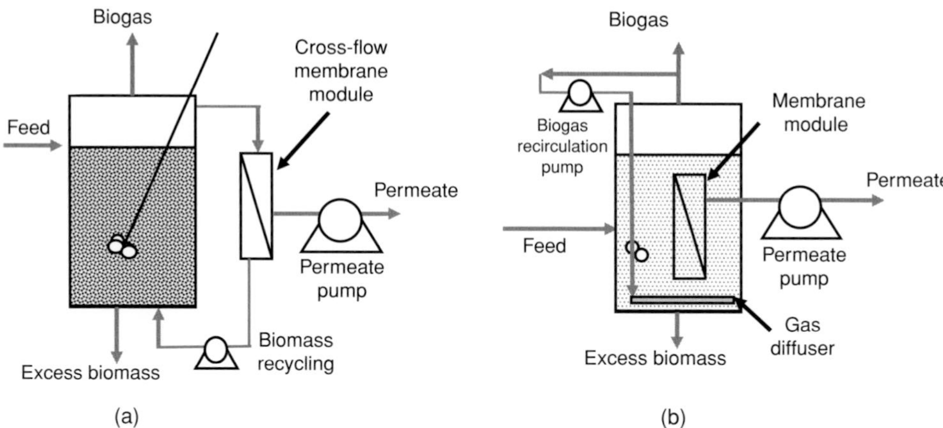

FIG. 5.13. Anaerobic membrane bioreactors: (a) membrane in an external loop and (b) membrane immersed in the reactor.

Coupling the membrane to the anaerobic reactor prevents the washout of slow-growing microorganisms, especially methanogens. Factors affecting membrane performance include membrane types and modules, transmembrane pressure, cross-flow velocity, feed characteristics, biomass levels, and membrane fouling.

5.4 Membrane Technology for Syngas Fermentation to Ethanol

Ethanol can be produced from fermentation of sugar-rich feedstocks. Another approach is through fermentation of gaseous feedstock known as synthesis gas (syngas), which comprises CO and H_2. One major challenge with syngas fermentation is the gas–liquid mass transfer of CO and H_2 due to their poor solubility in an aqueous phase. Different approaches such as high gas and liquid flow rates, large specific gas–liquid interfacial areas, and use of increased pressure or solvents have been examined to enhance the efficiency of gas solubility in liquid phase (Lee and Rittmann 2002). However, the shortcoming of these methods are loss of gaseous feedstock as waste bubbles and difficulties in process control along with unpredictable gas–liquid contact (Ahmed et al. 2004).

The use of hydrophobic hollow-fiber membrane (HFM) can improve transfer of syngas to the close proximity of biofilm (Fig. 5.14). In the HFM, syngas is diffused through the walls of membranes without loss of gas through bubble formation. Besides, this approach offers great flexibility in process control as the gas partial pressure in the membrane lumen and the surface area available for gas transfer can be manipulated independently (Lee and Rittmann 2002). The use of HFM in syngas fermentation has not been reported at the time of writing this chapter. The other applications of HFM, where gas–liquid mass transfer is critical, are discussed here.

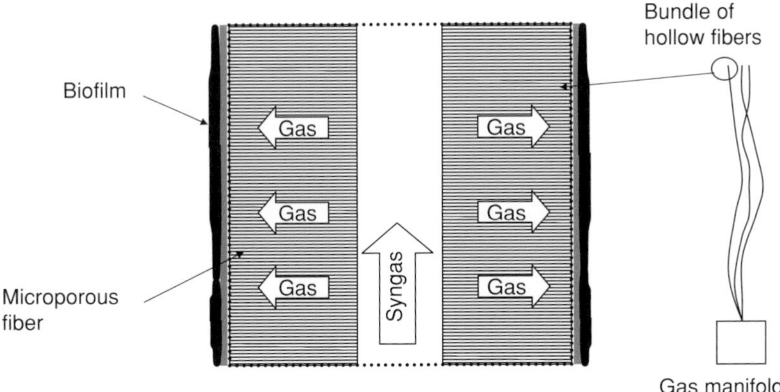

FIG. 5.14. Hollow-fiber membrane system for syngas fermentation.

Voss et al. (1999) employed polydimethylsiloxane (PDMS) coated polypropylene membrane for oxygen transfer. The gas pressure inside the non-porous-coated fibers increased up to 60 psi without bubble formation. Mitsubishi Rayon Corporation developed composite membranes that contain a nonporous, gas-permeable polyurethane layer (1 mm thick) embedded into the membrane wall as illustrated in Fig. 5.15. The polyurethane layer allows bubbleless operation at very high oxygen feed pressures similar to PDMS-coated membranes with a 100% oxygen transfer efficiency (Ahmed et al. 2004).

A hollow-fiber membrane bioreactor was developed to use hydrogen gas as an electron donor (Lee and Rittmann 2002; Nerenberg and Rittmann 2004). A bundle of sealed-end hollow fibers were housed in a hydrogen gas dissolution unit, and biofilm was developed on the surface of the hollow fiber. The bioreactor allowed 100% hydrogen transfer into the biofilm and achieved effective removal of nitrate, perchlorate, bromate, chlorate, chlorite, chromate, selenate, selenite, and dichloromethane.

The feasibility study of using HFM for syngas fermentation is currently being evaluated at the author's laboratory at the University of Hawaii.

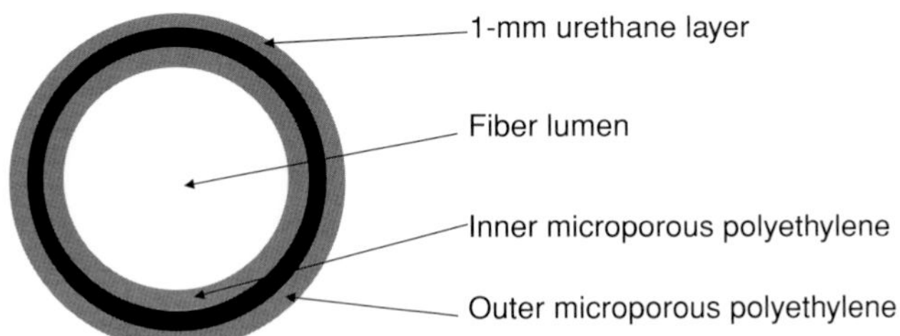

FIG. 5.15. Cross section of a polyethylene hollow fiber composite membrane with 1-mm polyurethane layer in the membrane wall.
Source: Ahmed et al. (2004). Reprinted with permission.

References

Ahmed, T., Semmens, M. J., and Voss, M. A. 2004. Oxygen transfer characteristics of hollow-fiber, composite membranes. *Adv. Environ. Res.* 8:637–646.

Dague, R. R., Habben, C. E., and Pidaparti, S. R. 1992. Intial studies on anaerobic sequential batch reactor. *Water. Sci. Technol.* 26(9–11):2429–2432.

Hall, E. R. 1992. Anaerobic treatment of wastewaters in suspended growth and fixed film processes. In *Design of Anaerobic Processes for the Treatment of Industrial and Municipal Wastes*, edited by J. F. Malina, Jr, and F. G. Pohland, pp. 1–40. Technomic Publishing Co. Inc., Lancaster, USA.

Henze, M., Harremoës, P., Jansen, J. C., and Arvin, E. 2002. *Wastewater Treatment: Biological and Chemical Processes*, 3rd edn, pp. 303–304. Springer-Verlag., Berlin, Germany.

Hulshoff Pol, L. W., Lopes, C. I. S., Lettinga, G., and Lens, L. N. P. 2004. Anaerobic sludge granulation. *Water. Res.* 38:1376–1389.

Kato, M. T., Field, J. A., Versteeg, P., and Lettinga, G. 1994. Feasibility of expanded granular sludge bed reactors for the anaerobic treatment of low-strength soluble wastewaters. *Biotechnol. Bioeng.* 44:469–479.

Lee, K.-C., and Rittmann, B. E. 2002. Applying a novel autohydrogenotrophic hollow-fiber membrane biofilm reactor for denitrification of drinking water. *Water Res.* 36:2040–2052.

Lettinga. G., Velsen, A. F. M., Hobma, S. W., de Zeeuw, W., and Klapwijk, A. 1980. Use of the upflow sludge blanket (USB) reactor concept for biological wastewater treatment, especially for anaerobic treatment. *Biotechn. Bioeng.* XXII:699–734.

Lettinga, L., and Hulshoff Pol, L. W. 1992. UASB design for various types of wastewaters. In *Design of Anaerobic Processes for the Treatment of Industrial and Municipal Wastes*, edited by J. F. Malina, Jr., and F. G. Pohland, pp. 119–145. Technomic Publishing Co. Inc., Lancaster, USA.

Mach, K. F. 2000. *Development of Static Granular Bed Reactor.* M.S. thesis, Iowa State University, Ames, IA, USA.

Manual of Practice (MOP) # 16 1987. *Anaerobic Sludge Digestion*, 2nd edn. Water Pollution Control Federation, Alexandria, VA, USA.

Metcalf and Eddy 2003. *Wastewater Engineering*: *Treatment and Reuse*, 4th edn. McGraw-Hill Companies, Inc., New York, USA.

Nerenberg, R., and Rittmann, B. E. 2004. Hydrogen-based, hollow-fiber membrane biofilm reactor for reduction of perchlorate and other oxidized contamitants. *Water Sci. Technol.* 49(11–12):223–230.

Puñal, A., Ménal, R., and Lema, J. M. 1998. Multi-fed upflow anaerobic filter: Development and features. *J. Environ. Eng. ASCE* 124(12):1188–1192.

Qureshi, N., Annous,B. A., Ezeji, T. C., Karcher, P., and Maddox, I. S. 2005. Biofilm reactors for industrial bioconversion processes: Employing potential of enhanced reaction rates. *Microb. Cell Fact.* 4:24. Available at: http://www.pubmedcentral.nih.gov/picrender.fcgi?artid=1236956&blobtype=pdf (accessed December 16, 2007).

Sung, S., and Dague, R. R. 1995. Laboratory studies on the anaerobic sequencing batch reactor. *Water Environ. Res.* 67(3):294–301.

Visvanathan, C., Ben Aim, R., and Parameshwaran, K. 2000. Membrane separation bioreactors for wastewater treatment. *Crit. Rev. Environ. Sci. Technol.* 30(1):1–48.

Voss, M. A., Ahmed, T., and Semmens, M. J. 1999. Long term performance of parallel flow bubbleless hollow fiber membrane aerator. *Water Environ. Res.* 71(1):23–30.

Young, J. C., and McCarty, P. L. 1969. The anaerobic filter for waste treatment. *J. WPCF* 41(5):R160–R173.

CHAPTER

6

Molecular Techniques in Anaerobic Biotechnology: Application in Bioenergy Generation

Srisuda Dhamwichukorn

6.1 Background

Anaerobic microorganisms can serve as cell factories for converting biomass to biofuel and bioenergy, and to optimize this function, it is necessary to closely examine the microbial communities responsible for bioenergy production. Traditional methods of microbial analysis include culture-type techniques such as plating and counting, microscopic enumeration, biochemical analysis, and immunolabeling. These techniques, however, cannot provide detailed information about the structure, function, dynamics, and diversity of microbial communities. Molecular techniques that can measure cell DNA (deoxyribonucleic acid) and RNA (ribonucleic acid) are more direct and robust. Not including viruses, all cells contain DNA that can be monitored at any point in an organism's existence, and in fact even after the cell has expired. Molecular techniques can characterize genes, proteins (enzymes), and metabolic products associated with bioenergy production at extremely low concentrations without difficulties of culturability. Molecular analysis of microbial cells and communities can, therefore, furnish useful information about structure (who they are), function (what they do), and dynamics (how they change through space and time). This chapter focuses on the application of molecular techniques for microbial bioenergy conversion within anaerobic biotechnology.

6.2 Molecular Techniques in Anaerobic Biotechnology

As molecular techniques evolved in recent decades, applications for anaerobic biotechnology and bioenergy production (e.g., methane, hydrogen, butanol, and

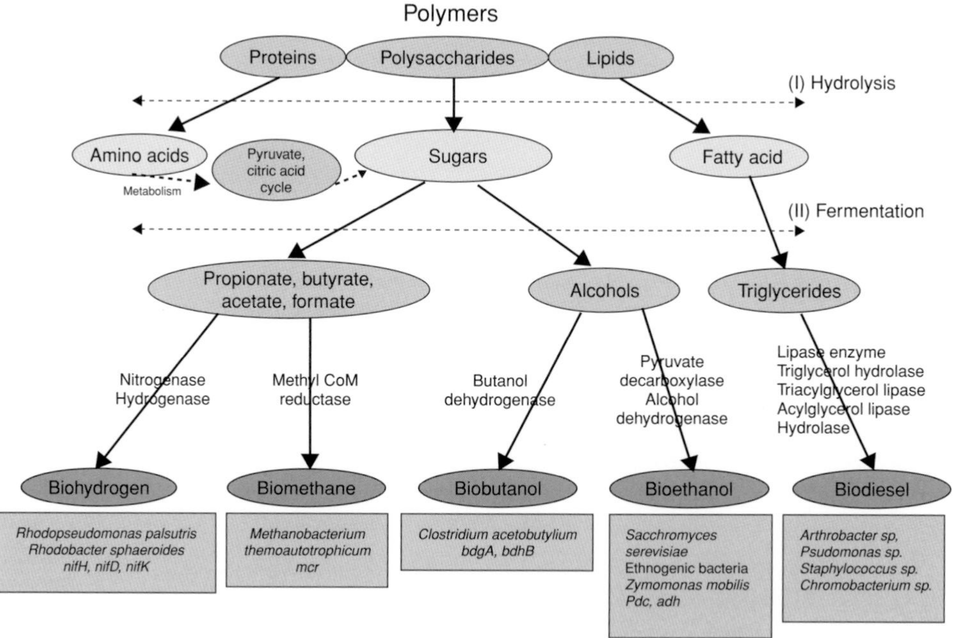

Fig. 6.1. Integrated bioconversion system for bioenergy production.

ethanol) have been developed to identify and quantify the responsible microbial communities.

All aspects of the biomass-to-bioenergy conversion process illustrated in Fig. 6.1 can be evaluated using molecular techniques. Anaerobes including hydrogen producers, clostridia, methanogens, sulfate reducers, and yeasts carry out the nonoxidative metabolism. Many bioenergy production systems using a nonoxidative metabolism similar to that shown in Fig. 6.1 are not functionally stable and often result in low yields. Numerous anaerobic studies using molecular techniques have demonstrated that characterization of the microbial community structure, function, and dynamics aids in better system design, process control, and stability. Successful systems also are those that successfully eliminate inhibiting microbial species while maintaining productive microbial species.

Concepts of using molecular techniques in anaerobic biotechnology and its application in bioenergy production, methane, hydrogen, butanol, and ethanol, are described in the following sections.

6.3 Fundamentals of Molecular Techniques

Molecular techniques are often used to examine genes–DNA sequences or DNA fragments encoded for certain cell functions such as cell structure and cell function for converting biomass to bioenergy. Culture-independent techniques that are commonly used include polymerase chain reaction (PCR), reverse transcription PCR (RT-PCR), denaturing gradient gel electrophoresis (DGGE), and clone-sequencing to target 16S rRNA structural and functional genes. Sequence and phylogenic analyses are often associated with the aforementioned techniques. PCR can be used as a tool for DNA replication and to selectively amplify a target gene sequence relating to bioenergy production. It is possible to amplify genes or DNA of the bacteria and archaeabacteria, and enhance key functional groups that are needed to assess changes in microbial community structure and abundance, prior to further analysis or characterization. Information is extracted from cell DNA by amplifying (PCR) and characterizing the DNA fragments. After amplification and characterization, the microbial community structure within the samples can be identified. Likewise, to obtain information from mRNA or protein, the mRNA fragments are amplified and characterized or the activity of the protein is monitored. The information gained should reveal the microbial community function and dynamics, for example, what genes are turned on or functioning (gene expression).

Replication or amplification of DNA by PCR requires the following:

1. DNA with target sequence
2. Two DNA primers
3. Nucleotide bases (A, T, G, C)
4. DNA polymerase enzyme
5. Salts, buffers, and a thermocycler

Amplification of specific sequences from complex mixtures traditionally has required the use of two specific and conserved primer sequences that correspond to structural (16s rRNA) or functional (e.g., nitrogenase (*nif*) for hydrogen production or methyl CoM reductase (*mcr*) for methane production) genes.

Using 16S rRNA gene sequences containing portions of both conserved and variable regions demonstrate the potential benefits for microbial diversity and community analysis. The highly conservative regions facilitate PCR amplification using only one universal primer set to amplify the 16S rRNA gene sequences for all eubacteria from environmental samples. The variable regions are different for each bacterial species, allowing accurate bacterial detection and identification. From these characteristics, the 16S rRNA genes help to determine the genetic diversity of a microbial community as well as to assist in identifying the phylogenetic affiliation of its members.

6.4 Phylogenetic Analysis

Phylogenetic tree analyses are used to taxonomically classify microbial gene sequences. These compare the DNA sequences or related 16S rDNA sequences of selected microbial groups documented in the National Center for Biotechnology Information nucleotide sequence databases and in Ribosomal Database Project by

Table 6.1. Enzymes and genes associated with bioenergy production.

Types of Bioenergy	Process Involved	Enzymes	Genes	Organisms
1. Methane (hydrogen as an intermediate)	Methanogenesis	Methyl CoM reductase	*mcr*	Anaerobic methanogens
2. Methane (acetate as an intermediate)	Methanogenesis	Methyl CoM reductase	*mcr*	Acetoclastic methanogens
3. Methane (formate as an intermediate)	Methanogenesis	Methyl CoM reductase	*mcr*	Methanogens
4. Methane (methanol as an intermediate)	Methanogenesis	Methyl CoM reductase	*mcr*	Methanogens
5. Hydrogen (acetic acid as an intermediate)	Dark fermentation	Hydrogenase	*hya, hup, hox*	Anaerobic bacteria, e.g., *Enterobactericeae, Bacillus,* and *Clostridium*
6. Hydrogen (butylic acid as an intermediate)	Dark fermentation	Hydrogenase	*hya, hup, hox*	Anaerobic bacteria, e.g., *Enterobacteria, Bacillus,* and *Clostridium*
7. Hydrogen	Photo fermentation	Nitrogenase	*nifH, nifD, nifK*	Nonsulfur photosynthetic bacteria, *Cyanobacteria*
8. Butanol	Fermentation	Butanol dehydrogenase	*bdh*	*Clostridium acetobutylicum* and *Clostridium beijerinckii*
9. Ethanol	Fermentation	Pyruvate decarboxylase, alcohol dehydrogenase	*pdc, adh*	*Saccharomyce* sp., *Kluyveromyce* sp., *Zymomonas mobilis, Clostridium* sp., *Thermoanaerobacter ethanolicus, Zymobacter palmae*

neighbor-joining and unweighted pair group method with arithmetic mean (UPGMA) (average linkage clustering) methods contained in the PHYLIP 3.5c software package (Altschul et al. 1990; Felsenstein 1989; Zhang and Madden 1997).

The enzymes and genes associated with nonoxidative metabolic bioenergy production are listed in Table 6.1.

6.5 Molecular Techniques for Microbial Community Structure Analysis: DNA Fingerprinting, Clone Library, and Fluorescent In Situ Hybridization

Bacterial abundance and diversity in anaerobic processes have been studied using various microbiological approaches. Traditional culturing approaches using counting techniques only provide a precursory estimate of bacterial diversity, disregarding the vast majority of unculturable bacteria because of limitations inherent to the

Table 6.2. Applications of molecular techniques for microbial community analysis in bioenergy production processes.

Types of Bioenergy	Molecular Techniques Utilized	Major Findings	References
Methane	PCR-DGGE and FISH	Microbial community changes with operational time	Diaz et al. (2006) and Zheng et al. (2006)
Methane	SSCP, qPCR, and FISH	Significant effect of volatile fatty acids concentrations on methanogenic community	Hori et al. (2006)
Hydrogen	PCR-DGGE, PCR-RISA, cloning, and T-RFLP	Identify dominant species and contaminant	O-thong et al. (2007)
Butanol	Genomic sequencing, recombinant DNA technology	Improvement in productivity	Ezeji et al. (2007)
Ethanol	Recombinant DNA technology, DNA microarrays	Improvement in productivity	Rogers et al. (2007) and Sedlak et al. (2003)

growth media. In contrast, molecular methods have the potential to provide a more complete assessment of microbial diversity because they provide a means to detect nucleic acids in microbial samples that may not be culturable.

6.5.1 DNA Fingerprinting

DNA fingerprinting techniques include denaturing gradient gel electrophoresis (DGGE), temperature gradient gel electrophoresis (TGGE), and terminal restriction fragment length polymorphism (T-RFLP). Examples of microbial community structure obtained from various molecular techniques are presented in Table 6.2. These data clearly show that PCR-DGGE has been employed more often than other fingerprinting techniques (Table 6.3), and this chapter subsequently focuses on DGGE.

6.5.1.1 Denaturing Gradient Gel Electrophoresis Analyses

DGGE is becoming the most widely employed molecular fingerprinting technique for investigating microbial diversity in anaerobic studies. The DGGE technique is usually accompanied by PCR-amplified partial 16S rDNA fragments, which requires conventional DNA extraction and PCR amplification of 16S rRNA gene.

DNA fragments of the same length but with different sequences can be separated using DGGE based on partial melting of double-stranded DNA molecules in a polyacrylamide gel containing a linear gradient of denaturant (a mixture of urea and formamide). As the DNA molecule begins to melt, its molecular volume increases, impeding its migration into the gel. Two molecules of DNA of the same length but differing sequences melt at different concentrations of denaturant. The molecule melting at a lower denaturant concentration does not migrate as far, thereby electrophoretically separating the molecules.

By using DGGE, 50% of the sequence variants can be detected in DNA fragments up to 500 base pair (bp). Detection can be increased to nearly 100% by the attachment of a GC-rich region (a GC-clamp) to one side of the DNA fragment. This is accomplished by adding a sequence of guanine and cytosine bases to the 5′ end of one of the PCR primers. The GC-rich region of the amplified DNA molecules acts as a high denaturant melting domain, preventing dissociation of DNA into single strands (Muyzer et al. 1993).

Since broad groups of microorganisms (i.e., multiple genera) are often targeted with each primer pair, substantial sequence variation in the amplified 16S rDNA is observed. The GC-clamp technology is used to improve separation of sequence variants, making it possible to effectively compare the microbial communities.

Major bands in the DGGE gels are excised, DNA extracted and reamplified by PCR, and rerun on a denaturing gradient gel to verify the purity of the PCR reamplification product. Usually, only those bands accounting for over 5% of the total fluorescent intensity of a lane are sequenced. Although the theoretical maximum number of bands sequenced from each lane is 20, the DGGE analysis of soil microbial communities has sequenced 5–10 bands, which generally produces the necessary fluorescence. PCR reamplification products are purified with a PCR purification kit. Sequencing is performed using an express DNA sequencer. DNA fragments with unlike sequences migrate differently in a DGGE gel, forming bands with unique positions relative to the loading wells. All major bands occupying unique positions are then sequenced. Positions and intensities of DGGE gel bands are recorded using an electrophoresis documentation and analysis system (Jena et al. 2006).

Numerous microbial diversity studies that have used DGGE to assess anaerobic bioenergy production have been conducted, with the outcome of an improved understanding of microbial diversity being the result (Muyzer et al. 1993).

6.5.1.2 Temperature Gradient Gel Electrophoresis Analyses

TGGE is very similar to DGGE, using the same concepts of DNA separation by differential melting of GC-rich segments in the amplified DNA molecules. Unlike DGGE, in which DNA is melted by urea salt, TGGE melts DNA with elevated temperature, as the name implies (Gilbride et al. 2006).

6.5.1.3 Terminal Restriction Fragment Length Polymorphism Analyses

T-RFLP is a useful and rapid method of comparing rDNA (rRNA genes) from bacterial community structures. The process begins with extraction of DNA from the microbial communities. Before DNA amplification, fluorescent tags are added to one or both primers to label the PCR products (amplicons). The PCR products are then digested by a restriction enzyme, and the fragments are separated by sequencing gel, a DNA sequencer or capillary electrophoresis (automated DNA sequencing instrument). Only the labeled terminal fragments are detected by the instrument and generate profiles unique to the microbial community (Liu et al. 1997, 1998). T-RFLP has been successfully used in community analysis of anaerobic methane and hydrogen production systems.

6.5.2 Clone Library

This technique follows extraction of DNA from the microbial community with PCR amplification of 16s rRNA genes. The PCR products of the 16s rRNA genes undergo a cloning reaction that inserts gene sequences into the synthetic plasmids (circular DNA containing specific DNA). The 16s rRNA genes of various species within plasmids are transferred into *Escherichia coli* cells. The colonies containing the plasmid DNA (i.e., 16s rRNA gene sequences) are then plated, selected, and analyzed. The plasmid DNA is then sequenced. The obtained DNA sequences are identified and characterized using phylogenetic analysis through the use of software and existing databases. Clone library identification of 16S rRNA genes has been used to determine the predominant microorganisms in biohydrogen and biomethane production (Sanz and Köchling 2007).

6.5.3 Fluorescent In Situ Hybridization

Fluorescent in situ hybridization (FISH) is a useful molecular technique to quantitatively identify a microbial community. A fluorescence probe specifically hybridizes its complementary target nucleic acid sequence and determines the microbial population within the community. FISH with 16S ribosomal RNA-targeted oligonucleotide probes can be used to observe the spatial distribution of archaeal and bacterial cells in biological systems (Sanz and Köchling 2007).

Examples of the various molecular techniques used for microbial community analysis in bioenergy production (e.g., methane, hydrogen, butanol, and ethanol) through nonoxidative metabolism are illustrated in Table 6.2.

6.6 Molecular Techniques for Functional Analysis

Using molecular techniques to understand and characterize the dominant species in a microbial community facilitates control of a microbial-based bioenergy production process. Equally important to defining the community structure is an understanding of function within the microbial community. It is important to examine temporal and spatial factors (detection of the expression of genes) and how the detected genes work (gene expression). It is possible to investigate the functional ability of genes through the detection of mRNA, in the forms of cDNA, which can be obtained from the reverse transcription-polymerase chain reaction (RT-PCR) of mRNA. For global gene expression or gene expression profiling, microarray techniques are employed.

6.6.1 Quantitative Real-Time PCR

Quantitative real-time PCR (qPCR) is used to quantify genes, which assists in the examination of microbial structure (using 16S rRNA genes) and function (using functional genes encoded for enzymes or functions). This technique has been used to examine microbial-mediated processes, for example, methane oxidation, sulfate reduction, and ammonium oxidation, to quantify gene expression in environmental samples.

6.6.2 Stable Isotope Probing

Stable isotope probing (SIP) has recently been developed and applied to recover the labeled DNA fraction from methane and methanol assimilation, ammonia oxidation, and anaerobic sludges. The technique allows for functional detection of microbial groups. In addition, microautoradiography (MAR), based on microscopic detection of microorganisms consuming radiolabeled biomass or substrates, can be used to measure the metabolic activity of members of diverse and complex microbial communities. MAR can be utilized concurrently with FISH (Lee et al. 1999; Wagner and Loy 2002).

6.6.3 Microarrays

DNA microarrays have been used almost exclusively for the study of genome-wide expression or transcriptional profiling within various organisms. Yet, due to their ability to utilize biological reactions in order to detect specific analytical targets (based on nucleic acid hybridization or recognition), microarrays can also be used as biosensors. Biosensors are fast, sensitive, and selective devices that use biological recognition to detect target analytes that provide either qualitative or quantitative results. The exploitation of microarrays as biosensors for bacterial detection and identification purposes, targeting either structural (e.g., 16S rRNA) or functional genes, has begun to produce noteworthy results. Successful studies have focused on cyanobacteria (Burja et al. 2003), methanogens, and many others. Microarrays have been developed to enhance understanding of microbial communities in regard to structure, function, and dynamics to enhance bioenergy production (Burja et al. 2003; Dhamwichukorn et al. 2004; Liu and Zhu 2005; Zhou 2003).

6.7 Nucleic Acid Extraction of Anaerobic Cells/Isolates and Sludge

Nucleic acid from anaerobic cells, biomass, or sludge can be extracted for further PCR amplification (DNA) or RT-PCR (mRNA), similar to soil and sediment samples. DNA/RNA extraction is a critical step in the sample preparation of all molecular techniques. Different extraction methods provide variable results during molecular analysis. There are many commercial kits available such as BIO101, Qiagen DNA Stool Mini DNA Extraction Kit, MoBio Ultraclean Soil DNA extraction Kit, MoBio Power Soil DNA extraction Kit, Epicentre, and the QBiogene FastDNA Spin Kit for Soil. Anaerobic sludge and cell DNA and RNA can be extracted by using one of these kits, the best choice being dependent on sample characteristics.

6.8 Molecular Techniques for Structure and Function Analysis

Concurrent utilization of multiple molecular techniques is a common practice for studying microbial community structure, function, and dynamics. For example, a combination of FISH and microautoradiography (FISH-MAR), FISH-SIP, and microarrays can be used to evaluate microbial structure and function. Figure 6.2 illustrates how different molecular techniques can be used to analyze microbial community structure, function, and metabolic transformation separately or together. FISH, DGGE, ribosomal intergenic spacer analysis (RISA), T-RFLP, clone libraries, DGGE, and T-RFLP are the monitoring techniques that have been used most frequently for microbial community structure analysis, with DGGE and T-RFLP being the procedures most commonly used concurrently with other techniques.

Table 6.3 summarizes information generated by molecular techniques for methanogens, fermentative hydrogen producers, butanol producers, and ethanol producers through nonoxidative metabolisms.

Table 6.4 presents a summary of currently available molecular approaches, including microarrays for microbial diversity and community analysis. Among the

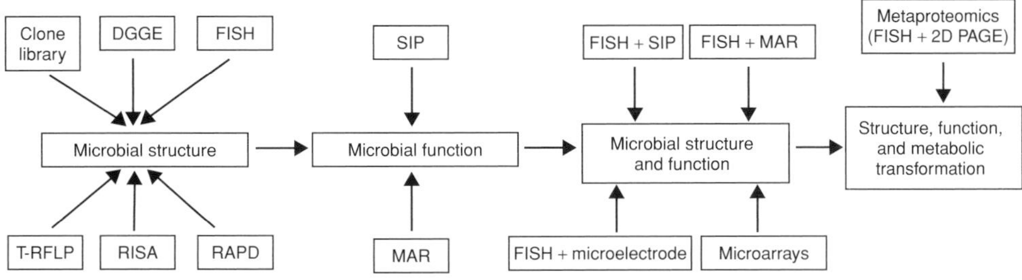

Fig. 6.2. Conceptualization of the role of currently available molecular biology and postgenomic techniques for analysis of microbial community structure, function, and metabolic transformation.

Table 6.3. Microbial findings of methane, hydrogen, butanol, and ethanol producers using molecular techniques.

Substrate/Product	Reactor Type	Dominated Microorganisms	Molecular Techniques Utilized	References
Brewery wastewater/methane	Upflow anaerobic sludge bed reactor (UASB)	*Deferribacteres*	Clone library	Diaz et al. (2006)
Sewage sludge/methane	Anaerobic sludge digester	*Euryarchaeota*	Clone library	Chouari et al. (2005)
Wastewater/methane	UASB, expanded granular sludge bed anaerobic filter (EGSB-AF), internal circulation reactor (IC)	*Mehanosaeta crenarchaeota*	FISH and clone library	Collins et al. (2005)
Municipal wastewater/methane	UASB	*Methanosaeta concilii*	FISH and PCR-DGGE	Zheng et al. (2006)
Synthetic wastewater/methane	Continuous stirred tank reactor (CSTR)	*Methanoculleus* sp., *Methanothermobacter* sp., *Methanosarcina* sp.	SSCP, qPCR, and FISH	Hori et al. (2006)
Swine manure/methane	Anaerobic membrane reactor (AnMBR)	*Methanosarcinaceae* sp., *Methanosaetaceae* sp., *Methanomicrobiales* sp.	T-RFLP	Padmasiri et al. (2007)
Olive mill solid wastes/methane	CSTR	*Methanosaeta concilii*	PCR-DGGE	Rincón et al. (2006)
Oleate-based effluent/methane	Anaerobic expanded granular sludge bed reactor (EGSB)	*Methanobacterium formicicum* DSM 1535^{NT}, *Syntrophomonas zehnderi* sp. nov.	PCR-DGGE	Souza et al. (2007)
Starch in wastewater/hydrogen	Batch	*Thermoanaerobacteriaceae, Saccharococcus* sp. clone ETV-T2.	Clone library FISH	Zhang et al. (2003)
Cellulose/hydrogen	Batch	*Thermoanaerobacterium, Clostridium thermoamylolyticum*	PCR-DGGE	Liu et al. (2003)
Carbohydrate-containing wastewater/hydrogen	Two-step process using dark fermentative CSTR and phototrophic bacteria (complete-mix cylindrical photoreactor)	Clostridia, *Rhodobacter capsulatus*	PCR-DGGE	Liu (2003)
Glucose/hydrogen	Anaerobic membrane reactor (AnMBR)	Clostridiaceae, Flexibacteracae, *C. acidisoli, Linmingia china, Cytophaga*	PCR-RISA	Oh et al. (2004a, b)

Food waste/hydrogen	CSTR	*Thermoanaerobacterium thermosaccharolytium, Desulfotomaculum geothermicum*	PCR-DGGE	Shin et al. (2004)
Food waste/hydrogen	CSTR	*Thermotogales, Bacillus* sp., *Prevotella* sp.	PCR-DGGE	Shin et al. (2004)
Sucrose/hydrogen	CSTR	*Clostridium* sp., *Bacillus* sp.	T-RFLP	Sung (2004)
Sucrose/hydrogen	CSTR	*Clostridium tyrobutyricum, Chryseobacterium ptoteolyticum, Clostridium acidisoli*	PCR-DGGE	Lin and Lay (2004)
Sucrose/hydrogen	CSTR	*Clostridium ramosum*	PCR-DGGE	Lin et al. (2006)
Rice slurry/hydrogen	Batch	*Clostridium* sp. 44a-T5zd	Clone library PCR-DGGE	Fang et al. (2005)
Glucose/hydrogen	Trickling biofilter reactor	*Thermoanaerobacterium thermosaccharolyticum*	PCR-DGGE	Ahn et al. 2005
Glucose/hydrogen	Batch	*Clostridium butyricum*	PCR-DGGE	Xing et al. (2005)
Palm oil mill effluent/hydrogen	Anaerobic-sequencing batch reactor	*Thermoanaerobacterium thermosaccharolyticum*	PCR-DGGE	O-Thong et al. (2007)
Liquefied cornstarch/butanol, ethanol	Batch	*Clostridium beijerinckii*	Genomic sequencing, recombinant DNA technology	Ezeji et al. (2007)
Carbohydrate-based feedstocks/ethanol	Batch	*Zymomonas mobilis*	Recombinant DNA technology, DNA microarrays	Rogers et al. (2007)
Lignocellulosic biomass/ethanol	Batch	*Saccharomyces* yeasts	DNA microarrays	Sedlak et al. (2003)

Table 6.4. Summary of molecular technologies for microbial diversity and community analysis.

Technology/ Approach	Nucleic Acid Extraction	PCR	Pattern Analysis/ Finger Printing	Advantages	Disadvantages	References
Fluorescence in situ hybridization (FISH)	No	No	NA	• Direct analysis and quantification • No heterogeneity problem • Suitable for targeting specific group/species	• Requires genes/RNA with high number of copies • Limit for total diversity mapping	Sanz and Köchling (2007)
Microarrays	Yes	No/yes	NA	• Suitable for mapping total diversity/pattern analysis • Further sequencing not required to identify group/species • High sensitivity • Fast and high throughput • Not laborious	• Do not reveal unexpected species, except using more probes	Burja et al. (2003), Dahllöf (2002), and Zhou (2003)
Denaturing gradient gel electrophoresis (DGGE)	Yes	Yes	Melting	• Suitable for pattern analysis or microbial community diversity • Further sequencing not required, except using a nonheterogeneous gene	• Can produce multiple bands from one template with bias PCR of heterogeneous genes (e.g., 16S) • Lacks resolution • Needs sequencing for species • May have PCR bias problems • Cannot actually indicate species richness	Dahllöf (2002), Jena et al. (2006), O-Thong et al. (2007), and Sanz and Köchling (2007)
Temperature gradient gel electrophoresis (TGGE)	Yes	Yes	Melting	• Suitable for pattern analysis	• Lacks resolution • Needs further sequencing • May have PCR bias problems • Cannot actually indicate species richness	Dahllöf (2002)
Terminal restriction fragment length polymorphism (T-RFLP)	Yes	Yes	Size	• Suitable for pattern analysis • One band per species • High resolution • Sensitive detection	• Restriction enzyme selection with unknown sequences • May have PCR bias problems • Cannot actually indicate species richness • Some species may show the same length of fragments	Dahllöf (2002)

Cloning and sequencing	Yes	Yes	NA	• Contains larger sequences • Provides more positive ID	• A large number of clones must be sequenced for positive diversity • Bias from PCR • Sequences need to compare with each other and libraries • Laborious and time consuming	Dahllöf (2002), Gilbride et al. (2006), and Sanz and Köchling (2007)
Amplified rDNA restriction anal (ARDRA)	Yes	Yes	Size	• Suitable for screening clone libraries or isolates before sequencing	• Not suitable for pattern analysis • Produces multiple bands for one species • Cannot actually indicate species richness	Dahllöf (2002)
Restriction fragment length polymorphism (RFLP)	Yes	Yes	Size	• Suitable for screening clone libraries or isolates before sequencing	• Not suitable for pattern analysis • Produces multiple bands for one species	Dahllöf (2002)
Single-stranded conformation polymorphism (SSCP)	Yes	Yes	Size	• Suitable for screening clone libraries or isolates before sequencing	• Suitable for pattern analysis • Produces multiple bands for one species	Dahllöf (2002)

several recent molecular techniques that have been used for microbial identification and detection, microarray is one of the most advanced and sophisticated techniques.

6.9 Postgenomic Approaches for Bioenergy Research

Although the success of the human genome sequencing projects has been most newsworthy, many important bacterial genomes related to bioenergy have also been sequenced successfully in recent years. The postgenomic technologies, that is, bioinformatics (computational biology), transcriptional profile studies (microarrays), translational study (proteomics), and metabolic pathway and products (metabolomics), and the potential benefits they could produce, have been acknowledged as cutting-edge technologies for future biofuel production.

Investigation of microbial biocomplexity as it relates to bioenergy production is possible with powerful new high-throughput monitoring of genes, proteins, and metabolites (Fig. 6.3). Understanding the complex interactions of organisms and biological systems can extend beyond the success of the genomic sequencing of organisms. In order to further expand and accelerate the accomplishments of these multidisciplinary research areas, interdisciplinary collaborative research is essential. The bacterial groups mentioned in this chapter are keys to carbon sequestration and bioenergy production. Examples of the global benefits produced by these approaches and the application of bacteria and yeasts are depicted in Fig. 6.4 (the illustration created by the U.S. Department of Energy (DOE) Genome Program's Genome Management Information System).

Postgenomics technologies

- Bioinformatics (computational biology, *in silico*)
- Microarrays (transcriptional profile studies)
- Proteomics (translational studies)
- Metabolomics (metabolic pathway and products)

The central dogma of molecular biology

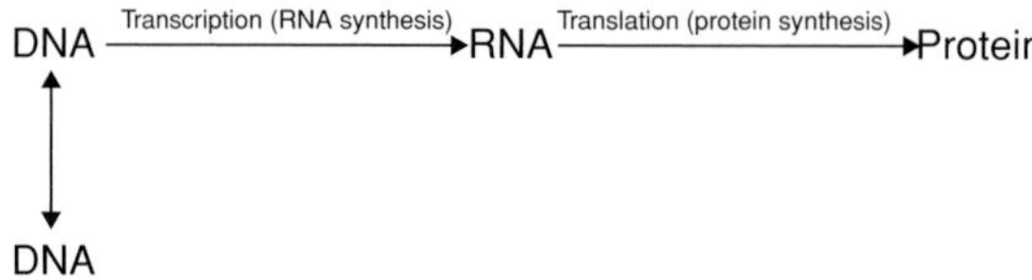

FIG. 6.3. Analogy of postgenomics technologies to the central dogma of molecular biology.

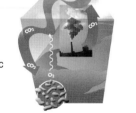

DOE GENOMICS:GTL
SYSTEMS BIOLOGY FOR ENERGY AND ENVIRONMENT
OFFICE OF SCIENCE
U.S. DEPARTMENT OF ENERGY

Payoffs for the Nation,
Grand Challenges for Biology
Develop biofuels as a major source for this century
Increase biological sources of fuels and electricity
• Efficient conversion of plant cellulose to ethanol
Contribute to U.S. energy security
• Biohydrogen-based industry in place

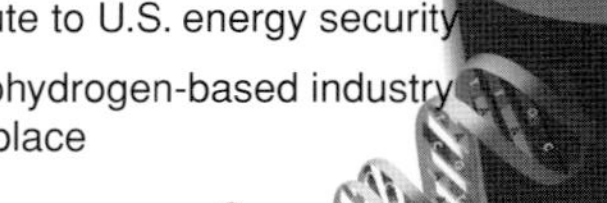

Bioreactor

Develop biology solutions for intractable environmental problems
Develop knowledge base for cost-effective cleanup strategies
Save billions of dollars in toxic waste cleanup and disposal

Understand biosystems climate impacts and assess sequestration strategies
Understand earth's natural carbon cycle and design strategies for enhanced carbon capture
Help stabilize atmospheric carbon dioxide to counter globel warming
YGG-01-0521R10

FIG. 6.4. DOE mission challenges require groundbreaking research for innovative solutions. The earth's microbial systems are the foundation for life and potential source of capabilities that we can put to use to these challenges; their study forms the core of the GTL program.
Source: Genomics: GTL Roadmap, U.S. Department of Energy Office of Science, August 2005, http://genomicsgtl.energy.gov/roadmap; The U.S. Department of Energy Genome Programs, http://genomics.energy.gov/gallery/gtl/detail.np/detail-06.html.

References

Ahn, Y., Park, E. J., Oh, Y. K., Park, S., Webster, G., and Weightman, A. J. 2005. Biofilm microbial community of a thermophilic trickling biofilter used for continuous biohydrogen production. *FEMS Microbiol. Lett.* 249(1):31–38.

Altschul, S. F., Gish, W., Miller, W., Myers, E. W., and Lipman, D. J. 1990. Basic local alignment search tool. *J. Mol. Biol.* 215:403–410.

Burja, A. M., Dhamwichukorn, S., and Wright, P. C. 2003. Cyanobacterial postgenomic research and systems biology. *Trends Biotechnol.* 21:504–511.

Chouari, R., Le Paslier, D., Dauga, C., Daegelen, P., Weissenbach, J., and Sghir, A. 2005. Novel major bacterial candidate division within a municipal anaerobic sludge digester. *Appl. Environ. Microbiol.* 71:2145–2153.

Collins, G., O'Connor, L., Mahony, T., Gieseke, A., de Beer, D., and O'Flaherty, V. 2005. Distribution, localization, and phylogeny of abundant populations of *Crenarchaeota* in anaerobic granular sludge. *Appl. Environ. Microbiol.* 71:7523–7527.

Dahllöf, I. 2002. Molecular community analysis of microbial diversity. *Curr. Opin. Biotechnol.* 13:213–217.

Dhamwichukorn, S., Bodrossy, L., and Kulpa, C. F. 2004. *Oligonucleotide Microarrays for Monitoring of Methane and Methanol Degrading Methylotrophs*, pp. 805–811. Proceeding in the JGSEE and Kyoto University Joint International Conference on Sustainable Energy and Environment. December 1–3, 2004, Hua Hin, Thailand.

Diaz, E. E., Stams, A. J. A., Amils, R., and Sanz, J. L. 2006. Phenotypic properties and microbial diversity of methanogenic granules from a full-scale upflow anaerobic sludge bed reactor treating brewery wastewater. *Appl. Environ. Microbiol.* 72:4942–4949.

Ezeji, T. C., Qureshi, N., and Blaschek, H. P. 2007. Bioproduction of butanol from biomass: From genes to bioreactors. *Curr. Opin Biotechnol.* 18:220–227.

Fang, H. H. P., Li, C., and Zhang, T. 2005. Acidophilic biohydrogen production from rice slurry. *Int. J. Hydrogen Energy.* 31(6):683–692.

Felsenstein, J. 1989. PHYLIP, phylogeny inference package (version 3.2). *Cladistics* 5:164–166.

Gilbride, K. A., Lee, D. Y., and Beaudette, L. A. 2006. Molecular techniques in wastewater: Understanding microbial communities, detecting pathogens, and real-time process control. *J. Microbiol. Methods* 66:1–20.

Hori, T., Haruta, S., Ueno, Y., Ishii, M., and Igarashi, Y. 2006. Dynamic transition of a methanogenic population in response to the concentration of volatile fatty acids in a thermophilic anaerobic digester. *Appl. Environ. Microbiol.* 72:1623–1630.

Jena, S., Jeanmeure, L. F. C., Dhamwichukorn, S., and Wright, P. C. 2006. Carbon substrate utilization profile of a high concentration effluent degrading microbial consortium. *Environ. Technol.* 27:863–873.

Lee, N., Nielsen, P. H., Andreasen, K. H., Juretschko, S., Nielsen, J. L., Schleifer, K., and Wagner, M. 1999. Combination of fluorescent in situ hybridization and microautoradiography—a new tool for structure-function analyses in microbial ecology. *Appl. Environ. Microbiol.* 65(3):1289–1297.

Lin, C. Y., and Lay, C. H. 2004. Carbon/nitrogen ratio effect on fermentative hydrogen production by mixed microflora. *Int. J. Hydrogen Energy* 29:41–45.

Lin, C. Y., Lee, C. Y., Tseng, I. C., and Shiao, I. Z. 2006. Biohydrogen production from sucrose using base enriched anaerobic mixed microflora. *Process. Biochem.* 41(4):915–919.

Li, H. 2003. Bio-hydrogen production from carbohydrate-containing wastewater. In *Civil Engineering*, pp. 113, University of Hong Kong.

Liu, H., Zhang, T., and Fang, H. H. P. 2003. Thermophilic hydrogen production from a cellulose containing wastewater. *Biotechnol. Lett.* 25:365–369.

Liu, W. T., Marsh, T. L., Cheng, H., and Forney, L. J. 1997. Characterization of microbial diversity by determining terminal restriction fragment length polymorphisms of genes encoding 16S rRNA. *Appl. Environ. Microbiol.* 63:4516–4522.

Liu, W. T., Marsh, T. L., and Forney, L. J. 1998. Determination of the microbial diversity of anaerobic-aerobic activated sludge by a novel molecular biological technique. *Water Sci. Technol.* 37:417–422.

Liu, W. T., and Zhu, L. 2005. Environmental microbiology-on-a-chip and its future impacts. *Trends Biotechnol.* 23(4):174–179.

Muyzer, G., de Waal, E. C., and Uitterlinden, A. G. 1993. Profiling of complex microbial populations by denaturing gradient gel electrophoresis analysis of polymerase chain reaction-amplified genes coding for 16S rRNA. *Appl. Environ. Microbiol.* 59:695–700.

Oh, S. E., Lyer, P., Bruns, M. A., and Logan B. E. 2004a. Biological hydrogen production using a membrane bioreactor. *Biotechnol. Bioeng.* 87(1):119–127.

Oh, Y. K., Kim, S. H., Kim, M. S., and Park, S. 2004b. Thermophilic biohydrogen production from glucose with trickling biofilter. *Biotechnol. Bioeng.* 88:690–698.

O-Thong, S., Prasertsan, P., Intrasungkha, N., Dhamwichukorn, S., and Birkeland, N. K. 2007. Improvement of biohydrogen production and treatment efficiency on palm oil mill effluent with nutrient supplementation at thermophilic condition using an anaerobic sequencing batch reactor. *Enzyme Microb. Technol.* 41:583–590.

Padmasiri, S. I., Zhang, J., Fitch, M., Norddahl, B., Morgenroth, E., and Raskin, L. 2007. Methanogenic population dynamics and performance of an anaerobic membrane bioreactor (AnMBR) treating swine manure under high shear conditions. *Water Res.* 41:134–144.

Rincón, B., Raposo, F., Borja, R., Gonzalez, J. M., Portillo, M. C., and Saiz-Jiminez, C. 2006. Performance and microbial communities of a continuous stirred tank anaerobic reactor treating two-phases olive mill solid wastes at low organic loading rates. *J. Biotechnol.* 121:534–543.

Rogers, P. L., Jeon, Y. L., Lee, K. J., and Lawford, H. G. 2007. *Zymomonas mobilis* for fuel ethanol and higher value products. *Adv. Biochem. Eng. Biotechnol.* 108:263–288.

Sanz, J. L., and Köchling, T. 2007. Molecular biology techniques used in wastewater treatment: An overview. *Process Biochem.* 42:119–133.

Sedlak, M., Edenberg, H. J., and Ho, N. W. Y. 2003. DNA microarray analysis of the expression of the genes encoding the major enzymes in ethanol production during glucose and xylose co-fermentation by metabolically engineered *Saccharomyces* yeast. *Enzyme Microb. Technol.* 33:19–28.

Shin, H. S., Youn, J. H., and Kim, S. H. 2004. Hydrogen production from food waste in anaerobic mesophilic and thermophilic acidogenesis. *Int. J. Hydrogen Energy.* 29(13):1355–1363.

Souza, D. Z., Smidt, H., Alves, M. M., and Stams, A. J. M. 2007. *Syntrophomonas zehnderi* sp. Nov., an anaerobe that degrades long-chain fatty acids in co-culture with *Methanobacterium formicicum. Int. J. Syst. Evol. Microbiol.* 57:609–615.

Sung, S. 2004. *Biohydrogen Production from Renewable Organics Wastes.* Technical Report (DE-FC36–00G010530). Iowa State University.

Wagner, M., and Loy, A. 2002. Bacterial community composition and function in sewage treatment systems. *Curr. Opin. Biotechnol.* 13:218–227.

Xing, D. F., Ren, N. Q., Gong, M. L., Li, J., and Li, Q. 2005. Monitoring of microbial community structure and succession in the biohydrogen production reactor by denaturing gradient gel electrophoresis (DGGE). *Sci. China C-Life Sci.* 48:155–162.

Zhang, J., and Madden, T. L. 1997. PowerBLAST: A new network BLAST application for interactive or automated sequence analysis and annotation. *Genome Res.* 7:649–656.

Zhang, T., Liu, H., and Fang, H. H. P. 2003. Biohydrogen production from starch in wastewater under thermophilic condition. *J. Environ. Manage.* 69(2):149–156.

Zheng, D., Angenent, L. T., and Raskin, L. 2006. Monitoring granule formation in anaerobic upflow bioreactors using oligonucleotide hybridization probes. *Biotechnol. Bioeng.* 94:458–472.

Zhou, J. 2003. Microarrays for bacterial detection and microbial community analysis. *Curr. Opin. Microbiol.* 6:288–294.

CHAPTER

7

Bioenergy Recovery from Sulfate-Rich Waste Streams and Strategies for Sulfide Removal

Samir Kumar Khanal

7.1 Background

Anaerobic processes are gaining popularity for both treatment and recovery of bioenergy from high-strength industrial waste streams. Some industrial effluents, however, are not amenable to anaerobic treatment, such as those from pulp and paper, molasses fermentation, seafood processing, potato-starch, tanneries, edible oil refineries, pharmaceutical and petrochemical production, and wine distillery slops. These waste streams contain elevated concentrations of sulfate and/or sulfide. Existing sulfide, and that produced by reduction of sulfate, is toxic to the organisms responsible for methanogenesis, a major pathway for methane production. Upset of methanogenesis due to sulfide toxicity often leads to process failure. Hydrogen sulfide also produces objectionable odors, causes corrosion, is toxic to humans, and adulterates the produced biogas by reducing its energy potential.

This chapter provides background information on anaerobic treatment of sulfate-rich waste streams with an emphasis on bioenergy production, sulfide toxicity, and suitable strategies such as selection of reactor design and operating conditions for treating high-sulfate wastewater. Various methods for sulfide removal from both aqueous and gaseous phases are also covered.

7.2 Sulfate-Reducing Bacteria

Sulfate-reducing bacteria (SRB) are strict anaerobic prokaryotes first discovered by Beijerinck in 1895 (Postgate 1984). SRB characteristically generate odorous hydrogen sulfide that can react with ferrous iron to form black iron sulfide. More importantly, SRB have many physiological and ecological similarities with methane-producing bacteria (MPB) (Smith 1993). The simplest sulfate-reducing reaction is illustrated by Eq. (7.1):

$$SO_4^{2-} + \text{organic matter} \rightarrow HS^- + H_2O + HCO_3^- \tag{7.1}$$

In the absence of sulfate, certain strains of SRB may act as fermentative bacteria, using lactate, ethanol, pyruvate, choline, malate, glycerol, and dihydroxyacetone as both electron donors and acceptors (Barton and Tomei 1995). They may also act as hydrogen-producing acetogens using lactate or ethanol in the presence of hydrogen-consuming MPB (Bryant et al. 1977). In the presence of sulfate, some SRB may oxidize volatile fatty acids (VFAs) (e.g., propionic and butyric acid) completely to CO_2 and generate sulfide during the process. Others may break down VFAs incompletely to acetate, also with sulfide production (Lens et al. 1998). In fact, SRB are able to metabolize a wider range of organic compounds when compared to MPB. SRB are especially effective in competing with MPB for acetate and hydrogen in the presence of sulfate. Sulfide generated from this metabolism is toxic to MPB. The degradation pathway of organic matter in the presence of sulfate is illustrated in Fig. 7.1.

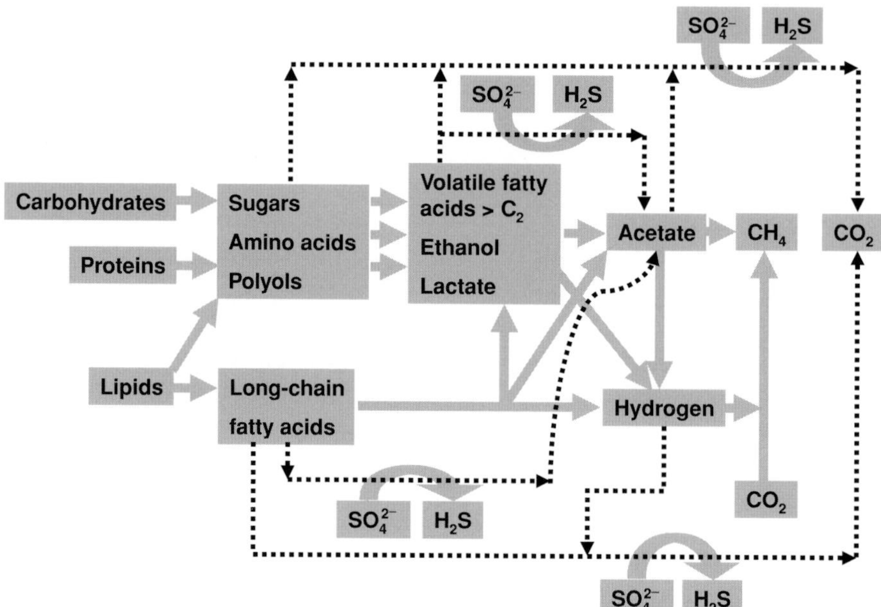

FIG. 7.1. Methane production pathway in presence of sulfate.
Source: Lens et al. (1998).

Table 7.1. Typical characteristics of high-sulfate industrial wastewater.

Wastewater Source	COD (mg/L)	Sulfate (mg/L)	References
Molasses fermentation	44,800–55,600	2,500–3,460	Carrondo et al. (1983)
Seafood processing	10,000–50,000	600–2,700	Mendez et al. (1995)
Potato-starch factory	17,500–18,000	317	Nanninga and Gottschal (1986)
Tannery	2,900–8,200	750–1,250	Genschow et al. (1996)
Distillery slops	95,000	6,000	Szendry (1983)
Edible oil refinery	1,010–8,200	3,100–7,400	Anderson et al. (1988)
Pharmaceutical plant	28,540	14,800	Mohanroa et al. (1970)
Citric acid	30,000	4,500	Svardal et al. (1993)

7.3 High-Strength Sulfate-Rich Wastewater

High-organic-strength sulfate wastewater is generated primarily by the agri- and food processing industries. Such waste is produced through the use of sulfuric acid, either to control pH in industrial processing or to extract starch from potatoes to recover protein (Nanninga and Gottschal 1986). Because of the high concentration of biodegradable organic matter in these types of wastes, anaerobic treatment is an economically attractive option, and it also provides an opportunity for bioenergy recovery in the form of methane gas. Some typical high-strength sulfate wastewaters are presented in Table 7.1.

Pulp and paper mills are other important sources of concentrated organic waste containing sulfate, sulfite, and other sulfur compounds, depending on the pulping processes employed. Thermomechanical pulping produces wastewater typically with chemical oxygen demand (COD) concentrations of 2,000–5,000 mg/L and sulfate concentrations of 200–700 mg/L (Habets and de Vegt 1991; Rintala et al. 1991). Chemothermomechanical pulping wastewater usually has a COD of 7,500–10,400 mg/L, with sulfide from 50 to 200 mg/L and sulfate from 1,200 to 1,500 mg/L (Habets and de Vegt 1991). The cooking liquor from a sodium-based sulfite pulp mill contains a mean COD of 20,000 mg/L, and its mean sulfate and sulfite concentrations are 1,050 and 1,125 mg/L, respectively (Särner 1990).

Acid mine drainage contains 1,980 mg/L of sulfate accompanied by low COD concentrations of less than 100 mg/L (Maree and Strydom 1985). Available carbon is barely sufficient for sulfate reduction. Such a waste stream is not suitable for bioenergy generation and is not considered further in this chapter.

7.4 Methane Recovery from High-Strength Sulfate-Laden Wastewater

For wastewater with a COD/SO_4 ratio of 0.67 (or $COD/S = 2.0$), organic carbon concentrations are just sufficient enough to biologically reduce the sulfate completely without external carbon supplementation. For wastewater with $COD/SO_4 \gg 0.67$, there is sufficient residual carbon for methane fermentation. Both MPB and SRB are able to degrade organic matter. Degradation of organic matter through sulfate-reducing pathways, however, has many ill effects, and it is actually more beneficial to completely eliminate SRB activity and allow only MPB to convert the organic matter into bioenergy. Two approaches for selective suppression of sulfate-reducing activities are discussed as follows:

Use of chemical inhibitors. Molybdate has been reported to selectively suppress sulfate-reducing activity (Anderson et al. 1988). A long-term application of this inhibitor in continuous operation, however, showed that SRB could develop resistance to it, thereby requiring higher doses (Lens et al. 1998). Karhadkar et al. (1987) also reported that at 20 mM of molybdate, both sulfate reduction and methanogenesis were inhibited. Similarly, transition metals (Cu, Zn, Mn, Co, Ni, and Cd) and antibiotics could also selectively suppress the sulfate-reducing activity.

Control of operating conditions. Hilton and Oleszkiewicz (1988) proposed a novel approach to enhance carbon flow toward the methanogenic pathway in the presence of sulfate. They reported that operating an anaerobic reactor at pH greater than 8.0 would cause a self-inhibition of SRB due to high total sulfides (>400 mg/L), thereby allowing MPB to utilize the carbon without competition.

In practical applications, it is difficult to achieve selective inhibition of sulfidogenesis, and both methanogenesis and sulfidogenesis continue to proceed during anaerobic treatment. Two types of anaerobic systems (single-stage and two-stage) can be employed for effective treatment of such waste streams.

7.4.1 Single-Stage Anaerobic System

In a single-stage system, both methanogenesis and sulfate reduction occur in the same reactor. The anaerobic biodegradation of organic matter in the presence of sulfate is shown in Fig. 7.1, and it is apparent from this illustration that acetate and hydrogen can be utilized by both SRB and MPB. Thus, the presence of sulfate will reduce the methane yield due to substrate diversion (Winfrey and Zeikus 1977). Furthermore, sulfide generated from sulfate reduction is toxic to MPB (Khanal and Huang 2006). It is also noteworthy that certain SRB species utilize long- and short-chain fatty acids, giving this group a competitive advantage over MPB.

7.4.2 Two-Stage Anaerobic System

A two-stage treatment system concept was originally proposed to treat high-solid organic waste, with the first stage used for acidification and the second stage for methanogenesis (Pohland and Ghosh 1971). A similar approach was then applied to the treatment of sulfate-rich wastewater. Because SRB can utilize a wide range of organic substrates (Hansen 1993) and tolerate a low pH with a high free H_2S (Mizuno et al. 1998), they function well in an acidogenic phase for sulfate reduction. Thus in a two-stage system with sulfate reduction occurring in the acidogenic (first) stage and methanogenesis in the second stage, sulfide toxicity to MPB can be greatly reduced. Moreover, sulfate reduction in the acidogenic phase generates the favorable intermediate, acetic acid, which augments methanogenesis by reducing carbon loss (Reis et al. 1991).

Särner (1990) developed an anaerobic trickling filter as an acidogenic reactor for maximum sulfite and/or sulfate reduction for the treatment of cooking liquor from a sodium-based sulfite pulp mill. The author reported inorganic sulfur (sulfite and sulfate) removal exceeding 85%. Mizuno et al. (1998) also conducted a series of chemostat studies using sucrose as a carbon source at various sulfate levels and hydraulic retention times (HRTs) to evaluate the role of SRB in the acidogenic phase. They observed sulfate reduction even at HRTs as short as 2 h and with sulfate levels of up to 2,400 mg/L. The authors proposed that the major role of SRB in the acidogenic phase with high sulfate concentrations and low HRTs could be the scavenging of hydrogen. An increase of either sulfate or HRT did not significantly affect acetate generation.

7.5 Important Considerations in Treatment and Methane Recovery from High-Strength Sulfate-Laden Wastewater

The presence of sulfate poses several challenges for both treatment and bioenergy recovery. Some important factors are discussed here.

7.5.1 Sulfide Toxicity to MPB

Several studies have reported impairment of methanogenic activity by sulfide. MPB impairment decreases with sulfur speciation in the following order: H_2S >total sulfide > sulfite > thiosulfite > sulfate (Khan and Trottier 1978). Table 7.2 summarizes the literature data on sulfide toxicity to methanogenic activity. The data show that a wide range of sulfide concentrations can deteriorate the process. Sulfide toxicity can be dependent on many factors including operating pH, the type of treatment system (attached or suspended growth system), substrate types (simple or complex substrate), experimental run time, and degree of acclimation (see Box 7.1).

Box 7.1

Mechanism of Sulfide Inhibition

The inhibitory effect of sulfide is mainly due to the unionized sulfide (H_2S) since only the neutral molecule can penetrate the cell membrane. Once across the cell wall, H_2S denatures native proteins or key metabolic enzymes through the formation of sulfide and disulfide cross-links between the polypeptide chains (Lens et al. 1998). Parkin et al. (1990) reported, however, that sulfide inhibition could be reversible. H_2S may also interfere with the assimilatory metabolism of sulfur (Vogels et al. 1988). Winfrey and Zeikus (1977) proposed the possibility of metal precipitation or other secondary effects responsible for the suppression of methanogenesis. This is because some trace metals like Fe, Ni, and Co are extremely important in anaerobic treatment (Callander and Barford 1983). Any precipitation of trace metals with sulfide will limit their availability to microorganisms. Isa et al. (1986a, b) proposed a similar explanation for the poor activity of MPB in the presence of sulfide.

7.5.2 Operating pH

Unionized sulfide is the most toxic form of sulfide for anaerobes, and its concentration is pH dependent. As operating pH varies from one study to another, so does the inhibitory level of sulfide. Aqueous sulfide is a weak acid illustrated by Eqs (7.2) and (7.3) as follows:

$$H_2S_{(aq)} \overset{K_1}{\longleftrightarrow} HS^- + H^+ \tag{7.2}$$

$$HS^- \overset{K_2}{\longleftrightarrow} S^{2-} + H^+ \tag{7.3}$$

At the near-neutral pH normally encountered in anaerobic treatment, only Eq. (7.2) is significant. The equilibrium constant for Eq. (7.2) is expressed by:

$$\frac{[H^+][HS^-]}{[H_2S]} = K_1 \tag{7.4}$$

where $K_1 = 1.49 \times 10^{-7}$ ($pK_a = 6.83$) at 35°C.

$$\text{Dissolved sulfide (DS)} = H_2S_{(aq)} + HS^- \tag{7.5}$$

From Eqs (7.4) and (7.5), $H_2S_{(aq)}$ is given by:

$$H_2S_{(aq)} = \left(1 + \frac{K_1}{10^{-pH}}\right)^{-1} \times DS \tag{7.6}$$

Table 7.2. Sulfide levels imposing toxicity to methanogenic activity.

Substrate	Operation	Sulfide Concentration[a]	Remarks	References
Cellulose	Batch studies	56 mg/L (FS)	Onset of inhibition	Khan and Trottier (1978)
		320 mg/L (FS)	Complete inhibition	
Synthetic w/w[b]	Continuous studies	160–200 mg/L (DS)	—	Choi and Rim (1991)
Seafood processing w/w		240 mg/L (DS)	No inhibition	
Synthetic distillery w/w	—	5% H_2S in gas phase	50% inhibition of methanogenic activity	Karhadkar et al. (1987)
Lactate	Batch studies	100 mg/L (FS)	50% inhibition of methanogenic activity	McCartney and Oleszkiewicz (1993)
Leather processing w/w	Batch/continuous	100 mg/L (FS)	Sulfide was directly used and 15% decrease in efficiency	Wiemann et al. (1998)
Lactate	Continuous	69 mg/L (FS)	—	Maillacheruvu et al. (1993)
Glucose		75.4 mg/L (FS)	—	
Benzoate	Continuous (UASB)	769 mg/L (DS) or 234 mg/L (FS)	No inhibition	Fang et al. (1997)
Propionate	Continuous (upflow anaerobic filter (UAF))	350 mg/L (DS) or 110 mg/L (FS)	—	Parkin et al. (1991)
—	Continuous (UASB)	250 mg/L (FS)	50% inhibition of methanogenic activity	Koster et al. (1986)
Glucose	Continuous (CSTR)	613 mg/L (DS) or 228 mg/L (FS)	Onset of inhibition	Khanal and Huang (2005)
Glucose	Continuous (UAF)	804 mg/L (DS) or 80 mg/L (FS)	Severe inhibition of methanogenic activity	Khanal and Huang (2006)

[a] DS, dissolved sulfide (H_2S + HS^- + S^{2-}); FS, free sulfide (H_2S).
[b] w/w, wastewater.

The equilibrium distribution of H_2S between the aqueous and gaseous phases is governed by Henry's law:

$$\frac{[H_2S_{(g)}]}{[H_2S_{(aq)}]} = K_H \tag{7.7}$$

where $K_H = 13$ atm/mol at 35°C.

A small pH variation within the range of 6.0–8.0 can significantly affect the H_2S concentration. Koster et al. (1986) found that sulfide inhibition in methanogenic granules was dependent on the free H_2S concentration between pH 6.4 and 7.2.

No such correlation was observed between pH 7.8 and 8.0. The free H_2S concentrations producing 50% inhibition were 250.0 and 90.0 mg S/L at pH 6.4–7.2 and pH 7.8–8.0, respectively, suggesting that the sulfide toxicity at acid-to-neutral pH (pH 6.4–7.2) was due to free sulfide, but at the basic pH (pH 7.8–8.0), toxicity was caused by total sulfide. Karhadkar et al. (1987) reported that as pH decreased, inhibition increased due to an increase in free sulfide concentrations.

On the other hand, McCartney and Oleszkiewicz (1993) found that MPB were equally sensitive to the free H_2S at pH 7.0 or 8.0, unlike SRB that were affected differently at the two pH levels. SRB were more sensitive to free H_2S at pH 8.0 than at pH 7.0. The authors asserted that at pH 8.0, more total sulfide was needed to maintain the same free H_2S observed at pH 7.0. Omil et al. (1996) observed that a pH decrease from 8.0 to 7.0 caused an increase of H_2S from 50.0 to 240.0 mg/L and reduced COD utilization from 41 to 5% by MPB in an upflow anaerobic sludge blanket (UASB) reactor fed with a mixture of VFAs at a COD/SO_4 ratio of 0.5. The authors also found that at pH 8.1–8.3, free H_2S was more toxic to MPB than at pH 7.2–7.4.

Example 7.1

A continuous stirred tank reactor (CSTR) treating sulfate-rich wastewater generates dissolved sulfide of 270 mg/L at sulfate concentration of 2,000 mg/L under mesophilic conditions. The reactor pH is 6.9. What is the free sulfide level? If the pH drops to 6.6, what would be the free sulfide level?

Solution

At a mesophilic condition (35°C), the ionization constant (K_1) is 14.9×10^{-8}. From Eq. (7.6), we have:

$$H_2S_{(aq)} = (1 + (K_1/10^{-pH}))^{-1} \times DS$$

At 2,000 mg/L, free sulfide is given by:

$$H_2S_{(aq)} = (1 + (14.9 \times 10^{-8}/10^{-6.9}))^{-1} \times 270 = 123.7\text{mg/L}$$

Free sulfide concentration at pH of 6.6 $= (1 + (14.9 \times 10^{-8}/10^{-6.6}))^{-1} \times 270 =$ 169.5 mg/L

Note: The drop in pH could result in a significant increase of free sulfide, thereby increasing MPB inhibition.

7.5.3 Types of Treatment System

Sulfide toxicity is dependent on the types of treatment system, that is, attached growth or a suspended growth. An attached growth system is more resistant to toxicity than the suspended growth system for two reasons (Parkin and Speece

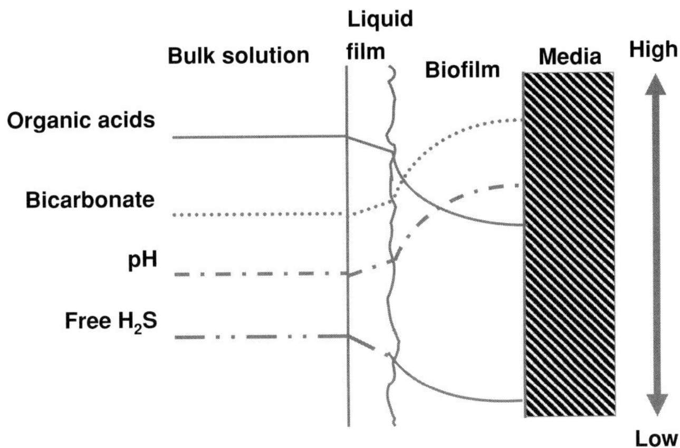

FIG. 7.2. Profiles of organic acids, HCO_3^-, pH, and free H_2S in an anaerobic biofilm. *Source:* Särner (1990). Reprinted with permission.

1983): (1) extremely high SRT at short HRT and (2) a quasi-plug flow hydraulic regime. Other possible reasons for higher tolerance of a biofilm (attached) system (or granules), especially for MPB, could be (a) inability of sulfide to penetrate deep into the biofilm and/or granules due to diffusion limitation, since MPB are believed to predominate at the inner part of the biofilm (or granules) because of their superior affinity to adhere to the carriers and/or aggregation (Santegoeds et al. 1999; Yoda et al. 1987); and (b) localized higher pH within the biofilm, granule, or floc because of the generation of bicarbonate from methanogenesis, which maintains a low free sulfide concentration (Särner 1990), as shown in Fig. 7.2. Therefore, the anaerobic filters were able to tolerate higher levels of dissolved and free sulfide than the CSTRs (Maillacheruvu et al. 1993). Fang et al. (1997) reported superior resistance of UASB granules to sulfide/H_2S toxicity because of their layered microstructure.

7.5.4 Reactor Configuration

High-rate anaerobic systems are ideal to treat high-strength, sulfate-rich wastewater. Two reactor configurations, the downflow anaerobic filter (DAF) and the static granular bed reactor (SGBR), best protect MPB against sulfide toxicity. In these configurations, sulfate reduction usually occurs in the upper zone with methanogenesis in the lower zone (Fig. 7.3). Rising methane gas bubbles generate turbulence that strips hydrogen sulfide gas from the upper zone of the reactor before it reaches the methane fermentation zone, enhancing conditions conducive to MPB. An example of a successful DAF system can be found in Puerto Rico, where the

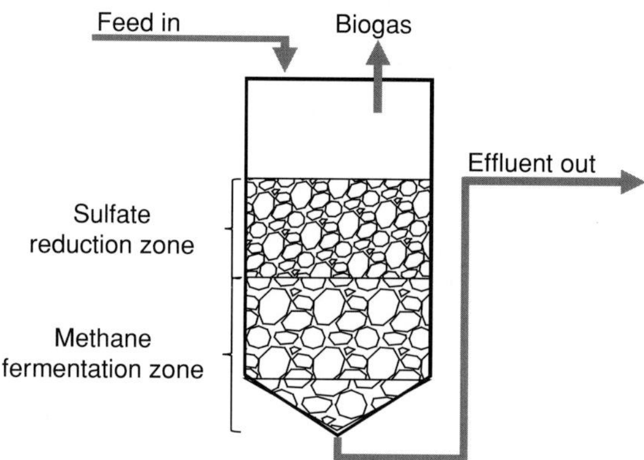

FIG. 7.3. Downflow bioreactor for enhanced methanogenesis.

reactor treats wastewater generated by a rum distillery. The influent COD and sulfate concentrations are 95 and 6 g/L, respectively (Speece 1996).

7.5.5 Degree of Acclimation

Methanogens can tolerate higher sulfide levels following a long acclimatization period. Isa et al. (1986b) reported that MPB were inhibited by free H_2S only at the very high concentrations exceeding 1,000 mg/L because of their adaptation to sulfides.

7.5.6 Organic Substrate Complexity

The type of substrate used as an electron donor affects both sulfide toxicity and methane recovery. An anaerobic system fed with complex substrates such as molasses, glucose, and lactate has more tolerance than that fed with simple substrates such as acetate and propionate. This is mainly due to the increased microorganism diversity supported by the more complex substrates (Maillacheruvu et al. 1993).

7.5.7 Alkalinity Generation

A major benefit of the anaerobic fermentation of sulfate-rich waste streams is the generation of in situ alkalinity. Alkalinity generation through sulfate reduction is illustrated by the following equations:

$$4H_2 + SO_4^{2-} + CO_2 \rightarrow HS^- + HCO_3^- + 3H_2O \quad (7.8)$$

$$CH_3COO^- + SO_4^{2-} \rightarrow 2HCO_3^- + HS^- \quad (7.9)$$

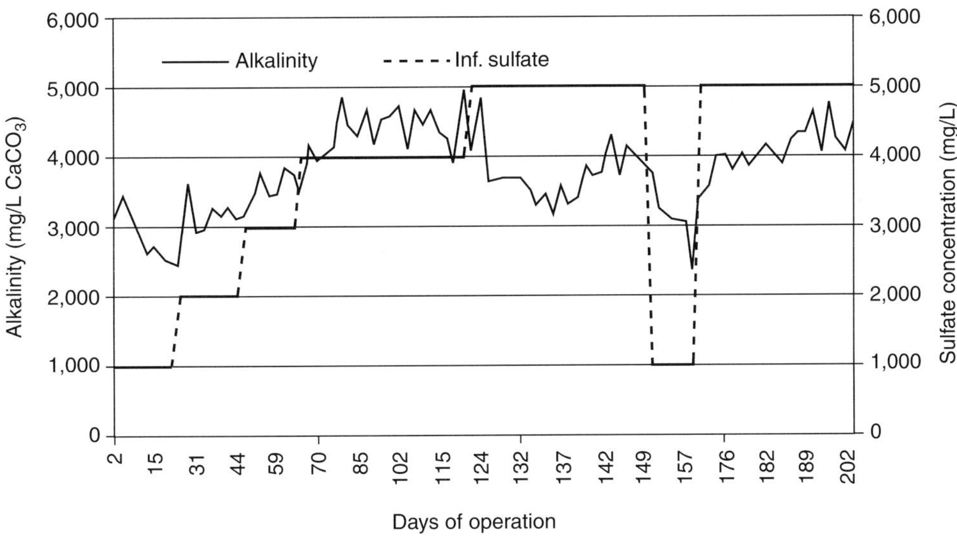

FIG. 7.4. Effluent alkalinity at various feed sulfate levels.

Theoretically, 2 mol (100 g) of alkalinity is generated through the complete reduction of 1 mol (96 g) of sulfate, that is, 1.04 g of alkalinity as $CaCO_3$ per gram of sulfate reduced (Greben et al. 2000).

Khanal and Huang (2005) observed progressively higher effluent alkalinity with increase in feed sulfate levels as shown in Fig. 7.4. Once sulfate concentrations exceeded 5,000 mg/L, however, alkalinity declined significantly. This was attributed to alkalinity consumption by the neutralization of excess VFAs, which accumulated because of sulfide toxicity to MPB. In this study, 0.38, 0.44, 0.51, 0.57, and 0.41 g alkalinity as $CaCO_3$ was generated per gram of influent SO_4 reduced at 1,000, 2,000, 3,000, 4,000, and 5,000 mg/L, respectively. In another study using an anaerobic filter, the alkalinity requirement was reduced by nearly half when the influent sulfate concentration was increased from 1,000 to 3,000 mg/L (Khanal and Huang 2006). It goes without saying that chemical costs can be significantly reduced when significant levels of sulfate are present in wastewater.

7.6 Interactions between MPB and SRB

MPB and SRB share many ecological and physiological similarities, such as their common presence in a wide variety of anaerobic ecosystems and the utilization of acetate and molecular hydrogen as electron donors (Fauque 1995). Understanding interactions between MPB and SRB is critical for successful treatment and bioenergy recovery from high-strength, sulfate-rich wastewater.

7.6.1 Coexistence

Coexistence refers to a condition where MPB and SRB utilize different electron donors and yet exist in the same microecosystems. In a sulfate-rich environment, methylamines and methanol are preferred by MPB (Smith 1993). In a nonlimiting substrate environment, both MPB and SRB can thrive because competition is greatly reduced (Smith 1993).

7.6.2 Synergism

The syntrophic relationship between MPB and SRB exists in an environment consisting of complex organic molecules in which the two groups of microorganisms rely on each other for their metabolic activities. The fermentation of lactate and ethanol by SRB in the absence of sulfate proceeds only when MPB exist in the system (Bryant et al. 1977). MPB utilize the H_2 produced during fermentation, thereby lowering the H_2 partial pressure, which enables SRB to effectively ferment lactate and ethanol. This synergistic relationship is known as interspecies hydrogen transfer. Another example of a synergistic relationship is the oxidation of methane by SRB (Fig. 7.5).

7.6.3 Competition

Competition is defined as the use or defense of a resource by one individual that reduces the availability of that resource to other individuals, whether of the same species or of other species (Smith 1993). It is well documented that both MPB and SRB readily use acetate and H_2 as electron donors. It follows that under a substrate-limiting condition, MPB and SRB compete for acetate and H_2. The outcome of the competition governs overall methane recovery. Some important factors that affect the competitions are discussed in the following sections.

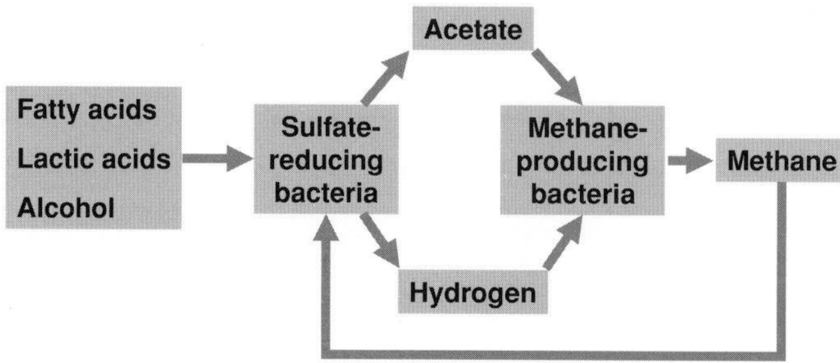

FIG. 7.5. Synergistic relationships between MPB and SRB.

Table 7.3. The standard free energy changes in sulfate-reducing and methanogenic reactions.

Reactions	ΔG° (kJ/reaction)
Sulfate-reducing reactions	
$4H_2 + SO_4^{2-} + H^+ \rightarrow HS^- + 4H_2O$	−152.4
$CH_3COO^- + SO_4^{2-} \rightarrow HS^- + 2HCO_3^-$	−47.6
Methanogenic reactions	
$4H_2 + HCO_3^- + H^+ \rightarrow CH_4 + 3H_2O$	−135.6
$CH_3COO^- + H_2O \rightarrow CH_4 + HCO_3^-$	−31.0

Source: Adapted from Lens et al. (1998).

7.6.3.1 Thermodynamics

Sulfate reduction thermodynamically yields more energy than the methanogenesis as evident from the Gibb's free energy changes (ΔG°) illustrated by the sulfate-reducing and methanogenic reactions in Table 7.3. It stands to reason then that MPB cannot effectively compete with SRB for substrates like acetate and hydrogen.

7.6.3.2 Kinetics

Kinetically, SRB are superior to MPB based on their higher affinity (or lower K_s value) for limiting substrates as illustrated in Eq. (7.10), where K_s is the half-velocity constant in the expression of the Monod kinetics as follows:

$$\mu = \mu_m \frac{S}{K_s + S} \tag{7.10}$$

where μ, μ_m, and S are specific growth rate, maximum specific growth rate, and limiting substrate concentration, respectively.

The apparent K_s values for hydrogen uptake were reported to be 0.012 mg/L for MPB and 0.002 mg/L for SRB (Kristjansson et al. 1982). Schönheit et al. (1982) reported the apparent K_s values for acetate uptake to be 177.0 mg/L and 11.8 mg/L, respectively, for MPB and SRB, while Gupta et al. (1994) reported K_s values of 6.0 mg/L and 0.84 mg/L for MPB and SRB, respectively, with acetate as the feed.

7.6.3.3 COD/SO_4 Ratio

The stoichiometry of sulfate reduction is given by:

$$8e + 8H^+ + SO_4^2 \rightarrow S^{2-} + 4H_2O \tag{7.11}$$

Each molecule of sulfate can accept eight electrons. On the other hand, oxygen can accept only four electrons. Therefore, the electron-accepting capacity of SO_4

is twice that of O_2; this then implies that the COD/SO_4 ratio $= 4/96 \sim 0.67$ (or COD/S = 2.0). As mentioned earlier, what this means for a wastewater with a COD/SO_4 ratio of 0.67 is that there is just enough carbon for sulfate reduction. When this ratio drops below 0.67, an external carbon source is needed for sulfate removal. As the ratio climbs above 0.67, residual organic matter becomes available for methane production. This illustrates the importance of COD/SO_4 ratio in the competition for available substrates.

Several researchers have investigated the degree of competitiveness between MPB and SRB relating to the COD/SO_4 ratio. Isa et al. (1986a, b) reported that electron flow attributed to SRB increased from 11 to 34% with a concomitant decrease in methane yield from 298.0 to 189.0 mL/g COD removed when the influent COD was decreased from 5.0 to 0.5 g/L. SRB predominated when the COD/SO_4 ratio dropped below 1.7. MPB dominated the system when the COD/SO_4 ratio was above 2.7 (Choi and Rim 1991).

McCartney and Oleszkiewicz (1993) found that when the COD/SO_4 ratio was decreased from 3.7 to $\leq$1.6, lactate degradation favored the sulfate-reducing pathway. Harada et al. (1994), using low-strength wastes in a UASB reactor, found that SRB contributed COD reductions of 4.8–5.8%, 23–34%, and 39–75%, respectively, at sulfate concentrations of 30, 150, and 600 mg/L. Annachhatre and Suktrakoolvait (2001) reported that at a high COD/SO_4 ratio of 5.0–6.7, 80% of the electron flow was diverted to methane generation and only 20% was utilized for sulfate reduction. At a lower ratio of 0.67, over 90% of the electron flow was removed via the sulfate-reducing pathway.

Example 7.2

A UASB reactor of 20 m^3 is employed to treat wastewater from a distillery plant. The COD loading rate is 30 kg/m^3·day. The wastewater contains 2 g/L sulfate. Assume the biogas contains 60% (by volume) methane gas. Compare the biogas production rate with and without sulfate.

Solution

(a) *Maximum biogas production rate without sulfate*:

Reactor volume = 20 m^3
COD loading rate = 30 kg/m^3·day
COD load = $30 \times 20 = 600$ kg COD/day
At STP, 1 kg COD produces 0.35 m^3 CH_4
Maximum CH_4 produced = $0.35 \times 600 = 210$ m^3/day
Biogas production rate = $210/0.6 = 350$ m^3/day

(b) *Biogas production in presence of sulfate*:

1 g sulfate consumes 0.667 g COD; thus total COD consumed in sulfate reduction = $2 \times 0.667 = 1.334$ g

COD consumed in sulfate reduction = 1.334 kg COD/m$^3 \times$ 20 m^3/day = 26.68 kg COD/day
COD available for methane production = 600 − 26.68 = 573.32 kg COD/day
Methane produced at STP = 573.32 × 0.35 = 200.7 m^3/day
Biogas production = 200.7/0.6 = 334.5 m^3/day

Note: The sulfate reduction resulted in a nearly 4.5% decrease in biogas production rate.

7.6.3.4 Substrate Type

The type of substrate also affects the competition between MPB and SRB since the two groups do not have the same affinity for all types of substrate. Isa et al. (1986a, b) reported that both sulfate reduction and H_2S content in the biogas were low for an acetate-fed, high-rate reactor with polyurethane media, implying that acetate was not a good substrate for SRB. Formate also did not enhance the sulfate-reducing activities. This group did observe, however, that addition of hydrogen or hydrogen precursors such as ethanol enhanced sulfate reduction.

Some studies contradict these findings. Gupta et al. (1994) reported that in an acetate-fed chemostat, 95% of the COD was metabolized through the sulfate-reducing pathway, indicating that SRB outcompeted MPB. In a formic acid-fed reactor, 24 and 62% of the COD was mineralized by MPB and SRB, respectively, illustrating the relative success each had for formic acid. In a methanol-fed chemostat, methanogenesis mineralized 87% of the total COD, suggesting that SRB were unable to use methanol, which supported other findings (Smith 1993). Yoda et al. (1987) reported that SRB could outcompete MPB for acetate in a long-term study under an acetate-limiting condition in an anaerobic fluidized-bed reactor. Harada et al. (1994) also concluded that although SRB competed poorly with MPB for acetate in a UASB reactor, SRB gradually gained an advantage over MPB in the long run because of their kinetic and thermodynamic advantages, and that a given substrate supports a more diverse group of SRB than MPB, thereby making the former more competitive.

7.6.3.5 pH

The competition between MPB and SRB also depends on environmental conditions such as pH. The optimum pH for acetotropic methane-producing bacteria (AMPB) and acetotropic sulfate-reducing bacteria (ASRB) is similar (Widdel 1988). Visser et al. (1996) reported that at pH below 6.9, AMPB outcompeted ASRB, while at pH above 7.7, the competition was favored by ASRB. They also reported that the growth of both ASRB and AMPB occurred over a wider pH range in granular sludge than in suspended sludge. The pH also has an indirect impact on the relative bioactivities between MPB and SRB. This is because pH affects speciation of sulfide and ammonia, both of which are considered more toxic in unionized forms. A decrease of pH from 8.0 to 7.0 resulted in an increase of H_2S from 50 to 240 mg/L. This reduced the fraction of COD used by MPB from 41 to 5% in a UASB reactor fed a VFA mixture with a COD/SO_4 ratio of 0.5. In thermophilic treatment of sulfate-laden wastewater, it was found that at neutral pH, equilibrium was established between SRB and MPB. However at pH $\geq$ 8.0, MPB became strongly inhibited, making SRB more competitive in utilizing this substrate (Visser et al. 1992).

7.6.3.6 Temperature

MPB and SRB thrive at similar temperatures during mesophilic operation. Visser et al. (1992), while batch testing using mesophilic (30°C) granular sludge, showed that SRB were less sensitive to high-temperature shocks (65°C for 8–9 h) compared to MPB. In continuous-mode reactors, a decrease in the reactor temperature from 30 to 25°C for a prolonged period of time (30 days) increased the SRB electron flow from 43 to 80% (Shin et al. 1996). This indicates that SRB is less sensitive than MPB to temperature reduction.

7.6.3.7 Treatment System

Anaerobic treatment of sulfate-laden wastewater can be conducted using a suspended growth system, an attached growth system, or a granulated sludge system such as UASB and SGBR. MPB and SRB have different degrees of immobilization or attachment capacity to the inert media or onto sludge granules, which may give one or the other a competitive advantage. MPB have been found to better colonize and adhere to polyurethane media (Isa et al. 1986b). Yoda et al. (1987) reported that MPB were able to form a biofilm faster than SRB in a fluidized-bed reactor at a higher acetate concentration, establishing an advantage in this situation as well. Visser et al. (1996) found that in granular sludge, a wider pH range supported growth of both ASRB and AMPB than in suspended sludge. Fang et al. (1997), using benzoate as the sole carbon source in a UASB reactor, reported a high resistance of granules to sulfide/H_2S toxicity because of the layered microstructure in the UASB granules.

7.6.3.8 Experimental Run Time and Acclimation

The experimental run time affects the degree of acclimation of MPB to sulfide concentrations and also to other environmental factors in the treatment system. The resistance of MPB to sulfide toxicity in a fixed-film reactor at a high concentration

Table 7.4. Ill effects of sulfide.

Sulfide toxicity	Sulfide in the range of 50–300 mg/L as free sulfide imposes toxicity to MPB
Biogas quality	Lowers biogas quality and reduces its value as an energy source. For boiler firing, sulfide levels $<1,000$ ppmv are recommended. For electricity generation by internal combustion, sulfide levels should be <100 ppmv, and for pipeline transfer of recovered biogas, H_2S should be <4 ppmv (Zicari 2003). Sulfide also interferes with metal catalysts used in thermocatalytic processes
Odor generation	Produces an objectionable odor detectable to humans at concentrations of 0.5 ppb to 0.2 ppm
Corrosion	Corrosive to metals and other materials due to the formation of sulfuric acid
Human health	Human respiratory toxin with a threshold limit value and short-term exposure limit of 10 and 15 ppm, respectively (Buisman et al. 1991)
Oxygen demand	Consumes up to 2 mol of oxygen per mole of sulfide

of free H_2S ($>1,000$ mg/L) is due to adaptation of MPB (Isa et al. 1986a, b). Fang et al. (1997) also reported a similar finding in a UASB reactor treating sulfate-laden wastewater. Harada et al. (1994) found that SRB outcompeted MPB during long-term operation of a UASB reactor when COD loading was increased from 0.5 to 3 g/L·day with an influent sulfate concentration of 600 mg/L. In this study, suppression of MPB was not due to sulfide toxicity since free H_2S was less than the threshold toxicity level, but was a result of competition between MPB and SRB.

7.7 Sulfide Removal

If a high-quality biogas that has value as a renewable energy source is to be recovered from a system, efficient removal of sulfide from both aqueous and gaseous phases is extremely important. Table 7.4 presents various nuisances caused by sulfide.

Methods for sulfide removal include physical (stripping), chemical (adsorption, precipitation, and oxidation) and biological, and combinations of all three. Removal of both aqueous and gaseous sulfides is discussed in the following sections.

7.7.1 Stripping

The equilibrium between aqueous-phase (H_2S_{aq}) and gaseous-phase (H_2S_{gas}) sulfides is governed by Henry's law (Eq. (7.7)). Stripping sulfide from the aqueous phase effectively reduces gas-phase sulfide. Sulfide in the biogas can be easily removed in an external unit, either chemically or biologically. In the chemical method, the sulfide-laden biogas is allowed to enter a tower of iron fillings/iron sponge, or

Fe_2O_3, where the sulfide is removed through adsorption and subsequent oxidation. The sulfide-free biogas is then recirculated back to the reactor to strip additional free sulfide from the reactor. The exhausted Fe_2O_3, now significantly Fe_2S_3, can be regenerated by heating in the presence of air/O_2. The dry-bed chemical process is most suitable for removal of low level of H_2S. For high levels of sulfide, the biological method is more appropriate. Sulfide-oxidizing bacteria convert the sulfide to elemental sulfur in presence of oxygen/air.

7.7.2 Chemical Precipitation

Divalent metals such as iron, zinc, and copper are capable of precipitating aqueous sulfide as insoluble metal sulfides (Eq. 7.12). Although iron sulfide is more soluble than sulfides of zinc, copper, nickel, and cobalt, iron salts (Fe^{2+} and Fe^{3+}) are widely used because of economic and toxicity considerations.

$$Fe^{2+} + HS^- \rightarrow FeS^- + H^+ \tag{7.12}$$

$$[Fe^{2+}][S^{2-}] = K_s, \quad \text{where } K_s = 6 \times 10^{-18}$$

Ferrous sulfide is essentially insoluble, and its continuous precipitation in the reactor can reduce the effective volume of the reactor, so the operator must be mindful of this and other potentially negative consequences.

7.7.3 Biological Sulfide Oxidation

Sulfide is primarily oxidized to elemental sulfur in an oxygen-limiting condition (<0.1 mg/L) by aerobic sulfide-oxidizing bacteria (Janssen et al. 1997), illustrated in Eq. (7.13). In a sulfide-limiting condition, sulfate is the main end product (Eq. 7.14).

$$2HS^- + O_2 \rightarrow 2S^0 + 2OH^- \left(\Delta G^{o'} = -\frac{29.50 \text{ kJ}}{\text{mol HS}^-}\right) \tag{7.13}$$

$$2HS^- + 4O_2 \rightarrow 2SO_4{}^{2-} + 2H^+ \left(\Delta G^{o'} = -\frac{772.43 \text{ kJ}}{\text{mol HS}^-}\right) \tag{7.14}$$

The bacteria responsible for sulfide oxidation belong to a group of colorless sulfur bacteria, of which *Thiobacillus* is the best known. *Thiobacillus* is mostly facultative autotrophic, utilizing reduced inorganic sulfur compounds as electron donors and carbon dioxide as a carbon source. Some heterotrophic *Thiobacilli* were, however, reported in a sulfide-oxidizing reactor, when the sulfide-laden wastewater contained organic matter (Janssen et al. 1997) (see Box 7.2).

Box 7.2

Merits of Biological Sulfide Removal

- The process requires no chemical catalyst or oxidant except air/oxygen.
- It does not generate chemical sludge.
- The process mainly converts sulfide to easily settleable elemental sulfur with low levels of sulfate and thiosulfate.
- The recovered sulfur can be used in several applications, such as an electron donor in autotrophic denitrification and bioleaching of heavy metals, as a soil amendment and as a fungicide.
- The process is less energy intensive.
- Overall, the process is less expensive.

7.7.3.1 Biological Sulfide Removal from Aqueous Phase

Some of the major findings concerning removal of aqueous sulfide are summarized in Table 7.5. Factors relevant to biological sulfide removal are discussed in the following sections.

Seed Inoculum and Start-Up

Sulfide-oxidizing bacteria are ubiquitous in nature, and they can be inoculated into a reactor with compost, marshland sludge, and other similar materials. One study employed ditch mud as an inoculum of sulfide-oxidizing bacteria (Buisman et al. 1990). Waste-activated sludge serves as an ideal seed inoculum, and an enriched culture of sulfide-oxidizing bacteria can be obtained by a continuous sulfide feed. The ideal pH for biological sulfide oxidation is in the alkaline range of 7.5–8.0; ideal temperatures range from 20 to 30°C. During start-up, the sulfide level should be kept at a relatively low level of 10–15 mg/L; sulfide levels of 5–30 mg/L have been reported toxic to *Thiobacilli*—a predominant sulfide-oxidizing bacteria (Buisman et al. 1991). Sulfide levels can be increased stepwise following acclimatization.

Operation of Sulfide-Oxidizing Bioreactor

For efficient bioconversion of sulfide to elemental sulfur, the sulfide-oxidizing bioreactor should be operated in an oxygen-limiting condition. The optimal molar oxygen to sulfide consumption ratio for maximum sulfur formation has been reported to be 0.7 (Janssen et al. 1997). *Thiobacilli* excrete sulfur particles, which are found to remain attached to the surface of bacterial cells. These particles eventually

Table 7.5. Summary of biological oxidation of sulfide in the aqueous phase.

		Operating Conditions				Sulfide				
Sulfide Source	Bioreactor Type	pH	Temperature (°C)	HRT (min)	Sulfide Loading Rate (kg $HS^-/m^3 \cdot$ day)	Influent (mg/L)	Effluent (mg/L)	Efficiency (%)	O_2 Level (mg/L)	References
Synthetic (Na_2S)	CSTR with BSP[a]	7.5	20	45	0.6	100	2	98	1.0	Buisman et al. (1990)
Anaerobically treated paper mill wastewater	Biorotor (46 rpm) with BSP[b]	7.7–8.2	27	13	NA	138	6.9	95	20% O_2 in gas phase	Buisman and Lettinga (1990)
				26		144	2.7	98		
Synthetic (Na_2S)	Expanded-bed	7.2–7.6	22 ± 2	NA	14	240	NA	NA	>0.1	Janssen et al. (1997)
Sulfide from anaerobic reactor	Fixed-film	7.0–7.5	25	NA	NA	109	ND	~100	NA	Khanal (2002)
						317	0.7	99.8		
						804	12.2	98.5		

[a] Recticulated polyurethane foam was used as biomass support particles (BSP).
[b] Raschig rings was used as BSP; NA, not available.

grow into an aggregate with diameter up to 3 mm. The sulfur-rich granules settle well with a velocity of 25 m/h (1.37 ft/min). For effective sulfur granule formation, the expanded-bed reactor/gas-lift reactor is more appropriate. Excessive turbulence can disintegrate the sulfur granules, and separation of the sulfide oxidation/granule formation zone from the aeration zone can mitigate this problem (Janssen et al. 1997).

End-Product Formation during Biological Sulfide Oxidation

Biological sulfide oxidation typically yields elemental sulfur and sulfate. However, thiosulfate formation also occurs if the biological system is overloaded. Thiosulfate formation is mainly driven through a chemical (or abiotic) route known as autoxidation. In an oxygen-limiting condition, that is, $[O_2/S^{2-}]_{consumption}$ of 0.5, thiosulfate ($S_2O_3^{2-}$) formation occurs as shown by Eq. (7.15) (Janssen et al. 1995):

$$2HS^- + 2O_2 \rightarrow H_2O + S_2O_3^{2-} \tag{7.15}$$

Thiosulfate formation is often observed during an initial phase of sulfide oxidation due to limited biological activity and an oxygen-limiting condition. At a molar ratio of $[O_2/S^{2-}] > 1.0$, sulfide is mainly oxidized to sulfate. This occurs when oxygen dosing is extremely high or the system is underloaded, and the elemental sulfur is further oxidized to sulfate, as shown by Eq. (7.16):

$$2S^o + 3O_2 \rightarrow 2SO_4^{2-} + 2H^+ \tag{7.16}$$

Elemental sulfur is a desired end product of sulfide oxidation, and maximum sulfur formation occurs at $[O_2/S^{2-}]$ of 0.7. Thus, aqueous oxygen plays a pivotal role in the dynamics of end-product formation, and a suitable oxygen dosing control method (e.g., oxidation–reduction potential (ORP)-based oxygenation) should be employed for selective formation of elemental sulfur (Khanal and Huang 2006).

Heterotrophic Sulfur-Reducing Activity

Studies reported in Table 7.5 were conducted using synthetic sulfide solutions that contained no organic matter. The sulfide stream from an anaerobic reactor often contains residual organic matter, which interferes with biological sulfide oxidation. The presence of abundant elemental sulfur coupled with a microaerobic environment provides a favorable condition for some sulfur particles to be reduced to sulfide again by heterotrophic sulfur-reducing bacteria. Microbes in the *Desulfuromonas* genus use acetate as an electron donor for sulfur reduction, as illustrated by the following reaction (Widdel and Bak 1991):

$$CH_3COO^- + 4S^o + 4H_2O \rightarrow 2HCO_3^- + 4H_2S + H^+ \quad (7.17)$$

Janssen et al. (1997) also observed the occurrence of heterotrophic sulfur-reducing activity within the sulfur sludge of a sulfide-oxidizing reactor, when the influent was supplemented with VFAs. Khanal (2002) also made a similar observation during sulfide oxidation of an anaerobic effluent. Such activity reduces the efficacy of a sulfide-oxidizing bioreactor. Formed elemental sulfur should be routinely recovered and the oxygen level should be kept adequate to avoid an anaerobic condition in a sulfide-oxidizing system.

7.7.3.2 Sulfide Removal from Gas Phase

The physical and chemical methods for gaseous sulfide removal are discussed earlier. For biological removal, the gaseous sulfide is first absorbed into the liquid phase using a scrubber (absorption tower) followed by oxidation of dissolved sulfide, as illustrated in Section 7.7.3.1. Thus, the gaseous sulfide removal is essentially a two-step process:

Step 1: Absorption and hydrolysis of H_2S

$$H_2S + OH^- \rightarrow HS^- + H_2O \quad (7.18)$$

Step 2: Biological sulfide oxidation

$$2HS^- + O_2 \rightarrow 2S^0 + 2OH^- \left(\Delta G^{o'} = -\frac{29.50 \text{ kJ}}{\text{molHS}^-}\right)$$

A typical schematic diagram of a gaseous H_2S removal process is shown in Fig. 7.6. The formation of elemental sulfur contributes (Eq. (7.13)) aqueous hydroxide. Therefore, recycling the liquid stream into the scrubber increases the pH of the aqueous solution and subsequently lowers chemical costs. The working principle of this process is patented and is known as Shell–Thiopaq process, developed by Paques Biosystems, The Netherlands. Based on a pilot-scale study, biogas sulfide was reduced from 1 to 2% (by volume) to less than 10 ppmv at a biogas flow rate of 400 N m^3/h from paper industry wastewater (Janssen et al. 2000).

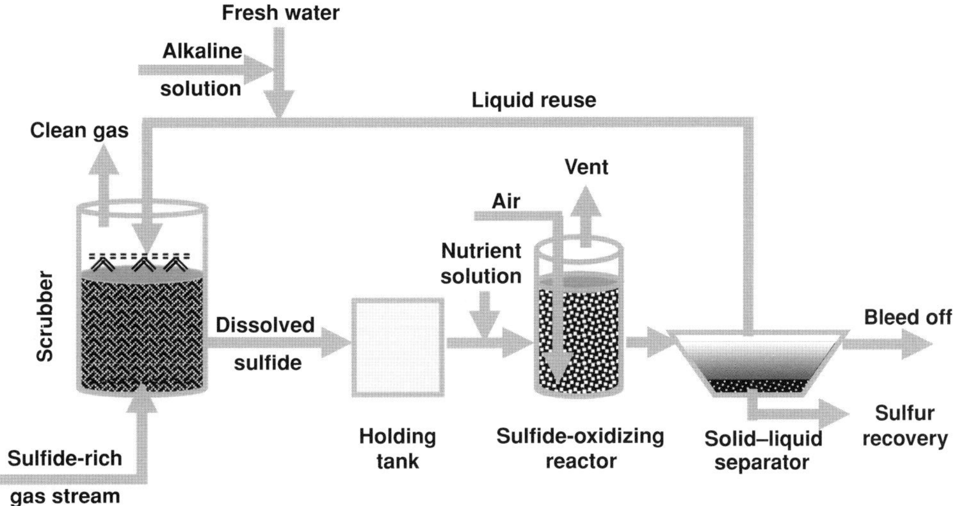

FIG. 7.6. Schematics of gaseous sulfide removal process.

Example 7.3

A seafood processing plant produces 2,500 m^3 of wastewater/day with a mean sulfate concentration of 6.0 g/L. How much elemental sulfur could be produced if sulfate and sulfide removal efficiencies are 75 and 90%, respectively?

Solution

Based on stoichiometry:

$$SO_4^{2-} + 8H^+ + 8e^- \rightarrow S^{2-} + 4H_2O$$

6 g/L will produce (32 × 6)/96 = 2 g sulfide (as S).
But with 75% efficiency, sulfide produced will be 0.75 × 2.0 = 1.5 g/L.
Similarly,

$$S^{2-} + \tfrac{1}{2}O_2 + H_2O \rightarrow S^\circ + 2OH^-$$

Thus, 1 g sulfide produces 1 g sulfur.
With sulfide oxidation efficiency of 90%, the sulfur production = 1.5 × 0.9 = 1.35 g/L.

$$S^\circ\text{production} = 2{,}500\ m^3/\text{day} \times 1.35\ kg/m^3$$
$$= 3{,}375\ \mathbf{kg/day\ (3.7\ ton/day)}$$

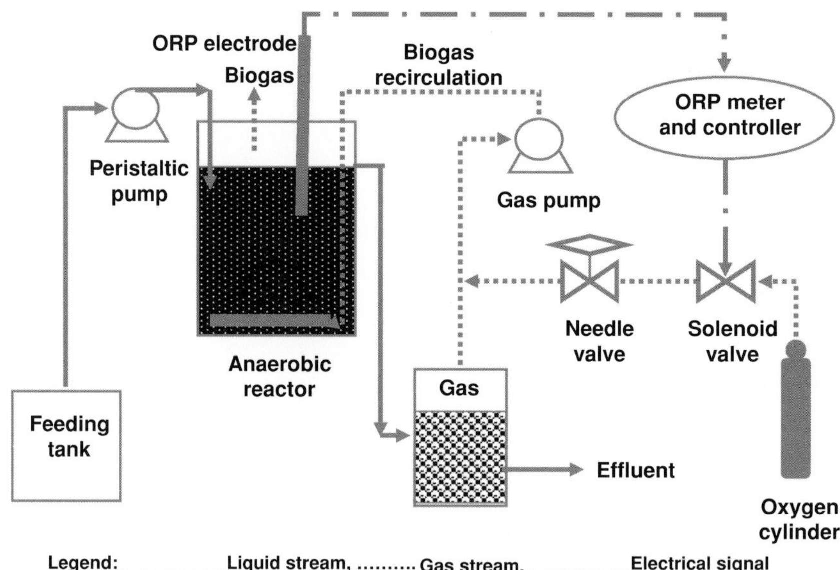

FIG. 7.7. Schematic of integrated process for sulfide control.

7.7.4 Integrated Approach for Sulfide Removal

Khanal (2002) developed an integrated approach for simultaneous elimination of aqueous and gaseous sulfides and sulfide toxicity through periodic oxygenation by controlling redox potential. As illustrated in Fig. 7.7, a small amount of oxygen

Table 7.6. Performance of an integrated system for sulfide removal.

	Without Aeration	With Aeration
Gas phase		
N_2 (%) (by volume)	0.5 ± 0.1[a]	6.8 ± 1.0
CH_4 (%)	65.6 ± 0.6	62.6 ± 2.4
CO_2 (%)	33.6 ± 1.1	29.8 ± 1.1
O_2 (%)	NT[b]	0.7 ± 0.1
H_2S (ppmv)	2450 ± 150	1.7 ± 1.7
Biogas production (L/day)	54.2 ± 4.5	59.8 ± 2.6
Methane production (L/day)	35.0 ± 0.6	37.8 ± 0.1
Liquid phase		
Sulfide (mg/L)	17.7 ± 1.7	ND
Sulfate (mg/L)	ND[c]	12.5 ± 8.3
Thiosulfate (mg/L)	ND	ND
ORP (mV)	−277 ± 8	−246 ± 3
pH	7.17 ± 0.01	7.23 ± 0.01

[a] Mean value ± standard deviation of seven data.
[b] NT, not tested.
[c] ND, not detected.

is periodically injected into an external sulfide-oxidizing unit, where aqueous and gaseous sulfides enter continuously. The sulfide-free biogas is recycled back to the reactor to strip freshly formed H_2S and is then returned to the sulfide-oxidizing unit for subsequent oxidation.

The process lowered gaseous sulfide levels from 2, 6, and 16% (by volume) to <0.1% at feed sulfate levels of 1,000, 3,000, and 6,000 mg/L, respectively. Similarly, the dissolved sulfide levels dropped from 109, 317, and 804 mg S/L to nondetectable 0.7 and 12.2 mg S/L, respectively, at sulfate levels of 1,000, 3,000, and 6,000 mg/L. The integrated process also eliminated the sulfide toxicity to MPB.

The process was extended to the removal of sulfide from biogas using simulated swine manure with a TS content of 2.1%. The 92-L working volume CSTR was connected to a 1-L sulfide-oxidizing unit (SOU). The CSTR was fed at an organic loading rate of 0.8 kg VS/m^3·day and operated at an HRT of 20 days at 25°C. The produced sulfide-rich biogas was recirculated to the SOU at a rate of 0.5 L/min. Treated effluent was periodically fed to the SOU to provide a liquid medium for sulfide oxidation. The major findings are summarized in Table 7.6 (Duangmanee et al. 2007). The data clearly show the effectiveness of an integrated system for efficient sulfide removal from both aqueous and gaseous phases. The sulfide was mainly oxidized to elemental sulfur, as apparent from deposition of white particles in the SOU.

References

Anderson, G. K., Sanderson, J. A., Saw, C. B., and Donnelly, T. 1988. Fate of COD in an anaerobic system treating high sulfate bearing wastewater. In *Biotechnology for Degradation of Toxic Chemicals in Hazardous Wastes*, edited by R. J. Scholze, Jr, E. D. Smith, J. T. Bandy, Y. C. Wu., and J. V. Basilico, pp. 504–531. Noyes Data Corporation, Park Ridge, NJ, USA.

Annachhatre, A. P., and Suktrakoolvait, S. 2001. Biological sulfate reduction using molasses as a carbon source. *Water Environ. Res.* 73(1):118–126.

Barton, L. L., and Tomei, F. A. 1995. Characteristics and activities of sulfate reducing bacteria. In *Sulfate Reducing Bacteria*, edited by L. L. Barton, pp. 1–32. Plenum Press, New York, USA.

Bryant, M. P., Campbell, L. L., Reddy, C. A., and Crabill, M. R. 1977. Growth of *Desulfovibrio* in lactate and ethanol media low in sulfate in association with H_2-utilizing methanogenic bacteria. *Appl. Environ. Microbiol.* 33:1162–1169.

Buisman, C. J. N., Geraats, B. G., Ijspeert, P., and Lettinga, G. 1990. Optimization of sulphur production in a biotechnological sulphide-removing reactor. *Biotechnol. Bioeng.* 50:50–56.

Buisman, C. J. N., and Lettinga, G. 1990. Sulphide removal from anaerobic effluent of a paper mill. *Water Res.* 24(3):313–319.

Buisman, C. J. N., Lettinga, G., Paasschens, C. W. M, and Habets, L. H. A. 1991. Biotechnological sulphide removal from effluent. *Water Sci. Technol.* 24(3–4):347–356.

Callander, I. J., and Barford, J. P. 1983. Precipitation, chelation, and the availability of metals as nutrients in anaerobic digestion. II. Applications. *Biotechnol. Bioeng.* 25(8):1959–1972.

Carrondo, M. J. T., Silva, J. M. C., Figueira, M. I. I., Ganho, R. M. B., and Oliveira, J. F. S. 1983. Anaerobic filter treatment of molasses fermentation wastewater. *Water Sci. Tech.* 15(8–9):117–126.

Choi, E., and Rim, J. M. 1991. Competition and inhibition of sulfate reducers and methane producers in anaerobic treatment. *Water Sci. Tech.* 23:1259–1264.

Duangmanee, T., Khanal, S. K., and Sung, S. 2007. *Micro-Aeration for Sulfide Removal in Anaerobic Treatment of High-Solid Wastewater: A Pilot-Scale Study.* In CD-ROM Proceedings of Water Environment Federation 80th Annual Conference & Exposition. October 13–17, 2007, San Diego, CA, USA.

Fang, H. H. P., Liu, Y., and Chen, T. 1997. Effect of sulfate on anaerobic degradation of benzoate in UASB reactors. *J. Environ. Eng. ASCE* 123(4):320–328.

Fauque, G. D. 1995. Ecology of sulfate-reducing bacteria. In *Sulfate Reducing Bacteria,* edited by L. L. Barton, pp. 217–241. Plenum Press, New York, USA.

Genschow, E., Hegemann, W., and Maschke, C. 1996. Biological sulfate removal from tannery wastewater in a two-stage anaerobic treatment. *Water Res.* 30(9):2072–2078.

Greben, H.A., Maree, J. P., and Mnqanqeni, S. 2000. Comparison between sucrose, ethanol and methanol as carbon and energy sources for biological sulphate reduction. *Water Sci. Technol.* 41:247–253.

Gupta, A., Flora, J. R. V., Gupta, M., Sayles, G. D., and Suidan, M. K. 1994. Methanogenesis and sulfate reduction in chemostats I. Kinetic studies and experiments. *Water Res.* 28(4):781–793.

Habets, L. H. A., and de Vegt, A. L. 1991. Anaerobic treatment of bleached TMP and CTMP effluent in the biopaq UASB system. *Water Sci. Tech.* 24:331–345.

Hansen, T. A. 1993. Carbon metabolism of sulfate reducing bacteria. In *The Sulfate-Reducing Bacteria: Contemporary Perspectives,* edited by J. M. Odom and R. Singleton, Jr, pp. 21–40. Springer-Verlag, Berlin, Germany.

Harada, H., Uemura, S., and Momonoi, K. 1994. Interaction between sulfate-reducing bacteria and methane-producing bacteria in UASB reactors fed with low strength wastes containing different levels of sulfate. *Water Res.* 28(2):355–367.

Hilton, B. L., and Oleszkiewicz, J. A. 1988. Sulfide-induced inhibition of anaerobic digestion. *J. Environ. Eng. ASCE* 114(6):1377–1391.

Isa, Z., Grusenmeyer, S., and Verstraete, W. 1986a. Sulfate reduction relative to methane production in high-rate anaerobic digestion: Technical aspects. *Appl. Environ. Microbiol.* 51(3):572–579.

Isa, Z., Grusenmeyer, S., and Verstraete, W. 1986b. Sulfate reduction relative to methane production in high-rate anaerobic digestion: Microbial aspects. *Appl. Environ. Microbiol.* 51(3):580–587.

Janssen, A. J. H., Dijkman, H., Janssen, G. 2000. Novel biological processes for the removal of H_2S and SO_2 from gas streams. In *Environmental Technologies to Treat Sulfur Pollution: Principles and Engineering,* edited by P. N. L. Lens and L. Hulshhoff Pol, pp. 265–279. IWA Publishing, London.

Janssen, A. J. H., Ma, S. C., Lens, P., and Lettinga, G. 1997. Performance of a sulfide-oxidizing expanded-bed reactor supplied with dissolved oxygen. *Biotechnol. Bioeng.* 53(1):32–40.

Janssen, A. J. H., Sleyster, R., Van Der Kaa, C., Jochemsen, A., Bontsema, J., and Lettinga, G. 1995. Biological sulphide oxidation in a fed-batch reactor. *Biotechnol. Bioeng.* 47(3):327–333.

Karhadkar, P. P., Audic, J. M., Faur, G. M., and Khanna, P. 1987. Sulfide and sulfate inhibition of methanogenesis. *Water Res.* 21(9):1061–1066.

Khan, A.W., and Trottier, T. M. 1978. Effect of sulfur-containing compounds on anaerobic degradation of cellulose to methane by mixed cultures obtained from sewage sludge. *Appl. Environ. Microbiol.* 35(6):1027–1034.

Khanal, S. K. 2002. Single-Stage Anaerobic Treatment of High Sulfate Wastewater with Oxygenation to Control Sulfide Toxicity. Ph.D. thesis, The Hong Kong University of Science and Technology, Hong Kong.

Khanal, S. K., and Huang, J.-C. 2005. Effect of high influent sulfate on anaerobic wastewater treatment. *Water Environ. Res.* 77(7):3037–3046.

Khanal, S. K., and Huang, J.-C. 2006. Online oxygen control for sulfide oxidation in anaerobic treatment of high sulfate wastewater. *Water Environ. Res.* 78(4):397–408.

Koster, I. W., Rinzema, A., de Vegt, A. L., and Lettinga, G. 1986. Sulfide inhibition of the methanogenic activity of granular sludge at various pH levels. *Water Res.* 20(12):1561–1567.

Kristjansson, J. K., Schönheit, P., and Thaur, R. K. 1982. Different Ks values for hydrogen of methanogenic bacteria and sulfate reducing bacteria for the apparent inhibition of methanogenesis by sulfate. *Arch. Microbiol.* 13:278–282.

Lens, P. N. L., Visser, A., Janssen, A. J. H., Hulshoff Pol, L. W., and Lettinga, G. 1998. Biotechnological treatment of sulfate-rich wastewaters. *Crit. Rev. Environ. Sci. Technol.* 28(1):41–88.

Maillacheruvu, K. Y., Parkin, G. F., Peng, C. Y., Kuo, W. C., Oonge, Z. I., and Lebduschka, V. 1993. Sulfide toxicity in anaerobic systems fed sulfate and various organics. *Water Environ. Res.* 65(2):100–109.

Maree, J. P., and Strydom, W. F. 1985. Biological sulfate removal in an upflow packed bed reactor. *Water Res.* 19(9):1101–1106.

McCartney, D. M., and Oleszkiewicz, J. A. 1993. Competition between methanogens and sulfate reducers. Effect of COD:sulfate ratio and acclimation. *Water Environ. Res.* 65(5):655–664.

Mendez, R., Lema, J. M., and Soto, M. 1995. Treatment of seafood-processing wastewaters in mesophilic and thermophilic anaerobic filters. *Water Environ. Res.* 67(1):33–45.

Mizuno, O., Li, Y. Y., and Noike, T. 1998. The behavior of sulfate-reducing bacteria in acidogenic phase of anaerobic digestion. *Water Res.* 32(5):1626–1634.

Mohanroa, G. J., Subrahmanyan, P. V. R., Deshmukh, S. B., and Saroja, S. 1970. Waste treatment at a synthetic drug factory in India. *J. Water Pollut. Control Fed.* 42(8):1530–1543.

Nanninga, H. J., and Gottschal, J. C. 1986. Anaerobic purification of wastewater from a potato-starch producing factory. *Water Res.* 20(1):97–103.

Omil, F. Lens, P., Hulshoff Pol, L., and Lettinga, G. 1996. Effect of upward velocity and sulfide concentration on volatile fatty acid degradation in a sulfidogenic granular sludge reactor. *Process Biochem.* 31:699–710.

Parkin, G. F., Lynch, N. A., Kuo, W. C., Van Keuren, E. L., and Bhattacharya, S. K. 1990. Interaction between sulfate reducers and methanogens fed acetate and propionate. *J. Water Pollut. Control Fed.* 62(6):780–788.

Parkin, G. F., Sneve, M. A., and Loos, H. 1991. Anaerobic filter treatment of sulfate-containing wastewaters. *Water Sci. Technol.* 23:1283–1291.

Parkin, G. F., and Speece, R. E. 1983. Attached versus suspended growth anaerobic reactors: Response to toxic substances. *Water Sci. Technol.* 15:261–289.

Pohland, F. G., and Ghosh, S. 1971. Developments in anaerobic stabilization of organic wastes. The two-phase concept. *Environ. Lett.* 1:255–266.

Postgate, J. R. 1984. *The Sulphate-Reducing Bacteria*, 2nd edn. Cambridge University Press, Cambridge, UK.

Reis, M. A. M., Lemos, P. C., Martins, M. J., Costa, P. C., Goncalves, L. M. D., and Carronda, M. J. T. 1991. Influence of sulfates and operational parameters on volatile fatty acids concentration profile in acidogenic phase. *Bioprocess Eng.* 6:45–151.

Rintala, J., Sanz Martin, J. L., and Lettinga, G. 1991. Thermophilic anaerobic treatment of sulfate-rich pulp and paper integrate process water. *Water Sci. Technol.* 24:149–160.

Santegoeds, C. M., Damgaard, L. R., Hesselink, G., Zopfi, J., Lens, P., Muyzer, G., and de Beer, D. 1999. Distribution of sulfate-reducing and methanogenic bacteria in anaerobic aggregates determined by microsensor and molecular analysis. *Appl. Environ. Microbiol.* 65(10):4618–4629.

Särner, E. 1990. Removal of sulphate and sulphite in an anaerobic trickling (ANTRIC) filter. *Water Sci. Technol.* 22(1–2):395–404.

Schönheit, P., Kristjansson, J. K., and Thaur, R. K. 1982. Kinetic mechanism for the ability of sulfate reducers to out-compete methanogens for acetate. *Arch. Microbiol.* 133:285–288.

Shin, H. S., Oh, S. E., and Bae, B. U. 1996. Competition between SRB and MPB according to temperature change in the anaerobic treatment of tannery wastes containing high sulfate. *Environ. Technol.* 17:361–370.

Smith, D. W. 1993. Ecological actions of sulfate-reducing bacteria. In *The Sulfate-Reducing Bacteria: Contemporary Perspectives*, edited by J. M. Odom and R. Singleton, Jr, pp. 161–188. Springer-Verlag, Berlin, Germany.

Speece, R. E. 1996. *Anaerobic Biotechnology for Industrial Wastewaters.* Archae Press, Nashville, TN, USA.

Svardal, K., Götzendorfer, K., Nowak, O., and Kroiss, H. 1993. Treatment of citric acid wastewater for high quality effluent on the anaerobic—aerobic route. *Water Sci. Technol.* 28:177–186.

Szendry, L. M. 1983. *Startup and Operation of the Bacardi Corporation Anaerobic Biofilter*, pp. 365–377. Proceedings of the 3rd International Symposium on Anaerobic Digestion. Evans and Falkner Inc., Watertown, MA, USA.

Visser, A., Gao Y., and Lettinga, G. 1992. Anaerobic treatment of synthetic sulfate-containing wastewater under thermophilic conditions. *Water Sci. Technol.* 25(7):193–202.

Visser, A., Hulshoff Pol, L. W., and Lettinga, G. 1996. Competition of methanogenic and sulfidogenic bacteria. *Water Sci. Technol.* 33(3):99–110.

Vogels, G. D., Kejtjens, J. T., and Van Der Drift, C. 1988. Biochemistry of methane production. In *Biology of Anaerobic Microorganisms*, edited by A. J. B. Zehnder. John Wiley & Sons, New York, USA.

Widdel, F. 1988. Microbiology and ecology of sulfate- and sulfur-reducing bacteria. In *Biology of Anaerobic Microorganism*, edited by A. J. B. Zehnder, pp. 469–585. John Wiley & Sons, New York, USA.

Widdel, F., and Bak, F. 1991. Gram-negative mesophilic sulfate-reducing bacteria. In *The Prokaryotes*, 2nd edn, edited by N. R. Krieg and J. G. Holt, pp. 3352–3378. Springer-Verlag, New York, USA.

Wiemann, M., Schenk, H., and Hegemann, W. 1998. Anaerobic treatment of tannery wastewater with simultaneous sulphide elimination. *Water Res.* 32(3):774–780.

Winfrey, M. R., and Zeikus, J. G. 1977. Effect of sulfate on carbon and electron flow during microbial methanogenesis in freshwater sediments. *Appl. Environ. Microbiol.* 33(2):312–318.

Yoda, M., Kitagawa, M., and Miyaji, Y. 1987. Long term competition between sulfate-reducing and methane-producing bacteria for acetate in anaerobic biofilm. *Water Res.* 21(12):1547–1556.

Zicari, S. M. 2003. Removal of Hydrogen Sulfide from Biogas Using Cow-Manure Compost. MS thesis, Cornell University, Ithaca, New York, USA.

CHAPTER

Bioenergy Generation from Residues of Biofuel Industries

Samir Kumar Khanal

8.1 Background

The dependence of the global economy on fossil-derived fuels, coupled with political instability in oil-producing countries, has pushed petroleum prices near all-time highs. Demand for energy is expected to increase more than 50% by 2025, mostly due to emerging economies—India and China. Increased use of fossil fuels will also increase atmospheric carbon dioxide, hastening the global warming crisis. Thus, there is an ongoing quest to develop sustainable, affordable, and environmentally sound energy from renewable resources. Biofuels derived from plant-based feedstocks are considered renewable and are an environmentally clean energy source, and have potential to significantly reduce consumption of fossil fuels.

Bioethanol and biodiesel are the most promising clean and alternative renewable fuels. These can be used in the form of a gasoline/diesel blend. Bioethanol is currently produced mainly from corn (United States) and sugarcane (Brazil). There is also growing interest in developing commercially viable cellulose-to-ethanol technology in the United States. Biodiesel is produced from soybeans (United States), rapeseed (European Union), and palm oil (Malaysia). Biofuel production processes also generate lower-value residues and wastes such as stillage and crude glycerin.

Anaerobic biotechnology converts organic-rich waste streams into methane gas, hydrogen, or ethanol, with methane generation the focus here. The produced methane can replace natural gas, coal, and electricity, currently used in ethanol and biodiesel plants. It is possible that biogas can be processed to a quality comparable to conventional natural gas and distributed in a natural gas pipeline system. The residues resulting from digestion are rich in nutrients (N and P) and can be land-applied, reducing use of fossil-fuel-based fertilizer. In addition, the treated effluent can be recycled as process water for in-plant use.

This chapter covers important feedstocks, biofuel production processes from these feedstocks, stillage and glycerin generation, and anaerobic digestion of these residues for bioenergy production. Also covered are water reclamation/reuse and biosolids disposal issues in biofuel industries.

8.2 Bioethanol Feedstocks

Feedstock for ethanol production can be broadly classified into the following:

8.2.1 Monomeric Sugars

These are substrates in which the carbohydrate is present in the form of simple, directly fermentable six- and twelve-carbon sugar molecules such as glucose, fructose, and maltose. Such feedstocks include sugarcane, sugar beets, fruit (fresh or dried), citrus molasses, and cane sorghum. Brazil produces its ethanol almost exclusively from sugarcane, and much of the Caribbean and other areas of South America also utilize this feedstock. Use of these feedstocks is not economically viable in the United States.

8.2.2 Starch

Starchy materials contain slightly complex carbohydrates and need to be enzymatically processed to yield simple sugars. Examples of this category are starch and inulin that can be broken down into the simpler six- and twelve-carbon sugars through the action of enzymes. The starch-based feedstocks include grains such as corn, grain sorghum, barley, and wheat and root/tubular crops such as cassava, potatoes, sweet potatoes, Jerusalem artichokes, cacti, and arrowroot. Ethanol is mainly produced from corn grain in the United States. There is also an increasing interest in sorghum-grain-based ethanol in areas of the United States where this crop is widely cultivated (Kansas, Oklahoma, and Texas). Similarly, in Thailand, where cassava is widely cultivated, the economic viability of cassava-based ethanol plant is being investigated.

8.2.3 Cellulosic Biomass

Lignocellulosic biomass, such as that obtained from forest and agricultural residues, also serves as feedstock for ethanol production. This type of feedstock, however, needs aggressive pretreatment with acid, alkali, or steam to break down the lignin or solubilize the hemicellulose component. The hydrolyzed biomass then undergoes enzymatic hydrolysis to produce fermentable sugars. The United States currently leads the effort to develop a viable cellulosic ethanol process, largely because of the

1.3 billion dry tons of cellulosic biomass that is available each year (USDA and USDOE joint report 2005).

Ethanol production from corn is currently very economically competitive in the United States because of political backing, but also significantly because value-added products are recovered from the process residues, which accounts up to 20% of the total revenue. Recovery of higher-value resources is a key to the future use of corn and all other biomass feedstocks used for ethanol production.

8.3 Biodiesel Feedstocks

Biodiesel is primarily produced from vegetable oil. Animal fats and waste grease are also used for biodiesel production in small facilities. Biodiesel is second to ethanol among biofuel in the United States and is mainly produced from soybean oil. Biodiesel produced from rapeseed oil is the major biofuel in the European Union (EU). China, Brazil, Malaysia, and India are other countries with significant biodiesel production. Along with soybean and rapeseed oils, palm, coconut, peanut, and sunflower oils are the primary feedstocks for biodiesel production.

8.4 Ethanol Production

The ethanol production process varies with feedstock types. Depending on substrate complexity, various pretreatment methods are needed. For production from sugarcane, sugar beets, and sorghum stalks, all of which contain simple sugars such as glucose and sucrose, no pretreatment is needed except size reduction and pressing. For starchy substrates such as corn, sorghum, and cassava, grinding (or milling) followed by enzyme hydrolysis is needed to obtain fermentable sugar. Lignocellulosic biomass requires more comprehensive physical and chemical pretreatments to break down the protective lignin layer. The cellulose/hemicellulose is then subjected to enzymatic hydrolysis to release simple sugars. Subsequent discussion will focus on ethanol production from starch- and lignocellulose-based feedstocks.

8.4.1 Corn-Based Ethanol Production

United States produced nearly 18.5×10^6 m^3 (4.9 billion gal) ethanol in 2006 from 110 biorefineries using primarily corn (Fink 2007). There are two commercial processes: dry-grind and wet milling.

Wet milling accounts for 21% of the total U.S. ethanol production (USDA 2006). Various parts of the corn kernel (i.e., starch, protein, fiber, and oil) are separated prior to fermentation. The waste stream from the wet milling process is relatively dilute (COD: 2–3 g/L) (Jasti et al. 2006) and is not economically attractive for methane production.

Dry milling accounts for the large majority of U.S. ethanol production because of its lower capital and operating costs, and is discussed in greater details in this section. In this process, whole-grain corn is subjected to fermentation following milling. Modified dry milling is a relatively new development that incorporates some aspects of both wet and dry mill technologies.

A dry-grind process is illustrated in Fig. 8.1. The whole corn is ground in a hammer mill or roller mill and then mixed with water to form a mash. The mash is cooked in a jet cooker at 80–90°C (215–220°F) for 15–20 min. A small amount of the enzyme α-amylase is added during jet cooking to assist liquefaction. Additional α-amylase is added during secondary liquefaction, which occurs for 90 min at 95°C (220°F). The cooked mash is then cooled to 60°C (140°F) and mixed with the enzyme glucoamylase to convert the starch to fermentable sugars, a process known as saccharification. This saccharified mash is fermented with yeast (*Saccharomyces cerevisiae*) to produce ethanol. In most plants, saccharification and fermentation occur simultaneously (simultaneous saccharification and fermentation (SSF)) to minimize inhibition of enzyme activity and the yeast cells by the product (sugar). Fermentation is usually conducted at pH of 4.8–5.0 and a temperature of 37°C (90°F) for 48 h. The fermented mash, often referred to as beer, is distilled to produce a 95% ethanol product by volume (or 190 proof). Dehydration of the 95% ethanol using molecular sieves, which preferentially retain the water while allowing the ethanol to pass, further purifies the product to 99.5% (~200% proof). The fermentation residues are referred to as whole stillage, which is centrifuged to obtain wet cake. The wet cake is passed through a series of dryers to obtain distillers dried grains (DDG). Thin stillage is the liquid portion obtained from centrifugation. A portion of the thin stillage is dehydrated by evaporation to obtain syrup. The syrup is blended with DDG to form distillers dried grains with solubles (DDGS). The remainder of the thin stillage is often recycled as process water.

Recent modifications of the dry-grind process include processing at reduced temperatures and prefractionation for recovery of germ and fiber prior to fermentation. This process has undergone extensive investigation by researchers at the University of Illinois (Singh et al. 2001). The reduced temperature or noncooking process uses a special enzyme that works effectively at low temperatures. STARGEN™ 001 is such enzyme, developed by Genencor International. STARGEN™ 001 contains *Aspergillus kawachi* α-amylase expressed in *Trichoderma reesei* and a glucoamylase from *Aspergillus niger* that functions synergistically to hydrolyze starch into glucose. There are several full-scale modified dry-grind plants currently in operation in the United States. Since there is no cooking step involved in the process, the thin stillage is not directly recycled upstream. It is sterilized by boiling to eliminate any possibility of bacterial contamination.

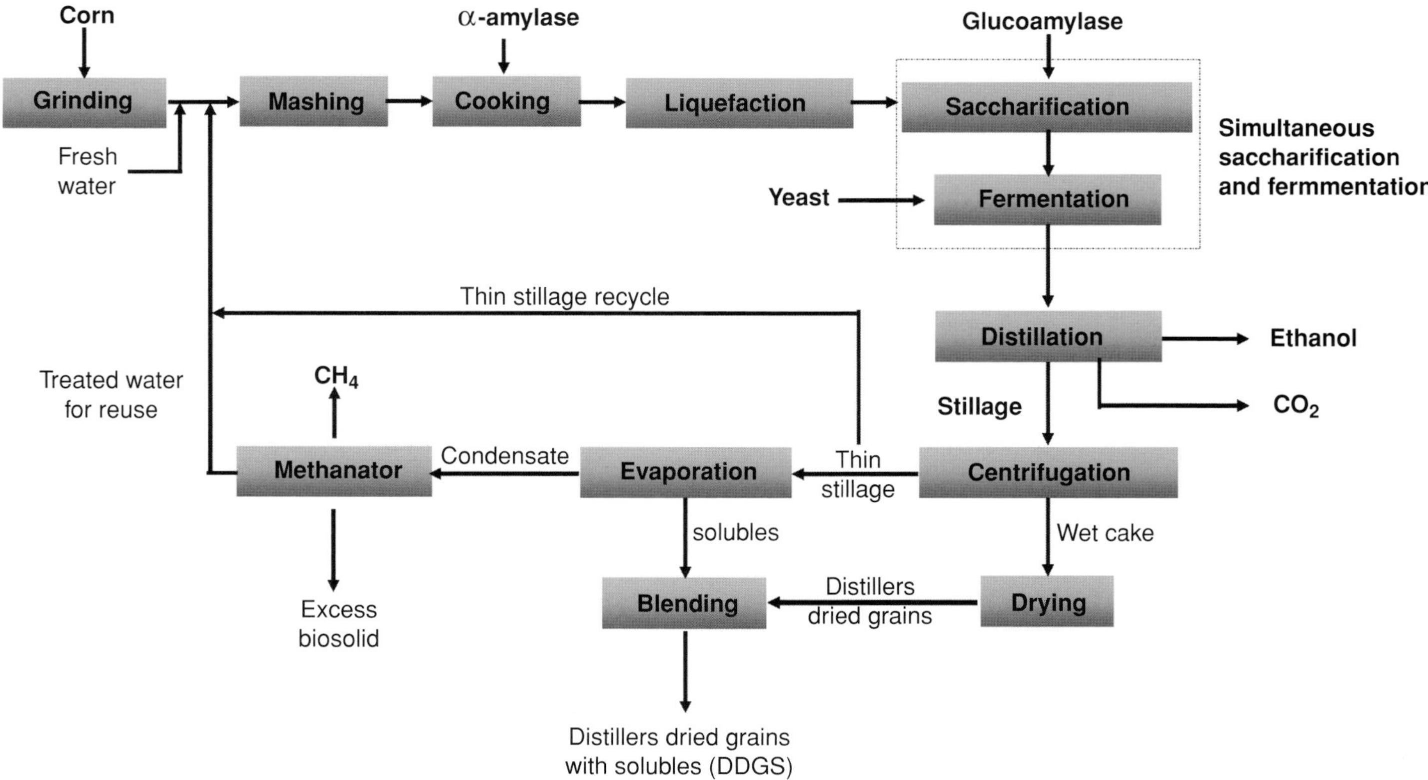

FIG. 8.1. Schematics of the conventional dry-grind ethanol plant.

8.4.2 Generation and Characterization of the Waste/Residue Streams

Despite the name, dry-grind plants require large amounts of water. Annual water consumption of a typical 50-million gal/year (MGY) dry-grind ethanol plant ranges from 150 to 300 million gal (Stanich 2007), equivalent to 3–5 gal/gal of ethanol. Water consumption varies from plant to plant with production capacity, plant age, intake water quality, process efficiency and control (e.g., water treatment, cooling tower, chiller, boiler, distillation, and centrifugation), and housekeeping practices. Water consumption also varies with season, with larger volumes being used in the summer due to greater evaporative losses. Water drawn into the plant and used for the above-described processes is termed "consumptive use" water, that is, the water consumed in the process, loss during evaporation/drying, reject from reverse osmosis (RO) units, and water that leaves with the products such as DDGS and ethanol. Water used for cooling tower and boiler blowdowns, effluent/liquid discharges, and internal recycling is not considered part of the total water demand.

Lanting and Gross (1985) examined a 230,000 m^3/year (~60 MGY) dry-grind ethanol plant in southern Ohio as part of an investigation focused on upgrading an aerobic trickling filter to an upflow anaerobic sludge blanket (UASB) system. The ethanol plant produced an average of 4,900 m^3/day (~1.3 million gal/day) of wastewater, with a peak flow of 5,500 m^3/day (~1.45 million gal/day). Evaporation and drying processes accounted for nearly 90% of the ethanol plant process water, with the remainder from distillation, hydrocarbon stripping, and miscellaneous plant wash water.

Many dry-grind ethanol plants claim to be closed loop (zero discharge) with regard to water use, and in fact most process wastewaters do get recycled in the plant. However, many plants use reverse osmosis (RO) as part of the production process, and the RO waste stream contains a high level of total dissolved solids (TDS). Significant dilution of RO reject is usually necessary prior to discharge.

Major liquid streams from a dry-grind plant include thin stillage, evaporative condensate, clean-in-place (CIP) wash water, blowdowns from the cooling tower and boiler, RO rejects, and miscellaneous wash waters. A brief discussion of these streams follows:

Thin stillage: Thin stillage contains very high levels of total solids (TS) and COD. Its characteristics are dependent on the fermentation and distillation processes, and the type and efficiency of solid–liquid separation. About 40–60% of thin stillage gets directly recycled in mash preparation. A remaining portion is evaporated to form syrup. Some plants do evaporate the entire thin stillage stream. A detailed discussion of thin stillage is described later.

Condensate: Condensate is the liquid stream recovered from evaporation of thin stillage. Condensate volume ranges from 200 to 250 gal/min for a 50-MGY plant when 50% of thin stillage is evaporated. Condensate contains significant levels of dissolved organics, primarily weak organic acids such as acetic, lactic, propionic, and butyric acid. COD varies from 2 to 8 g/L with a pH of 4–5. For example, in one condensate sample set, the volatile fatty acids (VFAs) and their concentrations (mg/L) were acetic acid, 1,664; propionic acid, 13; iso-butyric acid, 11; *n*-butyric acid, 32; and *n*-valeric, 8 mg/L. Acetic acid appears to be the major acid in most condensate samples. The evaporated condensate lacks many essential nutrients needed for successful biological treatment.

Wash water: This stream is generated from cleanup, pipe flushing, and rinsing of process vessels. CIP is one of the major components of the stream that mainly contains alkali solution (4–5% NaOH) and fermented mash. Because this waste stream has a high pH (10–12), it is often used to adjust pH in a UASB reactor or to optimize yeast propagation.

Blowdowns from boiler and cooling tower: Boiler blowdown is relatively clean and is directly recycled in the process. Cooling tower blowdown may contain some microbes due to atmospheric exposure, but is relatively free from organic matter and is discharged to a receiving stream or lake.

8.4.3 Treatment of Wastewater/Liquid Streams

Dry grind is considered to be a zero-discharge process, and wastewater treatment is not a standard practice in the ethanol industry. Anaerobic removal of organic acids from condensate is common, but the effluent from this process is recycled back. Because of its low solids content and high organic acid concentration, the condensate is typically treated in a UASB reactor often known as methanator. Other reactor configurations such as an expanded granular sludge bed (EGSB), anaerobic filter, and static granular bed reactor (SGBR) can be effectively used to treat this and similar waste streams.

In most cases, part of the CIP stream is dumped in a beer well. Portions of the CIP along with other wash water are discharged into a waste tank that feeds the methanator. CIP helps maintain the pH of the condensate near the neutral range needed for anaerobic treatment. Blowdowns from the cooling tower and boiler are relatively clean and do not require any specific treatment. Literature describing the treatment of wastewater and other liquid waste streams from ethanol plants is scarce. Data from one published study are summarized here.

8.4.3.1 Case Study 1: South Point Ethanol Plant, OH, USA

One of the early studies on anaerobic treatment of dry-grind corn ethanol plant wastewater was conducted using a 6-m^3 pilot-scale UASB reactor (proprietary name Biothane) by Lanting and Gross (1985). The wastewater appeared to be primarily condensate from thin stillage evaporation and process wash water. The goal of this

Table 8.1. Average wastewater characteristics and the UASB performance.

Parameters	Influent	Effluent	Efficiency
TCOD (mg/L)	3,627	874	76%
SCOD (mg/L)	2,889	416	86%
Total BOD (mg/L)	2,441	288	88%
Soluble BOD (mg/L)	1,910	181	90%
Volumetric organic loading rate (kg COD/m^3·day)		9.3	
HRT (h)		9.4	
Methane content (% by volume)		83	
Methane yield (m^3/kg $TCOD_{rem.}$)		0.33	

Source: Adapted from Lanting and Gross (1985).

study was to examine the effectiveness of an anaerobic system as a pretreatment step for post-aerobic treatment using the plant's existing trickling filters, which had difficulty meeting compliance requirements.

The average total chemical oxygen demand (TCOD) of the influent was 3,627 mg/L, which is within the range of condensate's organic strength. The influent was deficient in nitrogen and phosphorus and needed caustic addition for pH control. No data were available for wastewater pH and alkalinity. The important feed characteristics and details of the UASB reactor are presented in Table 8.1. UASB operating temperature was not indicated. Since the TCOD was significantly higher than the soluble chemical oxygen demand (SCOD), a significant amount of COD was contributed by particulate matter. The biogas methane content was unreasonably high at 83%.

The authors conducted shock loading tests by increasing the volumetric organic loading from 9.0 to 36.5 kg COD/m^3·day, nearly triple the normal loading rate. The UASB pilot plant was able to accommodate this variation without detrimental effects on system performance.

8.4.3.2 Case Study 2: Anaerobic Treatment of Condensate from Amaizing Energy, Denison, IA, USA

Amaizing Energy is a 189,250-m^3/year (50-MGY) dry-grind corn ethanol plant. The plant produces 114–136 m^3/h (500–600 gal/min) of thin stillage. About 68 $\pm$ 4.5 m^3/h (300 $\pm$ 20 gal/min) is sent to an evaporator tank to produce nearly 45 $\pm$ 2.3 m^3/h (220 $\pm$ 10 gal/min) of condensate. Condensate, along with part of CIP and wash water from the plant, is treated in a UASB reactor. The CIP stream is primarily used for pH control of the influent. The plant operates 2–113.6 m^3 (30,000 gal) UASB reactors to treat condensate at a total flow rate of 22.7 m^3/h (100 gal/min). Some important feed characteristics, UASB operating conditions, and digester performance results are presented in Table 8.2.

Table 8.2. Feed characteristics, UASB operating conditions, and performance.[a]

Parameters	Influent (mg/L)	Effluent (mg/L)	Efficiency (%)
$SCOD_{grab}$	2,259 ± 285	75 ± 18	96.7 ± 0.7
$SCOD_{composite}$	2,045 ± 426	31 ± 15	98.5 ± 0.8
TDS	92 ± 6	1,123 ± 72	—
Alkalinity as $CaCO_3$	NA	1,814 ± 265	—
Volumetric organic loading rate (kg COD/m^3·day)		9.8	
HRT (h)		10	
Methane content (% by volume)		NA	
Operating temperature (°C)		34 ± 0.5 (94 ± 1°F)	
Bioreactor pH		7 ± 0.05	
Biogas generation rate (m^3/day)		4,575 ± 55 (9,691,723 ± 115,609 cfm)	

[a] Mean value ± standard deviation of 20 steady-state data.

The influent lacks sufficient nutrients for anaerobic microbes. Inputs of urea (N) and phosphoric acid (P) are used to supplement deficient macronutrient. Trace mineral nutrients such as Ca (as $CaCl_2$), Mg (as $MgSO_4$), K (as KCl), and Fe (as $FeCl_3$) are also added. The produced biogas is purified using an iron sponge and is burned along with natural gas in a thermal oxidizer. The energy contribution of the produced biogas is less than 2% of the plant's total energy consumption. The treated effluent is recycled back to the mash preparation. The effluent COD of grab samples was higher than that measured in 24-h composite samples, possibly due to sulfide interference in the grab samples. The composite samples may have benefited from mixing and air exposure, enhancing sulfide oxidation. It is important to note that the feed COD varies considerably and is dependent on fermentor and distillation efficiency. Condensate COD spikes were observed as high as 9,000 mg/L when the fermentor was not running optimally. An equalization tank (also known as waste tank) provided some buffering of organic load fluctuations to the UASB.

8.4.4 Stillage from Dry-Grind Plant

Residues of yeast fermentation and ethanol recovery are termed stillage or whole stillage. The volume of stillage produced by a typical 50-MGY plant varies from 136 to 159 m^3/h (600–700 gal/min). The centrifuge centrate is known as thin stillage, and a 50-MGY plant typically produces 114–136 m^3/h (500–600 gal/min).

8.4.5 Why Biomethane Production from Stillage

Until recently, stillage was thought to be a high-value stream for animal feed production, representing about 20% of the total revenue in dry-grind plants, making

Table 8.3. Characteristics of whole stillage.

Parameters	Concentration
TS[a] (%)	11.4
VS[a] (%)	10.7
VS/TS ratio	0.93
Total suspended solids (TSS)[a] (%)	9.5
Volatile suspended solids (VSS)[a] (%)	9.4
TCOD (g/L)	203
SCOD (g/L)	48
COD/VS ratio	1.9
pH	4.46
Temperature (°C)	80–85
VFA (mg/L as acetic acid)	2,390
Alkalinity (mg/L as $CaCO_3$)	0
Carbohydrate[b] (mg/L as glucose)	10,700
Total Kjeldahl nitrogen (TKN) (mg/L as N)	4,020
NH_3–N[b] (mg/L)	18.5
Total phosphorus (TP) (mg/L as P)	1,331

Source: Adapted from Khanal (2005).
[a] 1% = 10,000 mg/L.
[b] Tested on soluble portion of the sample.

its use for biogas production economically questionable. The total revenue from DDGS has, however, declined by nearly 50% in recent years (Tiffany et al. 2007). The U.S. livestock market may have been unable to absorb all the DDGS produced by the rapidly expanding ethanol industry. The higher fat and sulfur content of DDGS, along with a poor balance of amino acids (especially low lysine levels), limits the use of DDGS for monogastrics species such as swine and poultry, and the price of DDGS will likely continue to decline. Secondly, energy costs continue to increase and are second only to raw material costs for dry-grind ethanol plants. This has motivated ethanol producers to explore alternatives (i.e., bioenergy production) for utilization of coproducts that could generate a higher rate of return. Biomethane production through anaerobic digestion is one such option that could reduce energy costs and fossil fuel use.

8.4.6 Whole Stillage Characterization

Whole stillage characterization is important for evaluation of anaerobic digestibility, estimation of methane production potential, and selection of a suitable reactor configuration. Literature on stillage characterization is limited. The characterization study presented here was conducted at author's laboratory. The sample was obtained from a local 50-MGY dry-grind plant. Table 8.3 shows the important characteristics of the whole stillage.

As shown in Table 8.3, the whole stillage contains an extremely high level of TS, most of which is organic material (VS/TS > 0.93). The TCOD exceeds 200 g/L, very high for a waste stream. The low pH (4.5) and zero alkalinity suggest that the stillage is not easily amenable to anaerobic treatment unless sufficient alkalinity is supplemented. The whole stillage does contain sufficient nutrients for anaerobic digestion. In another study, Rosentrater et al. (2006) reported total solids and volatile solids (VS) of 116.6 g/L (11.66%) and 107.8 g/L (10.78%), respectively, for whole stillage.

8.4.7 Anaerobic Digestion of Whole Stillage

Data for systematic studies examining anaerobic digestion of whole stillage are currently unavailable. The unique characteristics of whole stillage present both challenges and opportunities for anaerobic digestion, discussed later, along with considerations related to the anaerobic digestion of thin stillage.

8.5 Thin Stillage Characterization

One of the first studies characterizing thin stillage and its digestion was conducted by Stover's research group at Oklahoma State University in the early 1980. Thin stillage samples were collected from the university's 757-m^3/year (0.2-MGY) research facility and a 11,355-m^3/year (3-MGY) plant at Hydro, OK, USA. The important characteristics are presented in Table 8.4.

Table 8.4. Characteristics of thin stillage.

Parameters	Values[a]
TS	32,200 (9,300)
TSS	11,800 (3,700)
VSS	11,300 (3,500)
TDS	18,600 (7,100)
TCOD	64,500 (12,600)
SCOD	30,800 (6,200)
Total BOD_5	26,900 (800)
Soluble BOD_5	19,000 (2,100)
Total organic carbon	9,800 (2,200)
Total TKN	755 (115)
Soluble TKN	480 (85)
NH_3–N	130 (60)
TP	1,170 (100)
Soluble P	1,065 (75)
Total protein	4,590 (650)
Soluble protein	2,230 (780)
Total carbohydrate	8,250 (750)
Soluble carbohydrate	2,250 (550)
Glucose	<750
pH	3.3–4.0

Source: Adapted from Stover et al. (1983).

[a]All units in mg/L, except pH; numbers in the parenthesis are standard deviation.

Table 8.5. Characteristics of thin stillage.

Parameters	Values[a]
TCOD	53,373
BOD_5	39,840
[a]Total nonfilterable residues	30,492
Volatile nonfilterable residues	25,392
VFAs	1,332
pH	4.1
Alkalinity	0

Source: Adapted from Hunter (1988).
[a] All units in mg/L, except pH; nonfilterable residues can be considered suspended solids.

In another study, Hunter (1988) reported the characteristics of thin stillage obtained from a dry-grind plant located at Colwich, KS, USA. The plant utilized a mixture of corn and milo (ratio unspecified) (Table 8.5).

Significant changes in ethanol production processes have occurred over the past 20–30 years. As might be expected, this has resulted in changes in thin stillage composition and its digestibility. One such production change is increased capacity. Most dry-grind plants built today have a capacity of at least 189,250 m^3/year (50 MGY). The important characteristics of thin stillage obtained from a 50-MGY ethanol plant are presented in Table 8.6.

As illustrated in these tables, thin stillage is a high-nutrient, high-strength acidic stream with nearly zero alkalinity and the majority of the COD contributed by particulate matter. The extremely high volatile fraction (~80–90%) suggests that thin stillage is readily amenable to anaerobic digestion.

Table 8.6. Characteristics of thin stillage.

Parameters	Values
TS[a] (%)	6.1
VS[a] (%)	5.3
VS/TS ratio	0.87
TSS[a] (%)	2.1
VSS[a] (%)	2.1
TCOD (g/L)	94
SCOD (g/L)	41
COD/VS ratio	1.8
pH	4.46
Temperature (°F)	180
VFA (mg/L as acetic acid)	1,310
Alkalinity (mg/L as $CaCO_3$)	0
Carbohydrate[b] (mg/L as glucose)	13,600
TKN (mg/L as N)	1,720
NH_3–N[b] (mg/L)	32.1
TP (mg/L as P)	1,292

Source: Adapted from Khanal (2005).
[a] 1% = 10,000 mg/L.
[b] Results from the soluble portion of the sample.

8.5.1 Anaerobic Digestion of Thin Stillage

In recent years there has been renewed interest in anaerobic digestion of thin stillage for bioenergy recovery. High levels of biodegradable organics make it an ideal feedstock for anaerobic digestion.

8.5.2 Important Considerations in Stillage Digestion

8.5.2.1 Reactor Configuration

The large amounts of suspended solids found in thin stillage are not compatible with many digester configurations. High-rate anaerobic reactors such as UASB, EGBR, anaerobic filter, SGBR, and hybrid systems work best for soluble wastewater and are not suitable for thin stillage digestion. Continuous stirred tank reactor (CSTR) is an ideal configuration for thin stillage digestion. Solids retention time (SRT) and hydraulic retention time (HRT) are nearly equal in a CSTR. The high solids level in thin stillage does, however, require a long detention time due to a rate-limiting hydrolysis step. Various modifications of a CSTR can reduce detention times and improve the digestibility.

A membrane-coupled anaerobic process can be employed for thin stillage digestion. Large-pore membranes (e.g., microfiltration and nonwoven filter media) can retain biomass and particulate matter, which are then recycled back to the reactor. A major drawback of an anaerobic membrane bioreactor (AnMBR) system is its propensity to foul. Advancements in membrane technology may improve the tolerance to high-solids streams and provide a tangible solution to their digestion. Table 8.7 shows various reactor configurations suitable for thin stillage digestion.

8.5.2.2 Temperature

Thin stillage exits from the ethanol production process at 80–85°C (176–185°F). Anaerobic digestion near this temperature range eliminates the need for cooling and enhances hydrolysis of particulate matter. Although several hyperthermophilic methanogenic archaea (*Methanococcus jannaschii*, *Methanococcus igneous*, and *Methanococcus infernus*) are able to produce methane from H_2/CO_2 at temperature range of 45–91°C (113–196°F), with 85–88°C (185–190.4°F) being optimal (Burggraf et al. 1990; Jeanthon et al. 1998; Jones et al. 1983), their methane production potentials in a mixed-culture environment have not yet been demonstrated. Thermophilic digestion (55°C) of sewage sludge with a mixed culture has been applied in several full-scale operations. The same concept can be applied to thin stillage digestion. A successful laboratory-scale study examining the thermophilic digestion of thin stillage was conducted at Iowa State University and is discussed

Table 8.7. Different reactor configurations for stillage digestion.

Reactor configurations	Schematics	Remarks
Continuous stirred tank reactor (CSTR)	Biogas; Feed in → Effluent out	SRT ~ HRT. The digester can be operated at mesophilic (35°C) or thermophilic (55°C)
Anaerobic contact process (ACP)	Degassing; Feed in → Effluent; Sludge recycle; Biomass wastage	SRT > HRT. Sludge settleability governs SRT. Mesophilic temperature is ideal for settleability
Anaerobic sequential batch reactor	Biogas; Effluent; Feed, React, Settle, Decant	SRT > HRT. Sludge settleability governs SRT. Mesophilic temperature is ideal for settleability
Acid-phase digestion	Biogas; Feed in → Acidogenic reactor → Methanogenic reactor → Effluent	SRT ~ HRT. Acid-phase digestion enhances hydrolysis of particulate matter
Temperature-phase digestion	Biogas; Feed in → Thermophilic reactor → Mesophilic reactor → Effluent	SRT ~ HRT. Thermophilic digestion enhances particulate hydrolysis
Anaerobic membrane reactor (AnMBR)	Biogas; Feed in → Membrane → Effluent; Biomass recycle	SRT > HRT. Membrane retains the particulate matter and biomass, which is then recycled back into the reactor. Membrane fouling is a major limiting factor
Covered lagoon	Biogas	Low cost and simple. Lack of mixing and temperature loss in winter may lead to poor digestion and less methane yield

later. The reader should be aware that thermophilic systems are less adaptable to shock loading, and meticulous process control is extremely important for stable digester operation.

Mesophilic digestion (35–37°C) (95–98.6°F) is highly stable and less prone to failure, but the digestion rate is only about half that of thermophilic digestion. To improve both digestibility and stability, a temperature-phase anaerobic digestion (TPAD) system can be employed for thin stillage digestion. The first stage of this process is a thermophilic digestion operated at a short HRT of 3–5 days to enhance hydrolysis. The second stage is a mesophilic digestion that converts the hydrolyzed products (organic acids) into methane gas. Without thermophilic stage, mesophilic digestion of thin stillage is practical for bioenergy generation.

8.5.2.3 HRT and SRT

In a completely mixed reactor, HRT cannot be controlled independently of SRT. Long detention times in the range of 15–30 days are therefore required to minimize washout of slow-growing methanogens. If the digested biomass in an anaerobic sequential batch reactor (ASBR) and anaerobic contact process (ACP) settles well, HRT can be controlled independently of SRT. This effectively lowers the operating HRT and the anaerobic digester footprint.

(AnMBR) system is another approach that could reduce the retention time of anaerobic system. Other approaches that could improve the hydrolysis of particulate matter include physical, chemical, thermal, and biological pretreatments of the feed, which could subsequently help lower the detention times significantly.

8.5.2.4 pH and Alkalinity

As discussed previously, acidity does not present challenge to anaerobic digestion, especially during start-up. High levels of organic nitrogen (protein) in stillage can create in situ alkalinity and maintain an approximately neutral pH during digestion, due to the formation of ammonium bicarbonate. This substantially reduces or even eliminates the need for alkalinity addition during digestion. The author's research supports this conclusion and is discussed later. During plant start-up, digester pH should be carefully maintained near neutral range. The CIP stream can be used for pH control of the anaerobic digester, eliminating chemical costs otherwise required for this purchase.

8.5.2.5 Nutrients and Trace Elements

Unlike condensate, thin stillage contains diverse nutrients and trace minerals. The COD/N/P ratio of thin stillage is around 350:6.5:4.8, which is near the theoretical minimum of 350:7:1 recommended for a highly loaded system (Pohland 1992).

Table 8.8. Nutrients and trace elements in thin stillage.

Parameters	Values[a]
Protein	0.83
Phosphorus	0.11
Calcium	0.0042
Magnesium	0.049
Iron	0.00072
Potassium	0.18
Sodium	0.044
Sulfur	0.03

Source: Data obtained from Midwest Grain Processors.
[a] All units in %; 1% is equivalent to 10,000 mg/L.

Minerals such as calcium, magnesium, potassium, sodium, iron, and sulfur are also present, and Table 8.8 illustrates typical levels found. These mineral and nutrient levels essentially preclude supplementation.

8.5.3 Stillage Digestion at Mesophilic Conditions

The Stover group at Oklahoma State University conducted one of the first thorough studies examining the digestion of thin stillage (Stover et al. 1983). The feed material was supernatant obtained following gravity settling. An ACP and batch-scale digesters were both used for digestion. Various SRTs were examined. Because anaerobic biodegradability was the major focus of this study, the influent was diluted to one-third full strength for all systems, the exception being the investigation of a 30-day SRT case, where the stillage was examined at 67 and 100% strengths. The 2- and 4-day SRT systems were operated without sludge recycling, while the digesters at other SRTs were run with sludge recycling. The authors reported that pH and temperature were easily controlled at longer SRTs and higher influent strengths. Alkalinity requirements were reduced from 1,500 to 3,000 mg/L $CaCO_3$ at the shortest SRTs to only 200 mg/L $CaCO_3$ at an SRT of 30 days. Of note was the 98% SCOD removal efficiency in a 30-day SRT system with full-strength thin stillage and a CO_2 content of 23.5%. Excessive accumulation of organic acids (up to 2,500 mg/L) was observed with the diluted stillage at SRTs <4 days. This study did not report methane yields due to problems with the biogas measurement system. The digesters' performance is summarized in Table 8.9.

The Stover group further examined anaerobic digestion of settled thin stillage produced from a 1-MGY ethanol plant with the objective of bioenergy recovery, using both suspended-growth and fixed-film reactor systems (Stover et al. 1984). The results are presented in Table 8.10.

The results show the suspended-growth and fixed-film systems performing similarly. The suspended-growth system was essentially an ACP that maintained a high sludge concentration due to biomass recycling.

Table 8.9. Summary of continuous anaerobic treatment system performance.

SRT (days)	HRT (days)	Mixed liquor volatile suspended solids (MLVSS) (mg/L)	Alkalinity (mg/L $CaCO_3$)		VFA (mg/L)	Influent pH	Gas CO_2(%)	SCOD		
			Added	Effluent				Influent (mg/L)	Effluent (mg/L)	Efficiency (%)
2	2.0	675	3,000	3,500	1,000	10–12	—	6,500	5,900	9.2
4	4.0	650	3,000	4,400	2,600	10–12	<5.0	5,200	1,200	76.9
6	5.3	730	3,000	3,200	ND	9–12	10.0	8,900	2,470	72.4
10	5.3	1,670	1,500–2,500	3,560	ND	8–10	15.1	9,300	850	90.9
20	5.3	2,300	750–1,500	4,800	ND	7–8	19.1	12,250	460	96.2
30 (a)	5.0	8,700	300	4,950	ND	5–6	20.0	16,790	1,190	92.9
30 (b)	5.0	11,740	200	4,880	ND	3–5	23.5	28,620	560	98.0

Source: Adapted from Stover et al. (1983).

Table 8.10. Performance comparison of suspended-growth and fixed-film reactors.[a]

Suspended-Growth System					
F/M Ratio	Influent COD (mg/L)	Effluent COD (mg/L)	MLVSS (mg/L)	Methane (%)	Methane Yield (m^3/kg $COD_{removed}$)
0.50	5,125	380	3,380	78	0.62
0.56	10,100	380	5,380	71	0.55
0.55	16,000	425	8,500	70	0.53

Fixed-Film Reactor System				
Loading Rate (kg COD/1000 m^2 · day)	Influent COD (mg/L)	Effluent COD (mg/L)	Methane (%)	Methane Yield (m^3/kg $COD_{removed}$)
4.6	2,512	131	77	0.78
10.3	5,696	215	75	0.57
19.6	10,100	271	70	0.60
37.6	18,445	756	60	0.54

Source: Adapted from Stover et al. (1984).
[a] Data based on SCOD.

8.5.4 Evaluation of Methane Yield

Stover et al. (1984) estimated the potential methane production from 227.1 m^3/day (60,000 gal/day) of thin stillage to be 4,247.5 m^3/day (150,000 ft^3/day) based on continuous digestion studies and influent characteristics from Table 8.4 (see Box 8.1). This is equivalent to a methane yield of 0.29 m^3/kg COD_{added}. With the influent SCOD/TCOD ratio of 0.48, the calculated methane yields from Table 8.11 range from 0.25 to 0.37 $m^3 \cdot CH_4/kg \cdot COD_{removed}$, with an average of 0.29 $m^3 \cdot CH_4/kg \cdot COD_{removed}$. This methane yield data could be used to estimate the bioenergy production potential of thin stillage. Readers must be aware that the methane yield data in this research were extrapolated based on SCOD concentrations.

Box 8.1

Impact of Thin Stillage Digestion on Net Energy Balance

With a heating value of 1,000 Btu/ft^3 of CH_4, 130 MBtu/day could be generated from 4,247.5 m^3/day (150,000 ft^3/day) of methane gas. Stover et al. (1984) estimated a total energy requirement of 220 Mbtu/day in the production processes. Thus, if stillage evaporation were replaced with an anaerobic digestion, the methane produced could comprise up to 60% of the daily energy requirement for an ethanol plant. Anaerobic digestion could eliminate the energy needed for evaporation of thin stillage (28,400 Btu/gal ethanol from the total of 97,850 Btu/gal ethanol) and produce an additional 36,000 Btu/gal ethanol in the form of methane gas.

Table 8.11. Anaerobic digester performance.

Stillage Type	Effluent pH	Effluent Bicarbonate Alkalinity (mg/L $CaCO_3$)	VFA (mg HAc/L)	CH_4 Yield (m^3/kg VS Added)	CH_4(%)
Screened	7.7	7,909	2,664	0.57	63
Pressed	7.8	6,426	2,780	0.54	64

One of the limitations of the Stover's work is that it disregards the hydrolysis of particulate matter for methane production. A portion of the total COD is converted into soluble COD during digestion and contributes to methane production. The methane yield data may not accurately represent the true yield during digestion. The study examined anaerobic digestion at an HRT of 9.1 days, which may not be applicable if the feed contains particulate matter. Also, the methane yield estimation considers an influent SCOD concentration of 60 g/L that is much higher than the 2–18 g/L present in the feed used in the digestion studies.

Seely and Spindler (1981) also examined the thin stillage digestion option for bioenergy production from a farm-scale ethanol plant at Schroder Farm Alcohol, Campo, CO, USA. The thin stillage samples were obtained following solids removal using a Sweco screen separator. The retained solids were also digested separately after pressing, using a diamond screw press. The anaerobic digesters were operated at a mesophilic temperature (37°C) and at an HRT of 15 days. The digesters were initially started with dairy cow manure. The manure was subsequently blended with various fractions (50, 75, and 100% as VS) of thin stillage. Interestingly enough, blends of stillage and manure (50/50 and 75/25) did not require external alkalinity supplementation, even with the thin stillage at pH 3.4 and with no inherent alkalinity. The thin stillage, however, needed pH adjustment (to 6.8) when digested without manure. The authors of this study did not detail thin stillage characteristics. Digester performance using 100% thin stillage is illustrated in Table 8.11.

Based on energy balance calculations, the plant consumed 29,020 Btu/gal ethanol, and the produced biogas could comprise up to 48% of this plant's total energy need.

8.5.5 Stillage Digestion at Thermophilic Conditions

The higher metabolic rate observed under thermophilic conditions results in nearly double methanogenic activity compared to mesophilic conditions. This translates into a thermophilic digester that is half the size of that required for mesophilic digester of a similar configuration. Other advantages of thermophilic digestion are the enhanced hydrolysis of particulate matter in the stillage, increased volatile solids destruction, and enhanced liquid–solid separation (Ahring 1995). The increased heat requirement for the feed of ambient temperatures and reduced supernatant quality are the major disadvantages (Buhr and Andrews 1977). Since stillage exits in ethanol plant at higher temperature, thermophilic digestion appears to be an

Table 8.12. Trace element composition.

Chemical	Concentration (mg/L)
$FeCl_3 \cdot 4H_2O$	10,000
$CoCl_2 \cdot 6H_2O$	2,000
Ethylenediaminetetraacetic acid (EDTA)	1,000
$MnCl_2 \cdot 4H_2O$	500
Resazurin	200
$NiCl_2 \cdot 6H_2O$	142
Na_2SeO_3	123
$AlCl_3 \cdot 6H_2O$	90
H_3BO_3	50
$ZnCl_2$	50
$(NH_4)_6Mo_7O_{24} \cdot 6H_2O$	50
$CaCl_2 \cdot 2H_2O$	38
HCl (mL/L)	1

Source: Adapted from Zehnder et al. (1980).

ideal choice. A two-stage temperature or acid-phase digestion could alleviate the problem of poor effluent quality. The first thermophilic thin stillage digestion study was conducted at Iowa State University (Schaefer 2006), and the major findings are summarized in the following section.

8.5.5.1 Reactor Start-Up and Operation

Two 10-L thermophilic CSTRs started using anaerobically digested sewage sludge from a local thermophilic digester. The digesters were initially started with an F/M (food/microorganism) ratio of 0.5 and were filled with deoxygenated tap water. The digester pH was maintained at 7.0. The initial start-up consisted of long acclimation periods in a batch-mode operation so as to cultivate a viable population of thermophilic microbes. The digesters were fed with additional thin stillage when biogas production either stabilized or dropped and VFAs were stable or declining. The substrate used was full-strength thin stillage, the composition of which is illustrated in Table 8.6. The stillage was periodically supplemented with 1 mL of a trace mineral solution per 20 g COD (Table 8.12). The digesters were operated at HRTs of 30, 20, 15, and 12 days and believed to have reached quasi-steady state when the working volume had been displaced at least three times.

8.5.5.2 Performance of Anaerobic Digester

The performance of anaerobic digesters measured by VS destruction, stability, and methane production under quasi-steady-state conditions is summarized in Table 8.13.

TS/VS Removal

The digesters achieved TS removal of 76, 84, and 75% at HRTs of 30, 20, and 15 days, respectively. At an HRT of 12 days, VFA accumulation indicated that

Table 8.13. Summary of anaerobic digester performance.[a]

Parameters			HRT (days) 30	20	15	12[b]
COD (g/L)						
	Influent	Total	97.1	121.0	96.1	90.7
		Soluble	59.0	76.0	51.0	NA
	Effluent	Total	17.5	14.0	18.0	26.4
		Soluble	2.0	2.1	5.9	13.0
Solids (%)						
	Influent	TS	6.89	9.03	6.59	7.22
		VS	6.19	8.35	5.91	5.23
		TSS	2.77	3.42	2.54	2.95
		VSS	2.67	3.29	2.48	2.71
	Effluent	TS	1.68	1.48	1.65	2.32
		VS	1.09	0.85	0.93	1.48
		TSS	1.16	1.02	1.13	1.33
		VSS	1.00	0.86	0.94	1.10
VFA (mg/L as acetic acid)			160	200	2,400	6,300
Alkalinity (g/L as $CaCO_3$)			4.5	4.0	3.9	4.4
pH			7.44	7.17	7.09	6.86
Loading rate (kg·COD/m^3·day)[c]			3.2	6.1	6.4	7.6
Methane yield (m^3/kg·VS_{fed})			0.617	0.567	0.621	0.462
Methane yield (m^3/kg·$VS_{removed}$)			0.748	0.631	0.737	0.644
Methane yield (m^3/kg·COD_{fed})			0.393	0.391	0.382	0.266
Methane percentage			60.3%	56.8%	57.3%	52.6%

Source: Adapted from Schaefer (2006).
[a] Mean value after a minimum of four consecutive days of quasi-steady state.
[b] Based on one set of data due to digester instability.
[c] Multiply by 62.4 for lb·COD/1,000 ft^3·day.

the digester was not performing satisfactorily. The TS removal rate was less than that observed with the other HRTs, and the digester never reached a steady-state condition. The VS removal also showed similar trend with efficiencies of 82, 90, and 84%, respectively, at HRTs of 30, 20, and 15 days. In spite of an unexpectedly high 9% TS in the feed at 20-day HRT, the digester performance remained fairly stable, an indicator that a well-adapted thermophilic digester is capable of adsorbing shock load.

Alkalinity and VFA

Although the thin stillage contained nearly zero alkalinity, supplementation was not needed during digestion except 12 days. The effluent alkalinity was consistently high around 3.9–4.5 g/L as $CaCO_3$ at HRTs between 30 and 15 days. VFA concentrations increased with decreasing HRTs. Alkalinity addition of 6 g/L as $CaCO_3$ was needed at the 12-day HRT to maintain a neutral pH, due to VFA accumulation. Effluent VFA concentrations up to 2,400 mg/L were observed at the 15-day HRT, indicating the onset of instability. The digester was still able to operate, however, without failure.

The presence of organic nitrogen in the thin stillage, particularly in the form of protein, helped contribute in situ alkalinity. Hydrolysis of organic nitrogen during digestion resulted in formation of organic acid and ammonia as shown in Eq. (8.1). The ammonia then combined with carbon dioxide to form ammonium bicarbonate (Eq. (8.2)).

$$\underset{\text{Organic nitrogen}}{CHNH_2COOH} + 2H_2O \rightarrow RCOOH + NH_3 + CO_2 + 2H_2 \tag{8.1}$$

$$NH_3 + H_2O + CO_2 \rightarrow \underset{\text{(Alkalinity)}}{NH_4^+ + HCO_3^-} \tag{8.2}$$

A two-phase acid/temperature digestion and pretreatment of stillage may also enhance hydrolysis. This strategy can help provide sufficient alkalinity and reduce times needed for digestion.

Methane Yield and Biogas Composition

Methane yield is an important indicator of bioenergy production. The methane yield varied from 0.567 to 0.612 $m^3/kg{\cdot}VS_{added}$ (or 0.382–0.393 $m^3/kg{\cdot}COD_{added}$) for HRTs greater than 15 days, but dropped significantly to 0.462 $m^3/kg{\cdot}VS_{added}$ (or 0.266 $m^3/kg{\cdot}COD_{added}$) at an HRT of 12 days due to digester upset. The methane content of the generated biogas varied from 56.8 to 60.3% at HRTs between 30 and 15 days.

In another thermophilic digestion study, Zhang et al. (2001) employed a full-scale EGSB reactor to treat thin stillage, obtained following dewatering using a sheet and frame pressurized filter (see Table 8.6). A COD removal efficiency exceeding 90% was reported at an organic loading rate of 29 kg $COD/m^3{\cdot}day$ and an HRT of 24 h. Lime was added to maintain a neutral pH during digestion. During 2 weeks of operation conducted without alkalinity supplementation due to an availability of lime shortage, the reactor's performance remained fairly stable. The effluent suspended solids varied from 2 to 20 g/L with an average of 6 g/L, depending on production conditions. The authors observed continuous washout of nondigestable inert solids due to a high upflow velocity of 5–6 m/h. The biogas produced was used for firing the boiler.

8.5.6 Performance Comparisons

Table 8.14 summarizes important anaerobic digestion studies conducted on corn thin stillage to date. The data clearly show significantly higher methane yields for thermophilic digesters.

Table 8.14. Performance comparison of anaerobic reactors for thin stillage digestion.

Organic Loading Rate				Methane Yield		
kg VS/m^3·day	kg COD/m^3·day	Temperature	Reactor Type	m^3/kg VS_{fed}	m^3/kg COD_{fed}	References
1.6		Mesophilic	CSTR	0.54		Seely and Spindler (1981)
		Mesophilic	CSTR		0.29	Stover et al. (1984)
	3.2	Thermophilic	CSTR		0.39	Schaefer (2006)
	6.1	Thermophilic	CSTR		0.39	Schaefer (2006)
	6.4	Thermophilic	CSTR		0.38	Schaefer (2006)

8.6 Cassava-Based Ethanol Production

Cassava (*Manihot esculenta*) tubers are nearly 80% starch and <1.5% protein, on a dry weight basis. Pretreatments such as cleaning, peeling, and chipping of cassava tubers are conducted prior to drying. Dried cassava chips are then used for ethanol production. The ethanol production process from cassava chips follows

Box 8.2

Energy Recovery from Methane Gas

Energy costs are second only to the raw material (corn) costs in a dry-grind ethanol plant. Pimental and Patzek (2005) have long been critics of the "net energy balance" for corn ethanol. Several investigators have adequately shown that the corn-to-ethanol process yields a net positive energy balance (Farrell et al. 2006). Recovery of bioenergy from low-value coproducts such as stillage can further improve the net energy balance. Based on anaerobic digestion data, the following calculation shows the bioenergy recovery potential of methane gas from thin stillage produced by a 50-MGY plant:

Mean methane yield = 0.50 m^3/kg·VS_{fed} (conservative assumption)
Stillage produced by a 50-MGY plant = 113.6 m^3/h (500 gpm)
Total VS produced = 113.6 (m^3/h) × 24 (day/h) 59.1 (kg/m^3)
= 161,130 kg/day
Total methane generation = 80,565 m^3/day
Energy value of 1 m^3 methane = 35,310 Btu
Total energy produced from methane gas = 2,845 MBtu/day
= 1,038,425 MBtu/year

The annual fuel consumption of a typical dry-grind ethanol plant is 1,616,500 MBtu for a natural gas-based plant, and nearly two-thirds (1,075,000 MBtu) of this is consumed as boiler fuel for steam generation (EPA CHP Partnership 2006). Thus, methane gas could entirely replace the boiler fuel in a dry-grind process.

Table 8.15. Characteristics of cassava stillage.

Parameters	Values
COD (g/L)	81.1
BOD_5 (g/L)	31.4
Total solids (g/L)	44.5
Soluble solids (g/L)	40.4
Settled solids (g/L)	4.1
Organic matter (g/L)	37.1
Carbohydrates (g/L)	20.1
Total nitrogen (g/L)	0.65
Total phosphate (g/L)	0.38

Source: Adapted from de Menezes (1989).

nearly the same step as that for cornstarch illustrated in Fig. 8.1. The volume of stillage produced varies from 10 to 16 L/L of ethanol (2.6–4.2 gal/gal of ethanol), which appears to be in the same range that reported for a dry-grind corn-ethanol process (see Box 8.2).

Cassava-based stillage is equally rich in organic content (Table 8.15), but it lacks a suitable nutrient balance as an animal feed. This requires an entirely different stillage processing and utilization scheme. For such a stream, bioenergy recovery through anaerobic digestion could be one of the practical approaches to improve the net energy balance for such a plant. Studies on the anaerobic digestibility of cassava stillage are currently unavailable. It is likely that addition of macro- and micronutrients may be necessary to improve the biomethane production potential. The fibrous nature of the cassava stillage could impede hydrolysis, requiring pretreatment.

Example 8.1

Calculate the total energy production potential of stillage from a 1 million L/year cassava ethanol plant at thermophilic conditions.

Assume:

- Stillage generation: 16 L/L ethanol produced
- Methane yield: 0.3 $m^3/kg\ COD_{rem.}$
- COD removal efficiency: 70%

Total daily stillage generation $= 16 \times 1 \times 10^6 = 16 \times 10^6$ L
COD concentration in stillage $= 81.1$ g/L (Table 8.15)
Total daily COD produced $= 81.1\ (g/L) \times 10^{-3}\ (kg/g) \times 16 \times 10^6\ (L)$
$= 1{,}297{,}600$ kg
Total daily COD removed $= 1{,}297{,}600\ kg \times 0.70 = 908{,}320$ kg
The daily methane production $= 0.3 \times 908{,}320 = 272{,}496\ m^3$
Energy value of 1 m^3 methane = 35,310 Btu
Total energy produced $= 272{,}496 \times 35{,}310 = 9{,}622$ MBtu/day

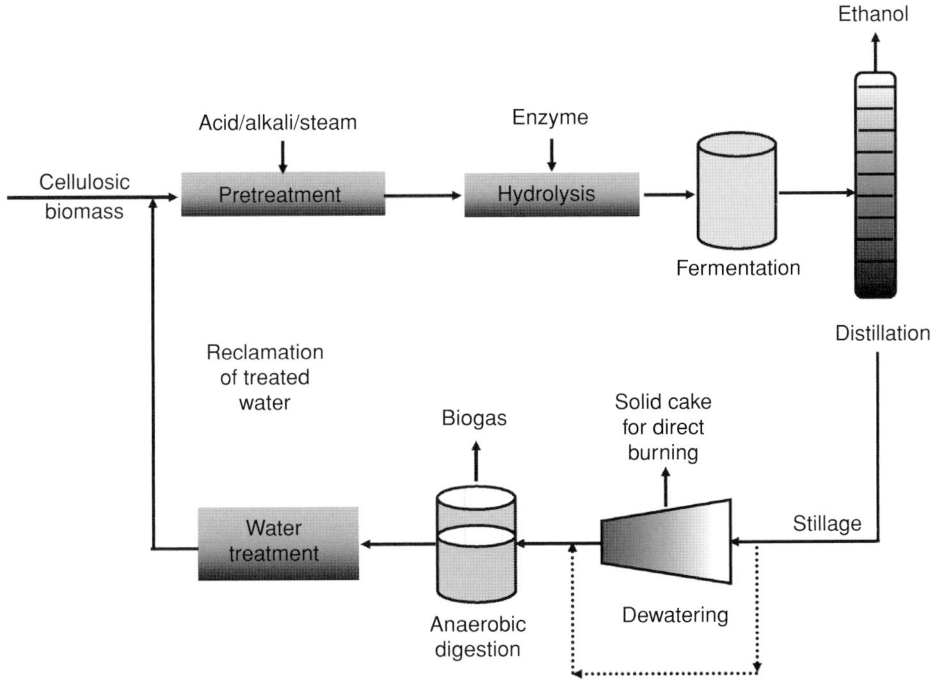

FIG. 8.2. Schematics of cellulosic ethanol production process.

8.7 Cellulose-Based Ethanol Production

There is great hope that lignocellulosic biomass can be used to produce liquid transportation fuel. However, due to its structural complexity (heterogeneity and crystallinity), direct utilization of biomass by microbes is extremely slow. It needs comprehensive pretreatment with acid, alkali, or steam explosion before enzymatic hydrolysis to release its underlying monomeric sugars for practical production of ethanol. The schematic diagram of a cellulose-based ethanol production process and stillage production step is shown in Fig. 8.2. A solid–liquid separation unit may be needed to remove recalcitrant cellulosic fibers from the stillage.

Although there is no full-scale cellulosic ethanol plant currently in operation, a laboratory-scale experiment did compare favorably in both quantity and stillage characteristics of stillage with respect to conventional feedstocks (Wilkie et al. 2000). The authors summarized the following data based on comprehensive literature review: stillage yield, 11.1 ± 4.1 L/L ethanol; COD, 61.3 ± 40 g/L; biochemical oxygen demand (BOD), 27.6 ± 15.2 g/L; N (total), 2.8 ± 4.6 g/L; P (total), 28 ± 30 mg/L; sulfate, 651 ± 122 mg/L; and pH, 5.35 ± 0.53. Stillage from a cellulosic ethanol plant could contain inhibitory compounds such as salts resulting from chemical treatment and biomass-derived phenolic compounds

(Wilkie et al. 2000). The authors further summarized the anaerobic digestibility of cellulosic stillage as follows: organic loading rate, 9.5 ± 5.4 kg COD/m^3·day; COD removal efficiency, 84 ± 7; and methane yield, 0.3 ± 0.1 m^3/kg COD_{fed}.

8.8 Bioenergy Recovery from Crude Glycerin

Glycerin or glycerol is a major byproduct formed during biodiesel production. As a rule of thumb, about 10 lb of glycerin is produced for every 100 lb of biodiesel. Increased biodiesel production may result in large amounts of glycerin that some may consider a "waste product." Crude glycerin has an extremely high organic strength and contains a substantial amount of methanol (~20–30% w/w). Ito et al. (2005) reported an organic strength of 570 g TOC/L with 41% (w/w) glycerin, 25% (w/w) methanol, and 8% (w/v) ash. In another study at Iowa State University, the COD of crude glycerine was 1,430 g/L with 70% glycerine and 30% methanol. Direct utilization of crude glycerin by microbes is extremely difficult due to substrate inhibition and salt toxicity. Dilution or codigestion with other waste streams could alleviate the inhibitory effects. A few strategies for renewable energy generation from crude glycerin are presented in the following sections.

8.8.1 Biomethane Production from Crude Glycerin

High-strength glycerin has the potential to produce methane gas. Theoretically, 0.43 L CH_4/g glycerin can be produced from 100% glycerin at standard temperature and pressure (STP).

A BMP test conducted at Iowa State University with crude glycerin at glycerin/swine manure (g/g VS) ratios of 1.2, 2.3, 4.6, and 9.3 produced methane yields (mL/g glycerin) of 232, 1, negative, and negative, respectively, compared to control. A negative yield indicates that the methane production was lower than the control, and at a glycerin/swine manure ratio of 4.6, methane generation was essentially zero. High glycerin levels appear to be toxic to methanogens due to a high salt and methanol level or due to substrate/product inhibition.

8.8.2 Hydrogen and Ethanol from Crude Glycerin

The hydrogen and ethanol production potential of biodiesel waste (crude glycerin) using *Enterobacter aerogenes* HU101 was examined by Ito et al. (2005). The authors reported molar hydrogen yields of 1.12, 0.90, 0.71, and 0.71 mol/mol glycerin at glycerin concentrations of 1.7, 3.3, 10, and 25 g/L, respectively, during batch testing. The respective yields for ethanol were 0.96, 0.83, 0.67, and 0.56 mol/mol glycerin. Interestingly, glycerin utilization was more efficient at the lower glycerin concentrations of 1.7 and 3.3 g/L, requiring just 4 h for complete utilization, whereas complete utilization of glycerin was not possible at higher concentration of 25 g/L during more than 48 h of incubation. A high salt level was the major cause of poor microbial activity. The crude glycerin needed dilution and supplementation of nutrients for efficient utilization by *E. aerogenes.*

References

Ahring, B. K. 1995. Methanogenesis in thermophilic biogas reactors. *Antonie Leeuwenhoek* 67(1):91–102.

Buhr, H. O., and Andrews, J. F. 1977. The thermophilic anaerobic digestion process. *Water Res.* 11:129–143.

Burggraf, S., Fricke, H., Neuner, A., Kristjansson, J., Rouviere, P., Mandelco, L., Woese, C. R., and Stetter, K. O. 1990. *Methanococcus ingeus* sp. nov., a novel hyperthermophilic methanogen from a shallow submarine hydrothermal system. *Syst. Appl. Microbiol.* 13:33–38.

de Menezes, T. J. B. 1989. The treatment and utilization of alcohol stillage. In *International Biosystems,* Vol. III, edited by D. L. Wise, pp. 1–14. CRC Press, Boca Raton, FL, USA.

EPA CHP Partnership. 2006. Available at: http://www.epa.gov/chp/documents/ethanol_energy_balance.pdf

Farrell, A. E., Plevin, R. J., Turner, B. R., Jones, A. D., O'Hare, M., and Kammen, D. M. 2006. Ethanol can contribute to energy and environmental goals. *Science* 311:505–508.

Fink, R. J. 2007. Ethanol Reality Check. In *Proceedings of the 5th International Starch Technology Conference – Energy Issues,* edited by K. Rausch, V. Singh, and M. Tumleson, pp. 56–67, University of Illinois at Urbana-Champaign, Urbana, IL.

Hunter, G. R. 1988. *Comparison of Anaerobic Systems for Treatment of Ethanol Stillage.* Ph.D. dissertation, University of Oklahoma, Oklahoma City, OK, USA.

Ito, T., Nakshimada, Y., Senba, K., Matsui, T., and Nichio, N. 2005. Hydrogen and ethanol production from glycerol-containing wastes discharges after biodiesel manufacturing process. *J. Biosci. Bioeng.* 100(3):260–265.

Jasti, N., Khanal, S. K., Pometto, A. L., and Van Leeuwen, J. (2006). Fungal treatment of corn processing wastewater in an attached growth system. *Water Sci. Pract.* 1(3):1–8.

Jeanthon, C., L'Haridon, S., Reysenbach, A. L., Vernet, M., Messner, P., Sleytr, U. B., and Prieur, D. 1998. *Methanococcus infernus* sp. nov., a novel hyperthermophilic lithotrophic methanogen isolated from a deep-sea hydrothermal vent. *Int. J. Syst. Bacteriol.* 48:913–919.

Jones, W. J., Leigh, J. A., mayer, F., Woese, C. R., and Wolfe, R. S. 1983. *Methanococcus jannaschii,* sp. nov., an extremely thermophilic methanogens from a submarine hydrothermal vent. *Arch Microbiol.* 136:254–261.

Khanal, S. K. 2005. *Anaerobic Digestion of Thin Stillage to Produce Methane and Class—A Biosolids.* Bioeconomy Conference, August 29–30, 2005, Ames, IA, USA.

Lanting, J., and Gross, R. L. 1985. *Anaerobic Pretreatment of Corn Ethanol Production Wastewater,* pp. 905–914. Proceedings of the 40th Industrial Waste Conference, Purdue University, West Lafayette, IN, USA. Butterworth, Boston, USA.

Pimental, D., and Patzek, T. W. 2005. Ethanol production using corn, switch grass, and wood, biodiesel production using soybean and sunflower. *Nat. Resour. Res.* 14(1):65–76.

Pohland, F. G. 1992. Anaerobic treatment: Fundamental concept, application, and new horizons. In *Design of Anaerobic Processes for the Treatment of Industrial and Municipal Wastes,* edited by J. F. Malina, Jr, and F. G. Pohland, pp. 1–40. Technomic Publishing Co. Inc., Lancaster, USA.

Rosentrater, K. A., Hall, H. A., and Hansen, C. L. 2006. *Anaerobic Digestion Potential for Ethanol Processing Residues.* In Proceedings an ASABE Annual International Meeting (Paper # 066167). July 9–12, 2006, Portland, OR, USA.

Schaefer, S. 2006. *Retooling the Ethanol Industry: Thermophilic, Anaerobic Digestion of Thin Stillage for Methane Production and Pollution Prevention.* M.S. thesis, Iowa State university, Ames, IA, USA.

Seely, R. J., and Spindler, D. D. 1981. *Anaerobic Digestion of Thin Stillage from a Farm-Scale Alcohol Plant.* In Proceedings of Energy from Biomass and Wastes V. January 26–30, 1981, Lake Buena Vista, FL, USA.

Singh V., Rausch, K. D., Yang, P., Shapouri, H., Belyea, R. L., and Tumbleson, M. E. 2001. *Modified Dry Grind Ethanol Process.* Publication of the Agricultural Engineering Department—University of Illinois at Urbana, Champaign UILU No. 2001–7021, July 18, 2001.

Stanich, T. 2007. Water efficiency as a result of sound water management. *Ethanol Producer Magazine* (February issue).

Stover, E. L., Gomathinayagam, G., and Gonzalez, R. 1983. *Anaerobic Treatment of Fuel Alcohol Wastewater by Suspended Growth Activated Sludge,* pp. 95–104. Proceedings of the 38th Industrial Waste Conference, Purdue University, West Lafayette, IN, USA. Butterworth, Boston, USA.

Stover, E. L., Gomathinayagam, G., and Gonzalez, R. 1984. *Use of Methane from Anaerobic Treatment of Stillage for Fuel Alcohol Production,* pp. 57–63. Proceedings of the 39th Industrial Waste Conference, Purdue University, West Lafayette, IN, USA. Butterworth, Boston, USA.

Tiffany, D. G., Morey, R. V., and DeKam, M. 2007. *Economics of Biomass Gasification and Combustion at Fuel Ethanol Plants,* pp. 76–85. In Proceedings of the 5th International Starch Technology Conference—Energy Issues, edited by K. Rausch, V. Singh, and M. Tumbleson, June 3–6, 2007. University of Illinois at Urbana-Champaign, Urbana, IL, USA.

USDA Report. 2006. The *Economic Feasibility of Ethanol Production from Sugars in the United States.* Available at: http://www.usda.gov/oce/EthanolSugarFeasibilityReport3.pdf (accessed June 18, 2007).

USDA and USDOE Joint Report 2005. *A Billion-Ton Feed Stock Supply for Bioenergy and Bioproducts Industry: Technical Feasibility of Annually Supplying 1 Billion Dry Tons of Biomass.* Joint Report, U.S. Department of Agriculture and U.S. Department of Energy, February 2005.

Wilkie, A. C., Riedesel, K. J., and Owens, J. M. 2000. Stillage characterization and anaerobic treatment of ethanol stillage from conventional and cellulosic feedstocks. *Biomass Bioeng.* 19:63–102.

Zehnder A. J. B., Huser, B. A., Brock, T. D., and Wuhrmann, K. 1980. Characterization of an acetate-decarboxylating, non-hydrogen-oxidizing methane bacterium. *Arch. Microbiol.* 124(1):111

Zhang, Z., Zhou, W., and Lin, R. 2001. Full-scale experiment on corn-ethanol production wastewater treatment with the thermophilic EGSB reactor. *Huangjing Kexue/Environ. Sci.* 22(4):114–116 (in Chinese).

CHAPTER

Biohydrogen Production: Fundamentals, Challenges, and Operation Strategies for Enhanced Yield

Samir Kumar Khanal

9.1 Background

The growing energy crisis, soaring fuel costs, and environmental concerns are major forces driving exploration of alternative energy sources. In addition, industrial nations wish to reduce their dependence on imported petroleum fuels. Hydrogen, an alternative energy carrier, can be produced renewably from domestic feedstocks such as organic wastes, biomass-derived sugars, and wastewaters.

Traditionally, methane recovery has been the major focus of anaerobic biotechnology. In recent years, the focus has shifted toward hydrogen production for several reasons. Hydrogen is a highly efficient energy source compared to methane, producing 143 kJ energy/g H_2 compared to 56 kJ energy/g of CH_4. Hydrogen can be directly used in fuel cell. In addition, it has other commercial values such as raw material for ammonia, ethanol and aldehyde synthesis, and hydrogenation of edible oil and fossil fuels (Hart 1997). From an environmental perspective, hydrogen is a clean energy source, producing water as its only byproduct when burned. This chapter describes biohydrogen production through dark fermentation, including the hydrogen production pathway, strategies for obtaining enriched cultures, important considerations in biohydrogen production, limitations of biohydrogen production, and possible remedial measures. In addition, a technoeconomic analysis of biohydrogen production is also covered.

Table 9.1. Different biological hydrogen production processes.

Biological Process	Microbial Group	Process Description	Disadvantages
Photosynthesis	Cyanobacteria	Cyanobacteria (or blue-green algae) are autotrophs and use CO_2 as a carbon source. They break down water into hydrogen and oxygen in presence of light energy	The process requires light energy. Carrier gas is needed to collect the evolved gas from the culture. Separation of oxygen and hydrogen is another limiting factor
Light fermentation	Phototrophic purple nonsulfur bacteria	These are heterotrophs and produce hydrogen using simple organic matter as a carbon source and light as an energy source under anaerobic conditions	Efficient light penetration and distribution in a highly turbid culture media is a major rate-limiting condition. The process can only use simple organic substrates
Dark fermentation	Nonphototrophic fermentative bacteria	Heterotrophs that produce hydrogen using complex organics as both carbon and energy sources under anaerobic conditions	The yield of hydrogen production is relatively low. Hydrogen partial pressure needs to be controlled at relatively low levels to enhance hydrogen yield

9.2 Biological Hydrogen Production

Hydrogen can be generated in a number of ways. Electrolysis electrically splits water into hydrogen and oxygen, chemical methods such as hydrocarbon cracking produce hydrogen, and biological methods such as dark and photo fermentation result in hydrogen production through microbial-mediated reactions. Although hydrogen production by electrolysis of water appears to be environmentally friendly, it requires large inputs of electricity derived from fossil fuel. Similarly, hydrogen production by thermocatalytic decomposition of hydrocarbon requires huge amounts of fossil-fuel-derived energy. Thus, both electrolysis and thermocatalytic approaches are not ideal methods for sustainable hydrogen production.

There are three microbial groups that have been studied for biological hydrogen production as shown in Table 9.1. The first group consists of the cyanobacteria that are autotrophs and directly decompose water to hydrogen and oxygen in the presence of light energy by photosynthesis (Hallenbeck and Benemann 2002). Since this reaction requires only water and sunlight and generates oxygen, it is attractive from an environmental perspective. However, the cyanobacteria examined thus far show rather low rates of hydrogen production and may have difficulty overcoming large Gibb's free energy (+237 kJ/mol hydrogen) requirements.

The second and third groups of bacteria are heterotrophs and use organic substrates as a carbon source for hydrogen production. The heterotrophic microorganisms produce hydrogen under anaerobic conditions, both in the presence or absence of light energy. Accordingly, the process is classified as either photofermentation or dark fermentation. Phototrophic purple nonsulfur bacteria produce hydrogen through photofermentation, and nonphototrophic fermentative bacteria produce hydrogen through dark fermentation. Thermodynamically, hydrogen production through photofermentation is not favorable unless light energy is supplied. In addition, only phototrophic bacteria are able to convert simple organic compounds such as organic acids to hydrogen, which thus limits the use of complex organic wastes.

Dark fermentation continuously produces hydrogen from renewable sources such as carbohydrate-rich wastes without an input of external energy, a considerable advantage. This is the main focus of this chapter.

9.3 Microbiology of Dark Fermentation

The hydrogen-producing microbes in an anaerobic fermentation process can be classified into two categories, namely (a) facultative anaerobes (or enteric bacteria, e.g., *Escherichia coli*, *Enterobacter*, and *Citrobacter*); and (b) strict anaerobes (clostridia, methylotrophs, methanogenic bacteria, and rumen bacteria). In a mixed-culture environment, these microbes may coexist and their abundance is governed by the relative competitiveness for the available substrates and the environmental conditions. Based on terminal restriction fragment length polymorphism (T-RFLP) analysis, about 70–80% of the total population identified in a mixed-culture hydrogen-producing bioreactor belonged to the genera *Clostridium* (Duangmanee et al. 2002). In another study using 16S rDNA-based technique, Fang et al. (2002) reported 65–69% *Clostridium* sp. in a mixed-culture community fermenting simple sugars such as glucose and sucrose. Among hydrogen-producing microbes, clostridia and enteric bacteria have been extensively studied. A brief discussion on these two microbial groups is presented here. A detailed discussion on microbiological aspects of hydrogen-producing bacteria including their phylogenic classification and identification using molecular techniques is presented in Chapter 6.

9.3.1 Enteric Bacteria

E. coli and *Enterobacter* are the two enteric bacteria commonly studied in biohydrogen production through dark fermentation. They are rod-shaped, gram-negative and facultative anaerobes. The enteric bacteria are less sensitive to oxygen and are able to recover following accidental air exposure (Nath and Das 2004). The presence of oxygen, however, causes degradation of formate—a major precursor for hydrogen production, without hydrogen formation.

9.3.2 Clostridia

Clostridia are generally obligate anaerobes and are rod shaped with round or pointed ends in some cases. Rod shape can be either straight or slightly curved with

0.5–2 μm in diameter and up to 30 μm in length. One of the important characteristics of clostridia is its ability to form endospore. Endospore is a survival structure developed by these organisms when the environmental conditions become unfavorable (e.g., high temperature, desiccation, carbon or nitrogen deficiency, and chemical toxicity). When favorable conditions return, the spores germinate and become vegetative cells (Doyle 1989).

Clostridia species are capable of using different organic substrates such as carbohydrates, amino acids, purines, and pyrimidines by a variety of fermentation pathways. Thus, clostridia are classified as proteolytic or saccharolytic depending on the types of organics they ferment (Cato and Stackebrandt 1989). Some clostridia are, however, both saccharolytic or proteolytic and some are neither. Proteolytic clostridia degrade proteins or amino acids. Saccharolytic clostridia ferment carbohydrate and are widely studied because of their ability to produce higher levels of hydrogen. One of the widely studied saccharolytic clostridia is *Clostridium butyricum* that produces butyric acid as the major fermentation product together with CO_2, acetate, and H_2 (Minton and Clarke 1989). This pathway was found in approximately 50% of all clostridia that have been isolated to date. Other fermentation pathways found in sacchrolytic clostridia are those leading to the production of propionate by *Clostridium arcticum* (Jones and Woods 1989), succinate by *Clostridium coccoides* (Kaneuchi et al. 1976), and lactate by *Clostridium barkeri* (Stadtman et al. 1972).

9.4 Hydrogen Production Pathway through Dark Fermentation

In most biological systems, hydrogen is produced by the anaerobic metabolism of pyruvate formed during the catabolism of various organic substrates. Simple sugars such as glucose are metabolized to pyruvate through various pathways that often involve the Embden–Meyerhoff–Parnas (also known as glycolysis) and the Entner–Doudoroff pathways. The pyruvate is broken down to acetyl coenzyme, from which adenosine triphosphate (ATP) can be derived, and hydrogen is generated from the products formate or reduced ferredoxin (Fe(red)) depending on the types of microbial groups present in the system. The enteric bacteria derive hydrogen from formate, whereas obligate anaerobes (e.g., clostridia) derive hydrogen from (Fe(red)). The various enzyme systems involved in breakdown of pyruvate are presented as follows (Hallenbeck 2004):

(a) Pyruvate:formate lyase (PFL)
 Pyruvate + CoA → acetyl-CoA + formate
(b) Pyruvate:ferredoxin (flavodoxin) oxidoreductase (PFOR)
 Pyruvate + CoA + 2Fe(ox) → acetyl-CoA + CO_2 + 2Fe(red)

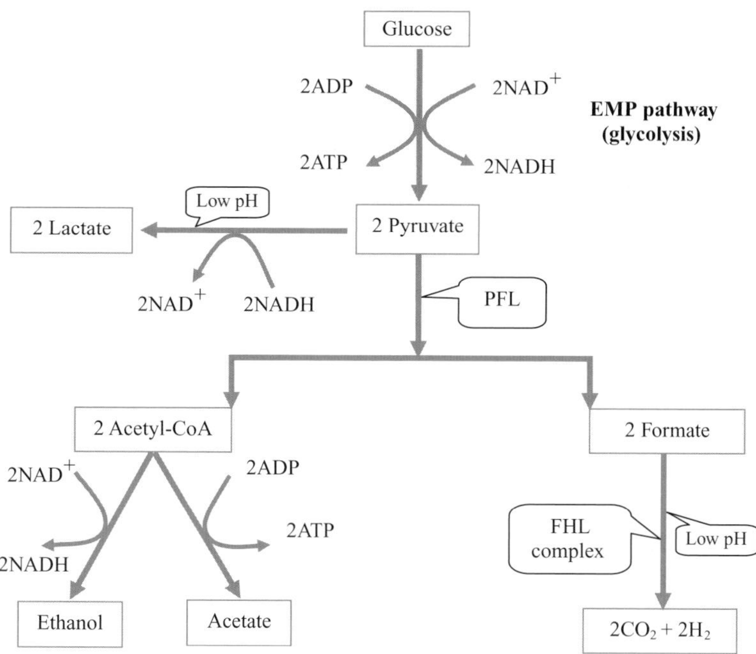

Fig. 9.1. Hydrogen production pathway of enteric bacteria during dark fermentation. *Source:* Adapted from Hallenbeck (2004).

A maximum hydrogen yield of 2 (enteric bacteria) or 4 (clostridia) mol/mol of glucose can be achieved. In the latter case, the hydrogen partial pressure must be extremely low for reduced nicotinamide adenine dinucleotide (NADH) to reduce protons to hydrogen. In reality, the yield is less than half the maximum as a portion of the reducing equivalents are diverted toward other reductive activities.

9.4.1 Fermentative Hydrogen Production by Enteric Bacteria

The hydrogen production pathway of dark fermentation by enteric bacteria is presented in Fig. 9.1. The carbohydrate-rich substrate is converted to pyruvate through glycolysis. The enteric bacteria generate acetyl coenzyme A (CoA) and formate through the enzyme PFL. Formate is then converted to hydrogen and carbon dioxide by enzyme formate hydrogen lyase (FHL) complex. An acidic condition causes induction of lactate dehydrogenase, thereby diverting some of the reducing power in pyruvate to lactate. This results in lower hydrogen yield. The enteric bacteria also produce other fermentation products such as lactate, ethanol, acetate, formate, carbon dioxide, succinate, and butanediol. The maximum yield by enteric bacteria during dark fermentation is 2 mol H_2/mol glucose.

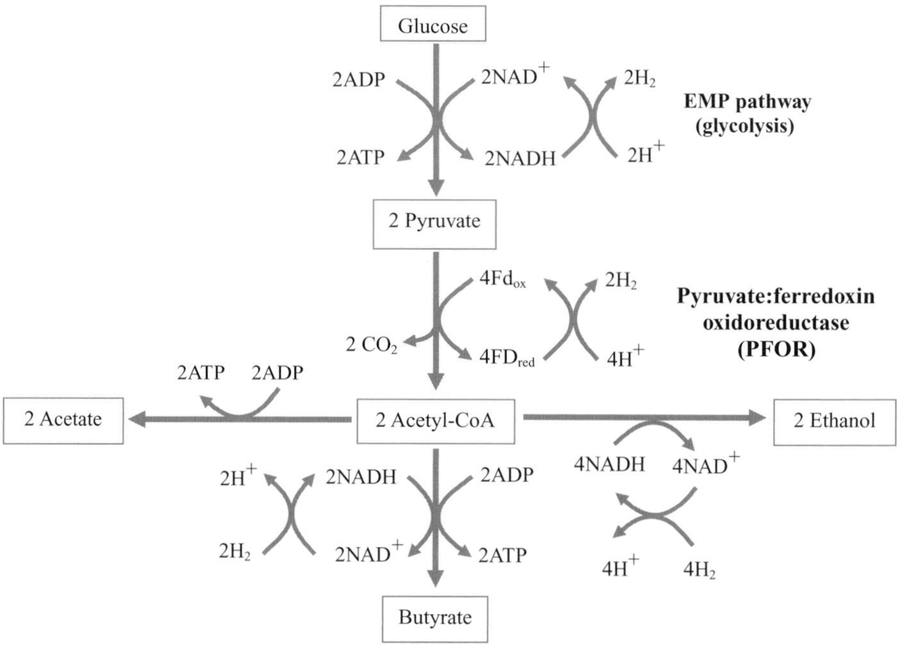

FIG. 9.2. Hydrogen production pathway of clostridia during dark fermentation. *Source:* Adapted from Hallenbeck (2004) and Li and Fang (2007).

9.4.2 Fermentative Hydrogen Production by Clostridia

The genus *Clostridium* has been widely studied for hydrogen production. The pathway of hydrogen production using glucose as a model substrate by clostridia is shown in Fig. 9.2. The organic substrate such as glucose is metabolized to pyruvate through glycolysis. Clostridia break down pyruvate into acetyl-CoA with the formation of Fe(red) using pyruvate:ferredoxin oxidoreductase (PFOR). The reduction of a proton by Fe(red) produces hydrogen through hydrogenase activity. The activity of hydrogenase, an iron-containing enzyme, is one of the most important factors in the overall hydrogen fermentation (Dabrock et al. 1992; Holt et al. 1988). There are a variety of end products (e.g., ethanol, acetate, acetone, butyrate, butanol, propionate, and propanol) that can be generated from acetyl-CoA depending on species and environmental conditions. The conversion of acetyl-CoA to ethanol or butyrate requires electrons that are supplied by NADH and oxidized to NAD^+. For generation of NADH, the electron is donated by hydrogen. Therefore, little or no hydrogen production occurs during the solvent (ethanol or butanol) production phase. Hydrogen production occurs mainly during the acid (acetate and butyrate) production phase. Hydrogen yield, however, is appreciably less with butyrate as a major fermentation product due to loss of hydrogen. The fermentation pathway is highly dependent on pH. Ballongue et al. (1987) reported that the large acid accumulations in the culture media triggered solvent production. The fermentation pathway shifts from acid production to solvent production as the pH drops below 4.5. It seems reasonable that in a natural environment, the shift to solvent production alleviates the inhibitory effect of acid accumulation. The increase in hydrogen partial pressure due to poor agitation has also been shown to be important in inducing solventogenesis (Doremus et al. 1985). The production of hydrogen and volatile fatty acids has been inversely correlated to alcohol production (Lay 2000).

Using glucose as a model substrate, hydrogen production with either acetate (Eq. (9.1)) or butyrate (Eq. (9.2)) formation can be given by (Miyake 1998):

$$C_6H_{12}O_6 + 2H_2O \rightarrow 2CH_3COOH + 2CO_2 + 4H_2, \quad \Delta G^\circ = -184 \text{ kJ} \tag{9.1}$$

$$C_6H_{12}O_6 \rightarrow CH_3CH_2CH_2COOH + 2CO_2 + 2H_2, \quad \Delta G^\circ = -255 \text{ kJ} \tag{9.2}$$

Thus, a maximum of 2 mol of hydrogen can be produced in the butyrate-type fermentation, whereas up to 4 mol of hydrogen can be obtained by the acetate-type fermentation. For acetate fermentation, the breakdown of pyruvate yields 2 mol H_2/mol glucose, and an additional 2 mol H_2/mol glucose is derived through NADH:ferredoxin oxidoreductase activity. The reduction of hydrogenase by NADH is energetically unfavorable under standard conditions unless an extremely low ($<10^{-3}$ atm) hydrogen partial pressure is maintained (Hallenbeck 2004).

Based on the Gibb's free energy change, butyrate fermentation is more energetically favorable and thus NADH is often used to promote butyrate fermentation. However, acetate and butyrate fermentations occur simultaneously during biological hydrogen production (Wood 1961). With butyrate as a predominant end product, the maximum hydrogen production may never exceed 2.5 mol H_2/mol glucose, as shown by Eq. (9.3):

$$C_6H_{12}O_6 + 0.5H_2O \rightarrow 0.75CH_3CH_2CH_2COOH + 0.5CH_3COOH + 2CO_2 + 2.5H_2 \tag{9.3}$$

The most likely reasons why clostridia favor the butyrate fermentation pathway during hydrogen production is the fact that the formation of one equivalent butyrate leads to less acidification of the organisms' environment than the two equivalents

Example 9.1

A biomass pretreatment plant produces 1,000 kg/h of mixed sugar stream. Calculate theoretical hydrogen production rate at STP if five-carbon sugars contribute 40% of the sugars.

Solution

1. *The theoretical molar yield of hydrogen from six carbons:*

   $C_6H_{12}O_6 + 2H_2O \rightarrow 2CH_3COOH + 8CO_2 + 4H_2$

   180 g 8 g

   Hydrogen production contributed by six-carbon sugars
   $= (8/180)\ g/g \times (1,000 \times 0.6)\ kg/h = 26.7\ kg/h$

2. *The theoretical molar yield of hydrogen from five carbons:*

   $3C_5H_{10}O_5 + 5H_2O \rightarrow 5CH_3COOH + 5CO_2 + 10H_2$

   150 g 20 g

   Hydrogen production contributed by five-carbon sugars
   $= (20/150)\ g/g \times (1,000 \times 0.4)\ kg/h = 53.3\ kg/h$

3. *Total hydrogen production rate:*
   $= 26.7 + 53.3 = 80\ kg/h$
   Or $(11.2)\ m^3/kg \times (80)\ kg/h = 896\ m^3/h$

Note: It is important to point out that hydrogen gas needs to be compressed significantly to obtain a desired mass. This is especially important when hydrogen is used in fuel cell.

of acetate. Thus, the generation of higher amount of butyrate may consume excess reducing equivalents (Ljungdahl et al. 1989).

9.5 Suppression of Hydrogen Consumers

In a traditional anaerobic digestion, where the main goal is to maximize methane yield, the produced hydrogen is quickly consumed by hydrogenotrophic methanogens. A part of the produced hydrogen is also consumed in the synthesis of acetate by homoacetogens. The produced hydrogen may also be consumed as an electron donor in sulfate reduction, autotrophic denitrification, and/or iron reduction, depending on whether sulfate, nitrate, iron is present in the feed. For sustainable anaerobic hydrogen production, therefore, the growth of hydrogen consumers needs to be repressed. Some of the strategies for obtaining an enriched culture of hydrogen-producing bacteria in a mixed-culture environment are use of chemicals (e.g., acetylene, 2-bromoethanesulfonate, chloroform, and limited oxygenation), acid/base treatment, heat treatment, kinetic control, and electric current.

9.5.1 Selective Inhibition of Hydrogen Consumers Using Chemicals

Most studies on biohydrogen production involve the inhibition of hydrogen consumers, especially methanogens. To date, acetylene and 2-bromoethanesulfonate (BES) have been used to inhibit methanogens. Acetylene has been known to inhibit the hydrogenase—a key enzyme in hydrogen production. It was previously mentioned that an acetylene concentration of 5% inhibits 50% of the hydrogenase activity in some cultures. BES could be used but there are mutants resistant to this chemical. In any long-term continuous study, however, microorganisms often develop resistance, increasing the require dose. And at high dosing levels, inhibitors can be bacteriostatic not only to hydrogen consumers but also to hydrogen producers. Therefore, from engineering perspective, such a strategy is not suitable and cost prohibitive.

9.5.2 Heat Treatment of Seed Sludge

Heat treatment is often employed to obtain enriched cultures of hydrogen-producing microbes. Heat treatment inactivates hydrogen-consuming microorganisms and therefore selects for hydrogen-producing bacteria. Part of the rationale for this idea is that many of the hydrogen-producing bacteria (e.g., *Clostridium* and *Bacillus* species) form endospores, which are "survival structures" developed by these organisms when unfavorable environmental conditions are encountered. When favorable conditions return, the spores germinate and become vegetative cells (Doyle 1989). A similar idea (heat application at 80°C for 10 min) has been used to eliminate non-spore-forming bacteria during isolation procedures of spore-forming bacteria, such as *Clostridium* species (Doyle 1989). Several studies employed heat treatment as a method to inactivate or eliminate these microorganisms. Lay (2000) and Okamoto et al. (2000) employed wet heat treatment (boiling for 15 min) of anaerobic digester sludge, whereas Van Ginkel et al. (2001) used dry heat treatment (baking at 104°C for 2 h) of compost and soils.

In addition to eliminating most vegetative cells (including hydrogen-consuming microorganisms), heat treatment activates spore germination. Germination of a spore involves three steps: activation, germination, and outgrowth, and heat treatment (or heat activation) is one way to initiate spore germination. Several researchers have studied heat activation of *Clostridium botulinum* and *Clostridium perfringens*, two human pathogens (Doyle 2002). The optimal temperature/time combination for heat activation of spores of *C. perfringens* strains T-65 and S-45 were 80°C/10 min and 75°C/120 min, respectively (Doyle 1989; Hui et al. 1994). The spores of some heat-resistant strains of *C. perfringens* can be activated by heat treatment at 100°C (Doyle 2002). In most cases, heat activation at 75–80°C for 15–20 min is employed to inactivate vegetative cells and activate germination of spores (Doyle 1989).

Based on number of studies conducted in the author's laboratory, a higher hydrogen production was achieved only for the first few days of reactor operation with initial heat-treated seed sludge; thereafter, it started to decline rapidly. In one study, the drop in hydrogen production was as high as 80%. Duangmanee et al. (2002) conducted a study using sucrose as an organic substrate in which the reactor biomass was repeatedly heat treated at 70°C for 20 min. Such heat treatment improved hydrogen production by as much as 70%. It is likely that heat treatment selectively eliminated hydrogen-consuming vegetative cells and facilitated the germination of hydrogen-producing spores. One study did report that heat treatment may not completely eliminate the hydrogen-consuming bacteria (Oh et al. 2003). The study reported the survival of homoacetogenic bacteria that consumed hydrogen for acetate production.

9.5.3 Acid/Base Treatment

Endospores are not only resistant to heat but also resistant to harmful chemicals including acids and bases (Brock et al. 1994). Maintenance of reactor pH beyond the optimal growth range of methanogens (6.8–7.4) would also discourage their growth. Therefore, applying low/high pH treatment to select spore formers is a more practical approach. Chen et al. (2002) demonstrated that anaerobic sludge pretreated with acid and base at pH 3.0 and 10.0, respectively, has been able to produce hydrogen.

9.5.4 pH Control

Since pH is of supreme importance for a *Clostridium*-rich bioprocess, biohydrogen production is dependent on its operating pH. Hydrogen production generates organic acids such as acetic, propionic, and butyric acids in the first (acidogenic) phase of anaerobic fermentation (Jones and Woods 1989). This implies that biohydrogen production is more favorable in the acidic range. Several researchers found that optimum operating pH for hydrogen production is around 5.5 with sucrose as the carbon source (Khanal et al. 2004; Liu and Fang 2003). Mizuno et al. (2000) found that a pH of 6.0 is optimal for hydrogen production when using glucose as an organic substrate. For hydrogen consumers, such as methanogens, the optimum pH is in the neutral range of 6.8–7.4. Thus, by operating a hydrogen-producing reactor at low pH range, the growth of hydrogen consumers, especially methanogens, could be suppressed. For each waste stream, however, the optimum pH needs to be evaluated individually. This is because the optimum pH for hydrogen production is governed by many factors, such as hydraulic retention time (HRT), solids retention time (SRT), substrate types, the mode of reactor operation (continuous or batch-mode feeding), experimental run times, and temperature, among others.

9.5.5 Kinetic Control

The specific growth rate (μ) and the yield coefficient ($Y_{X/S}$) of hydrogen-producing clostridia are much higher than the slow-growing hydrogen-consuming methanogens. Therefore, by selecting a shorter HRT in a continuous stirred tank reactor (CSTR), the slow-growing hydrogen consumers such as methanogens could be washed out from the system and selective growth of hydrogen-producing bacteria could be achieved. In a biokinetic study, Chen et al. (2001) reported maximum specific growth rate (μ_{max}) of 0.172 h^{-1} for sucrose utilizing hydrogen-producing cultures. The corresponding value for hydrogen-consuming bacteria is 0.055 h^{-1} (Rittmann and McCarty 2001). Similarly, the yield coefficient for hydrogen-producing cultures was found to be around 0.1 g VS/g chemical oxygen demand (COD) (Chen et al. 2001), whereas for hydrogen-consuming methanogens, the corresponding yield coefficient is approximately 0.056 g VSS/g COD (Rittmann and McCarty 2001). This apparently suggests that hydrogen consumers, especially methanogens, are slow growers, and by selecting a dilution rate greater than 0.055 h^{-1}, the hydrogen consumers could be washed out from the system. Thus, the kinetic data can be a useful tool in designing a hydrogen-producing bioreactor.

9.5.6 Electric Current

The use of low-voltage (3.0–4.5 V) electric current has been successfully used to suppress the growth of hydrogen consumers (Roychowdhury 2000). The author found no traces of methane gas when cellulosic landfill sludge was subjected to electric shock.

9.6 Hydrogen Yield

Chemically, 1 mol of glucose can produce up to 12 mol of hydrogen according to Eq. (9.4):

$$C_6H_{12}O_6 + 6H_2O \rightarrow 6CO_2 + 12H_2 \tag{9.4}$$

Dark fermentation, however, produces a maximum of only 4 mol of hydrogen (Eq. (9.1)). Thus, the maximum conversion efficiency (actual vs theoretical) for dark fermentation is 33% ((4/12) × (100%)). Therefore, the reader should note that literature values for hydrogen yield are a percentage of the hydrogen production with 33% being the baseline. Often, researchers express hydrogen yield as moles of hydrogen produced per mole of substrate fed. Some studies also use hydrogen yield as liters of hydrogen per gram of COD (L H_2/g COD). Based on Eq. (9.1), the maximum hydrogen yield through dark fermentation is 0.47 L H_2/g COD ((4 mol) × (22.4 L/mol))/((180 g glucose) × (1.07 g COD/g glucose)) at standard temperature and pressure.

Liu and Fang (2003) reported a maximum yield of 0.27 L H_2/g sucrose (i.e., 0.24 L H_2/g COD) using acidogenic granular sludge at an HRT of 13.7 h, 26°C, and pH 5.5 in a CSTR. The corresponding molar yield was approximately 3.76 mol H_2/mol sucrose. Thus, the hydrogen conversion efficiency was about 47% ((3.76/8) × 100%). Kataoka et al. (1997) studied continuous hydrogen production in a chemostat using a pure culture of *C. butyricum* SC-E1 with glucose as an organic substrate at an HRT of 8 h, 30°C, and pH 6.7. The authors reported a maximum yield of 1.3–2.2 mol H_2/mol glucose, which corresponded to a hydrogen conversion efficiency between 33 and 55%. The hydrogen production potential of cellulose was investigated using two types of natural inocula: anaerobically digested sludge and sludge compost in batch cultures at 60°C (Ueno et al. 1995). The results showed yields of 0.9 mol H_2/mol hexose for digested sludge and 2.4 mol H_2/mol hexose for compost. The respective hydrogen conversion efficiencies were 23 and 60%. At the time of this writing, there were over 150 publications related to biohydrogen production, and obviously it is not practical to include results from all these studies.

9.7 Important Considerations in Biohydrogen Production

Sustainable hydrogen production in a mixed-culture system requires suppression of hydrogen consumers, for example, hydrogenotrophic methanogens, homoacetogens, and sulfate reducers, and enhancement of hydrogen-producing activity by controlling operating parameters. Some of these parameters are discussed in the following section.

9.7.1 Types of Inocula and Enrichment Techniques

Hydrogen-producing bacteria can be readily obtained from nature. The hydrogen-producing potentials of different inocula such as compost, landfill sediment, and potato and soybean soils have been studied in depth. Many researchers employed anaerobically digested municipal sludge and animal manure as inocula in seeding hydrogen-producing bioreactors (Khanal et al. 2006; Lay et al. 1999, 2003). The hydrogen-producing capabilities of bioreactors seeded with natural inocula such as soil, sediment, digested biosolids, or manure evidently suggest an abundance of hydrogen-producing bacteria in nature. *Clostridium* is one of the most widely reported hydrogen-producing bacteria.

Heat treatment (especially boiling) of anaerobic sludge for 10–20 min helps obtain an enriched culture of *Clostridium* (Lay 2000). Baking compost or soil is another approach for obtaining an inoculum rich in hydrogen producers from solid matrices (Khanal et al. 2004; Lay et al. 2005). In addition to heat treatment, acid/base treatment has been studied as a selection pressure for hydrogen-producing *Clostridium*. Acid/base treatments at pH of 3.0 and 10.0 effectively inhibited methanogens (Chen et al. 2002). Alkaline treatment at pH 12.0 suppressed hydrogen consumers in sewage sludge (Cai et al. 2004). Kawagoshi et al. (2005) employed acid treatment to eliminate hydrogen consumers from various inocula including activated sludge, anaerobic sludge, refuse compost, watermelon soil, kiwi soil, and lake sediment.

9.7.2 Operating pH

As stated earlier, biohydrogen production is highly dependent on the pH of the system. To review, reasons for this include the following:

1. pH directly affects the hydrogenase activity that governs the hydrogen production.
2. Metabolic shift in favoring hydrogen production versus solvent production, and vice versa, are pH dependent.
3. Growth of hydrogen consumers can be inhibited by proper pH control.

Hydrogen production commonly occurs during the acid production phase of anaerobic fermentation (Jones and Woods, 1989). This apparently suggests that an acidic pH range favors hydrogen production. Batch studies indicated that the ideal initial pH with sucrose as a carbon source was in the range of 5.5–5.7 (Khanal et al. 2004; van Ginkel et al. 2001). Zhang et al. (2003) reported that the optimal initial pH for converting starch to hydrogen was around 6.0 under a thermophilic condition. Another study also showed that an initial pH 6.0 favored hydrogen production from cheese whey (Ferchichi et al. 2005). Fang et al. (2006) reported improved hydrogen production from rice slurry at an initial pH of 4.5. Based on these studies, the reader could conclude that an initial pH slightly less than neutral helps enhance hydrogen production. Somewhat contrarily, Yokoyama et al. (2007) reported that an approximately neutral pH was optimal for hydrogen production from dairy cow manure slurry at temperatures of 60 and 75°C.

Although there have been several investigations examining the hydrogen production in batch studies, operating pH should be determined based on continuous studies. Fang and Liu (2002) examined the effect of pH (ranging from 4.0 to 7.0 at 0.5 unit intervals) on hydrogen yield in a CSTR using glucose as a carbon source. The authors found that a pH of 5.5 was optimal for hydrogen production in this study, which produced a yield of 2.1 mol H_2/mol glucose. Lay (2000) optimized hydrogen production from synthetic starch wastewater by controlling the pH at 5.2. For beer processing wastes, Lay and coworkers determined an optimum pH of 5.8 in a CSTR, based on a statistical contour plot analysis (Lay et al. 2005). Thus, the reader can safely conclude that pH of 5.0–6.0 is optimal for hydrogen production in a continuous process.

9.7.3 Feed Composition

To date, most previous studies evaluating hydrogen production potential have examined carbohydrate-rich substrates (synthetic and real). These studies have included the examination of simple sugars (e.g., glucose and sucrose) for fermentative hydrogen production (Khanal et al. 2004; Lin and Jo 2003; Mizuno et al. 2000; van Ginkel et al. 2001). Other investigators exploring complex substrates have evaluated food wastes (Chen et al. 2006), cellulose-containing waste (Okamoto et al. 2000), and municipal solid wastes (Ueno et al. 1995).

Lay et al. (2003) compared hydrogen production from synthetic carbohydrate, protein, and fat-rich organic solid wastes. This batch study showed significantly higher hydrogen yield from a carbohydrate-rich substrate than from the other two substrates. Similar results were also achieved for bean curd manufacturing waste (protein-rich waste), and rice and wheat bran (carbohydrate-rich wastes) in which the brans performed more favorably for hydrogen fermentation (Noike and Mizuno 2000). Lactate (Logan et al. 2002), a mixture of peptone (40%) and glucose (60%) (Cheng et al. 2003), and filtrate from sewage sludge (Wang et al. 2003) were also examined for their hydrogen production potentials. The yields were, however, significantly low compared to carbohydrate-rich substrates.

9.7.4 Nutrients

All microbial-mediated processes require nutrients. Nitrogen is essential for cell synthesis, and hydrogen fermentation is affected by nitrogen concentration. Liu and Shen (2004) examined the effect of nitrogen supplementation ranging from 560 to 11,280 mg/L of NH_4HCO_3 (99–1,999 mg N/L) on hydrogen production using synthetic starch wastewater (15 g starch/L). The authors observed the maximum hydrogen yield and the maximum specific hydrogen production rate at a NH_4HCO_3 concentration of 5,640 mg/L (999 mg N/L). Lay et al. (2005)

suggested that an NH_4^+ concentration of 537 mg/L (418 mg N/L) was beneficial for hydrogen production from food wastes. Effect of four different carbon-to-nitrogen (C/N) ratios of 130, 98, 47, and 40 (at nitrogen level of 900 mg N/L) on hydrogen yield was examined for synthetic sucrose wastewater by Lin and Lay (2004a). The authors reported maximum hydrogen yield and hydrogen production rate of 2.4 mol/mol hexose consumed and 270 mmol/L·day at C/N = 47. A C/N ratio of 10 (or 0.4 g N/L) was optimal for hydrogen production from glucose using yeast as a nitrogen source (Morimoto et al. 2004). The source of nitrogen also affects hydrogen yield. Ueno et al. (2001) found reduction in hydrogen yield by 50% when peptone was substituted with ammonium chloride as a nitrogen source.

Phosphorus is equally important nutrient for biosynthesis. Phosphorus in the form of phosphate is often used as a buffering agent. Phosphorus as Na_2HPO_4 at concentration of 600 mg/L enhanced hydrogen production by nearly 1.9 times (Lin and Lay 2004b) compared to acidogenic nutrient formulation. Several studies reported optimal C/P ratios ranging from 120 to 130 for biohydrogen production (Hawkes et al. 2002; Lin and Lay 2004b). Based on multivariate analysis, Lay et al. (2005) found PO_4^{3-} of 1,331 mg/L optimal for clostridia using high-solid food wastes.

Apart from the macronutrients (N and P), trace metals are also extremely important in cell synthesis and other microbial metabolism. Iron is the most widely studied trace element in biohydrogen research due to its direct involvement in hydrogenase activity. An iron-limited growth media often showed poor hydrogen production and favored the pathway shift from acid production toward solvent production (Lee et al. 2001). A pure culture study with *C. acetobutylicum* revealed predominant lactate fermentation under an iron-limiting condition (Bahl et al. 1982), which essentially diverts the reducing equivalents from the hydrogen-producing pathway. Published studies show considerable variation in the optimal iron dosage needed for hydrogen production. For example, Liu and Shen (2004) reported 10 mg Fe^{2+}/L as an optimal dose for hydrogen production from starch, whereas 589 mg Fe^{2+}/L was optimal for sucrose, and 132 mg Fe^{2+}/L for food wastes using mixed cultures (Lay et al. 2005). Such variability essentially is governed by the types of wastes, cultures used, and operating conditions (e.g., pH, HRT, and temperature).

In addition to iron, fermentative hydrogen production requires other trace elements. There is no single recipe that could be used universally for biohydrogen production. Lin and Lay (2005) designed an experiment based on Taguchi orthogonal arrays to obtain a recipe for biohydrogen production. The authors reported the optimum concentrations of the following nutrients (mg/L): Mg^{2+}, 4.8; Na^+, 393; Zn^{2+}, 0.25; Fe^{2+}, 1; K^+, 2.94; I^-, 9.56; Co^{2+}, 0.25; NH_4^+, 16.8; Mn^{2+}, 2.4; Ni^{2+}, 1.23; Cu^{2+}, 1.25; Mo^{6+}, 0.04; and Ca^{2+}, 544. The study further concluded that Mg^{2+}, Na^+, Zn^{2+}, and Fe^{2+} were important for hydrogen production, with Mg^{2+} most significant.

9.7.5 Hydrogen Partial Pressure

Hydrogen partial pressure plays an important role in biological hydrogen production. The electron transfer from reduced NADH to a proton and the subsequent hydrogen production are primarily governed by hydrogen partial pressure (Box 9.1). Several strategies such as mixing, gas sparging, applying vacuum, and use of a hydrogen-permeable membrane, among others, have been used to reduce the hydrogen partial pressure in a bioreactor. Lay (2000) reported a twofold improvement in the hydrogen production when the agitation speed was increased from 100 to 700 rpm in a laboratory-scale CSTR fermenting starch. Sparging inert gas (nitrogen) increased hydrogen yield by 60% compared to an unsparged system (Hussy et al. 2005). Vacuum application and use of a hydrogen permeable membrane did not improve hydrogen production (Li and Fang 2007).

Box 9.1

Hydrogen Partial Pressure and Its Implications on Hydrogen Production

Hydrogen production through dark fermentation is primarily governed by hydrogenase activity that is enzyme-catalyzed transport of electrons from the intracellular carriers to protons. Protons, however, are poor electron acceptors as evident from their extremely low redox potentials ($Eh_{H2} = -414$ mV). Thus, for effective electron transport, the electron donor must be a strong reducing agent. The electron carriers—reduced ferredoxin and NADH—have low redox potentials of –400 and –320 mV, respectively, and are able to reduce protons to hydrogen. Redox potential of the net reaction under actual conditions determines the ability of electron donors (reduced ferredoxin and NADH) to reduce protons. Based on equal intracellular concentrations of oxidized and reduced ferredoxin and NADH, the hydrogen production is thermodynamically unfavorable at hydrogen partial pressures:

$$P_{H_2,max} \geq \left[\frac{2F(E^{o'}_{H_2} - E_x{}^{o'})}{RT} \right]$$

where $E_x{}^{o'}$ is the redox potential of the electron donor, F is Faraday's constant, R is the ideal gas constant, and T is the absolute temperature. Hydrogen production with reduced ferredoxin as the electron donor will continue as long as the hydrogen partial pressure is <0.3 atm (or 3×10^4 Pa). For reduced NADH, the hydrogen partial pressure must be maintained at extremely low levels ($<6 \times 10^{-4}$ atm or 60 Pa) for protons to be reduced to hydrogen. These values were calculated based on equal concentrations of the oxidized and reduced forms of electron donors.

Adapted from Angelent et al. (2004).

9.7.6 Hydraulic Retention Time

HRT affects hydrogen production primarily through dilution effect on the substrate and product, and subsequent organic loading. HRT can also be used as a control parameter to selectively wash out slow-growing hydrogen consumers. Shorter HRTs are more favorable for hydrogen production because methane producers essentially wash out from the system due to their low specific growth rate of 0.0167 h^{-1} (van Haandel and Lettinga 1994) compared to hydrogen producers (μ_{max}: 0.140–0.330 h^{-1}) (Horiuchi et al. 2002). Short HRTs may, however, reduce the substrate utilization efficiency, thereby lowering the overall process efficiency.

Lin and Chang (1999) obtained a maximum hydrogen yield of 1.76 mol/mol hexose consumed in a CSTR at an HRT of 6 h using glucose as a substrate. Ueno et al. (1996) reported a maximum hydrogen yield at an HRT of 12 h from sugary wastewater. While treating rice winery wastewater in a CSTR, a hydrogen yield of 1.32 mol/mol hexose consumed was obtained at an HRT of 24 h, and the yield decreased to 1.04 hexose consumed at an HRT of 4 h (Yu et al. 2003). Assessing HRTs ranging from 12 to 48 h, Lay et al. (2005) found that an HRT of 32 h was ideal for fermenting beer processing waste. Reducing the HRT from 18 to 12 h improved hydrogen yield without affecting starch removal efficiency for insoluble wheat starch coproduct (Hussy et al. 2003). For more complex and insoluble substrates such as starch, food waste, and other high-solid wastes, longer HRTs are needed due to the rate-limiting hydrolysis step. Hydrogen production from food waste continued to increase with increases in HRT from 48, 72, and 120 h (Shin and Youn 2005).

Since varying the HRT also affects the organic loading rate, differentiation of the effects of HRT and the organic loading rate on hydrogen production is difficult. Liu and Fang (2003) examined the effect of HRT ranging from 4.6 to 28.8 h, with corresponding sucrose concentrations ranging from 4.8 to 29.8 g/L, at a constant organic loading rate of 25 g sucrose/L· day on hydrogen production using granular acidogenic sludge. The authors reported a maximum yield at an HRT of 13.7 h with a sucrose concentration of 14.3 g/L.

9.7.7 Reactor Configuration

Most studies on fermentative hydrogen production using mixed cultures have been conducted in a CSTR. Few studies have examined high-rate reactors, where HRT and SRT can be controlled independently.

Development of hydrogen-producing granules was first reported by Liu and Fang (2003). The seed sludge obtained from acidogenic reactor was adapted for hydrogen production under low pH conditions. The granules were enriched in hydrogen-producing communities (Fig. 9.3). One study also examined the

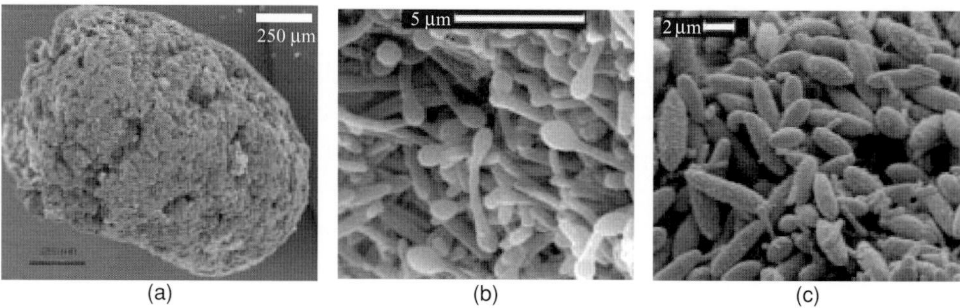

FIG. 9.3. Scanning electron microscopy images of (a) a typical hydrogen-producing acidogenic granule, (b) spore-forming bacteria, and (c) fusiform bacilli.
Source: Liu and Fang (2003). Reprinted with permission.

hydrogen production potential of UASB granules using glucose as an organic substrate under mesophilic conditions (Gavala et al. 2006). The hydrogen production rate of the UASB (19.05 mmol H_2/h·L) was significantly greater than that of the CSTR (8.42 mmol H_2/h·L) at a low HRT of 2 h. However, at longer HRT of 6 and 12 h, the CSTR outperformed the UASB. This implies that the low hydrogen production rate in CSTR at short HRT was due to biomass washout. Interestingly, the hydrogen yields at all HRTs for the CSTR were significantly higher (40–66%) than the UASB reactor. Since no methane was observed in the biogas, this suggests the possibility that the hydrogen produced in the UASB reactor was converted to acetate by homeacetogens. For rice winery wastewater, a hydrogen yield of 2.14 mol/mol hexose consumed was obtained for a 24-h HRT, with the yield diminishing to 1.74 mol/mol hexose consumed at a much lower HRT of 2 h (Yu et al. 2002). This study, however, did not examine the hydrogen yield using CSTR.

Immobilization in support media to form a biofilm is another approach of retaining biomass in a bioreactor. Wu et al. (2003) examined hydrogen production using biomass immobilized in alginate beads with acrylic latex and silicone in a three-phase fluidized-bed bioreactor. The best hydrogen yield of 2.67 mol/mol sucrose consumed was obtained at a short HRT of 2 h. In fixed-bed bioreactors using activated carbon (AC) and expanded clay (EC) as packed media for cell immobilization, Chang et al. (2002) reported the optimum hydrogen production rates of 1.32 (AC) and 0.42 L/h·L (EC) at HRTs of 1 and 2 h, respectively. For biomass immobilized in granular form, a carrier-induced granular sludge bed bioreactor can be operated at an HRT as low as 0.5 h, with a maximal hydrogen yield of 3.03 mol/mol sucrose consumed (Lee et al. 2004).

It is important to point out that reactor configurations such as UASB, biofilm, and membrane-coupled bioreactor may promote the growth of diverse microbial communities that include hydrogen consumers due to a long SRT. Thus, high-rate reactors may not always be an ideal option for hydrogen production. A comparative study of UASB versus CSTR clearly showed the superiority of CSTR in biohydrogen production (Gavala et al. 2006). However, if a granular or immobilized biomass is capable of maintaining an enriched culture of hydrogen producers for a considerable period of time, sustainable hydrogen production can be achieved using such reactor systems. As previously stated, two important factors that significantly impact hydrogen production are low-hydrogen partial pressure and suppression of hydrogen consumers, and should be considered when developing a reactor design.

9.7.8 Operating Temperature

Microbial-mediated processes such as hydrogen production are temperature dependent. Temperature has several effects on biohydrogen production, including microbial community selection, hydrogenase activity, mass transfer rate, hydrolysis of high-solid substrates, and hydrogen partial pressure. As a rule of thumb, reaction rate doubles for every 10°C increase in temperature up to its optimal value.

Biohydrogen studies have primarily been conducted at two temperature ranges, mesophilic (30–40°C) and thermophilic (50–60°C). Biohydrogen production at hyperthermophilic (70–80°C) conditions has merit—better reaction kinetics and operation at higher hydrogen partial pressure (van Groenestijn et al. 2002). In addition, higher temperatures allow nonaxenic operation of a bioreactor with no hydrogen-consuming activity (Yokoyama et al. 2007).

A batch hydrogen fermentation test using glucose as a substrate improved hydrogen yield by 72% when the temperature was raised incrementally (2°C) from 33 to 41°C (Mu et al. 2006). The specific hydrogen production rate also improved by nearly 86% with an increase in temperature from 33 to 39°C, but the rate decreased at 41°C. Based on the response surface method, Wang et al. (2005) examined the effect of temperature (from 20 to 45°C) on hydrogen production from sucrose. The result showed maximum hydrogen yield at 35°C. Temperatures from 25 to 40°C were found favorable for hydrogen production from sucrose, but production declined at temperatures greater than 45°C (Zhang and Shen 2006). Wu et al. (2005) also reported that 40°C was ideal for hydrogen production with anaerobic sludge immobilized by ethylene–vinyl acetate copolymer.

A temperature of 40°C was found optimal for a carrier-induced granular sludge reactor operating continuously and converting sucrose to hydrogen (Lee et al. 2006). Yu et al. (2002), however, reported that 55°C is the optimal temperature for hydrogen production from rice winery wastewater in a UASB reactor. Lin and Chang (2004) evaluated the simplicity and energy conservation of hydrogen production at ambient temperature in a CSTR. Yield and production rate were lower than desired at temperatures less than at 35°C.

Gavala et al. (2006) examined the effect of mesophilic and thermophilic temperatures on hydrogen production from glucose in CSTRs. The results show that the hydrogen yield for a thermophilic CSTR (~2.1 mol/mol glucose$_{consumed}$) was slightly higher than that for a mesophilic CSTR (~1.7 mol/mol glucose$_{consumed}$) at an HRT of 12 h. However, the specific hydrogen production rate (mmol/h·L·g VSS) for the thermophilic CSTR was 5–10 times higher than the mesophilic CSTR. Hydrogen production from cow waste slurry was examined at temperatures of 37, 50, 55, 60, 67, 75, and 85°C in a batch mode (Yokoyama et al. 2007). The results show maximum hydrogen production of 392 and 248 mL H_2/L slurry at temperatures of 60 and 75°C, respectively. The hydrogen production between 35 and 50°C was significantly lower due to consumption of hydrogen by methanogens. It can be concluded then that bioreactor operation at a temperature of at least 75°C eliminates the hydrogen-consumption activity.

9.7.9 Biokinetics of Hydrogen Production

Kinetic parameters play an important role in designing and operating a hydrogen bioreactor. Biokinetic studies of fermentative hydrogen production are very limited. Hydrogen-producing bacteria are generally classified as a group of bacteria responsible for acidogenesis, and the kinetic parameters such as maximum specific growth rate (μ_{max}: 0.140–0.330 h^{-1}) and biomass yield ($Y_{X/S}$: 0.20–0.25 g biomass/g COD) for acid-forming bacteria (Horiuchi et al. 2002) can be considered for hydrogen producers as well. Kumar et al. (2000) examined the growth kinetics of *Enterobacter cloacae* IIT-BT 08 and reported μ_{max} and $Y_{X/S}$ of 0.568 h^{-1} and 0.08 g biomass/g COD, respectively. For biohydrogen production using the extreme thermophile *Caldicellulosiruptor saccharolyticus*, μ_{max} of 0.130 h^{-1} was reported (van Niel et al. 2003). Apart from pure culture studies, Ueno et al. (2001) found $Y_{X/S}$ of 0.12 and 0.18 g biomass/g COD for a mixed microbial culture obtained from sludge compost-fermenting cellulose in both continuous and batch operation, respectively. Using sewage sludge for hydrogen fermentation from sucrose, μ_{max} of 0.172 h^{-1} and $Y_{X/S}$ of 0.09 g biomass/g COD were obtained (Chen et al. 2001).

In addition to μ_{max} and $Y_{X/S}$, hydrogen-producing microbes have a high half-saturation constant (K_s). Chen et al. (2006) examined the biokinetic parameters of hydrogen-producing bacteria using three different substrates. K_s values of 1.4, 6.6, and 8.7 g COD/L were obtained for sucrose, nonfat dry milk (NFDM), and food waste, respectively. Growth kinetics of hydrogen-producing bacteria from various cultures and substrates are summarized in Table 9.2. The data clearly show higher K_s values, suggesting that a high substrate concentration is essential for hydrogen fermentation. This finding illustrates the need for shorter HRTs for

Table 9.2. Summary of growth kinetics of hydrogen-producing bacteria.

Test Type	Culture	Temperature (°C)	Substrate	$\mu_{max}(h^{-1})$	R_{max} ($mL{\cdot}h^{-1}$)	K_S(g COD/L)	$Y_{X/S}$(g biomass/ g COD)	References
Continuous	*Citrobacter intermedius*	37	Glucose	0.220	121.4	NA[a]	0.11	Brosseau and Zajic (1982)
Continuous	Mixed	35	Sucrose	0.172	NA	0.068	0.09	Chen et al. (2001)
Continuous	Mixed	37	Glucose	0.140–0.330	NA	NA	0.20–0.25	Horiuchi et al. (2002)
Batch	*Enterobacter cloacae*	37	Glucose	0.568	NA	3.914	0.08	Kumar et al. (2000)
Batch	*Caldicellulosiruptor saccharolyticus*	70	Sucrose	0.130	67.2	0.801	NA	van Niel et al. (2003)
Batch	Mixed	35	Sucrose	ND[a]	13.9	1.446	ND	Chen et al. (2006)
Batch	Mixed	35	NFDM	ND	25.6	6.616	ND	Chen et al. (2006)
Batch	Mixed	35	Food waste	ND	29.9	8.692	ND	Chen et al. (2006)

Source: Chen et al. (2006). Reprinted with permission.

[a]NA, not available; ND, not determine.

continuous hydrogen fermentation. High substrate concentrations and/or high organic loading rates are key for successful operation of a hydrogen-producing reactor.

9.8 Limitations of Dark Fermentation and Potential Remedial Options

A major limitation of dark fermentation is its low hydrogen yield. Chemically, there is enough energy in glucose to produce up to 12 mol of hydrogen according to Eq. (9.4). However, no single microbe is known to carry out this reaction pathway. Because of the positive free energy change associated with this reaction ($\Delta G^{o'} = +3.2$ kJ/mol), essentially no cell growth can occur (Thauer 1976). To facilitate faster cell growth, microbes produce lesser amounts of hydrogen and acetic acid while producing an array of other waste products such as ethanol and lactic and butyric acids. These acids and alcohols normally accumulate in the growth medium as dead-end metabolites, since under an anaerobic dark condition in a pure culture system, the conversion of these acids into additional hydrogen is thermodynamically unfavorable (Classen et al. 1999). Nonetheless, energy considerations for these byproducts must be addressed so that several technical challenges be overcome. First, acid accumulation represents energy lost, which could have generated additional hydrogen with suitable microbes, if indeed they exist. Second, these acids lower the pH of the culture medium, causing a redirection of the cellular metabolic pathways toward solvent production (solventogenesis), which is unfavorable for hydrogen production. Third, the acids in the spent culture medium must be treated further so that the resultant liquid can be either recycled or released safely. Wastewater treatment adds additional cost. If additional energy can be captured from these waste acids, then the economics of fermentation would be much more favorable. Some of the strategies are discussed in the following section.

9.8.1 Hydrogen Fermentation Followed by Methane Production

The effluent from a hydrogen-producing bioreactor is rich in organic acids and solvents. Methanogenesis is the final step that converts these end products into methane. Han and Shin (2004) employed a UASB reactor to treat organic acids and alcohols generated from hydrogen fermentation of food waste. "Hy-Met" (hydrogen–methane) process, so named by Japanese researchers, also employed a similar concept to produce hydrogen from high-solid waste streams (e.g., bread and beer wastes) followed by methane production from the volatile organic acids in a UASB reactor (Nishio and Nakashimada 2007). Operation of the first-stage reactor at an elevated temperature range between 60 and 75°C minimized the activity of hydrogen consumers and enhanced hydrogen production. The methane generated from the second stage could be used to produce heat/steam needed for in-plant use.

9.8.2 Dark Fermentation Followed by Photofermentation

The integration of (light) photofermentation into dark fermentation essentially aims to enhance the molar yield of hydrogen. Purple nonsulfur photosynthetic bacteria belonging to the family of Rhodospirillaceae are most versatile in their various growth modes and the wide variety of organic carbon substrates they can use to support cell growth (Beatty and Gest 1981). In the presence of light, organic acids, including acetic, butyric, formic, lactic, malic, propionic, and succinic, can donate electrons to the photosynthetic electron transport chain to generate Fe (red) as the low-redox-potential electron donor and ATP as the energy source. With acetic and butyric acids as the main products of dark hydrogen fermentation, the stoichiometry of hydrogen production via photofermentation is given by Eqs (9.5) and (9.6):

$$\underset{\text{(Acetic acid)}}{CH_3COOH} + 2H_2O + \text{light} \rightarrow 4H_2 + 2CO_2 \tag{9.5}$$

$$\underset{\text{(Butyric acid)}}{CH_3CH_2CH_2COOH} + 6H_2O + \text{light} \rightarrow 10H_2 + 4CO_2 \tag{9.6}$$

The amount of hydrogen that can be theoretically produced from acetic and butyric acids is very encouraging for the economics of the overall integrated fermentation process. That said, not much is known about the hydrogen conversion efficiencies and rates from both acetic and butyric acids by the various photosynthetic bacteria. Several reports have documented conversion efficiencies from acetic acid ranging from low of 11% to high of 73% in two untyped species of *Rhodobacter* and *Rhodopseudomonas*, respectively (Barbosa et al. 2001; Mao et al. 1986). Hydrogen yields from butyric acid were approximately 8.4% in an untyped *Rhodopseudomonas* sp. (Barbosa et al. 2001) to 75% in *Rhodobacter sphaeroides* RV (Miyake et al. 1984). Although these values are scattered, photofermentation has a potential when integrated with dark fermentation to convert waste organic acids from dark fermentation into additional hydrogen. This integrated strategy could realistically convert 1 mol of glucose to nearly 12 mol of hydrogen, approaching the theoretical maximum yield.

9.8.3 Dark Fermentation Followed by Microbial Fuel Cell

The soluble organic end products of dark fermentation can be directly converted to electricity using a microbial fuel cell (MFC). The MFC operates similarly to a hydrogen fuel cell, which oxidizes hydrogen at the anode surface and transfers electrons to the cathode, where molecular oxygen is reduced to water. In an MFC, the driving force is the redox reaction of substrates (organic acids) mediated by anaerobic microorganisms. Thus, dark fermentation coupled with an MFC could improve the overall economics of hydrogen production. A detailed discussion on MFC is presented in Chapter 10.

9.8.4 Dark Fermentation Followed by Bioelectrochemically Assisted Microbial System

This is a modified version of an MFC in which the protons generated from the soluble substrate (in dark fermentation) at the anode chamber are converted to hydrogen gas at the cathode chamber. A schematic of the process is shown in Fig. 9.4. In order to produce hydrogen, no oxygen is supplied to the cathode side of the cell and a small voltage is applied to the circuit to make hydrogen evolution possible.

Liu et al. (2005) examined the hydrogen production potential of a bioelectrochemical-assisted microbial system using acetate as a carbon source. Over 90% of the protons and electrons produced from acetate were recovered as hydrogen at the cathode at an applied voltage of 250 mV. The authors further reported that with a coulombic efficiency of 78% and an electron recovery of 92% as hydrogen, up to 2.9 mol H_2/mol acetate could be recovered (assuming a maximum

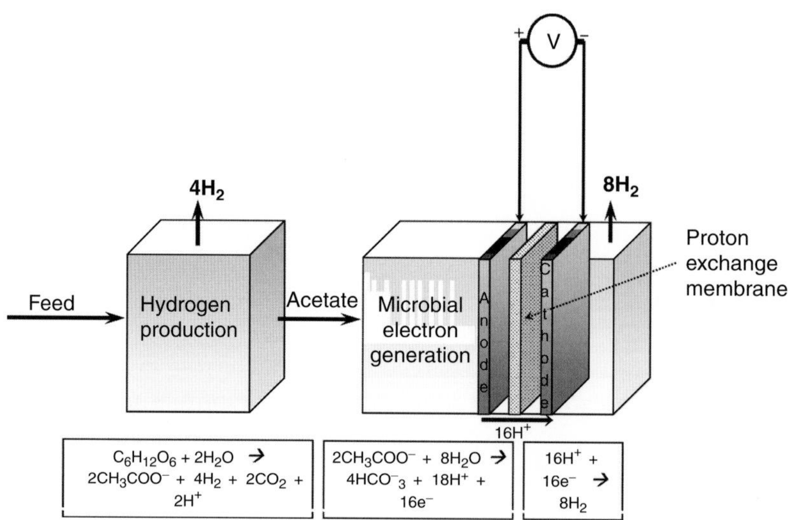

FIG. 9.4. Bioelectrochemical-assisted microbial system for enhanced hydrogen production.

of 4 mol H_2/mol acetate could be obtained). Thus, with 2–3 mol H_2/mol glucose generated through dark fermentation, an overall hydrogen yield of 8–9 mol H_2/mol glucose is obtainable.

9.8.5 Redirection of the Metabolic Pathway

Hydrogen production is always associated with acid production, which lowers the culture pH, creating an unfavorable environment for hydrogen producers. Redirecting the metabolic pathway by blocking acid production might enhance hydrogen production (Kumar et al. 2001). Dark fermentation also produces other reducing byproducts such as ethanol, propanol, and lactate, which divert hydrogen atoms away from hydrogen gas. The catabolism of glucose to pyruvate via glycolysis generates a reducing equivalent (NADH). Further conversion of pyruvate to ethanol, butanediol, butyric acid, or lactic acid will oxidize NADH to NAD^+, and this reduces hydrogen production. By blocking the formation of these reducing solvents and acids, the amount of NADH could be increased, improving hydrogen yield (Nath and Das 2004). Kumar et al. (2001) obtained an enhanced yield of 3.8 mol H_2/mol glucose for *E. cloacae* by blocking the organic acid production pathway by using proton-suicide technique with NaBr and $NaBrO_3$.

9.9 Technoeconomic Analysis of Hydrogen Fermentation

The technical feasibility of biological hydrogen production through dark fermentation has been widely studied. The major challenge associated with hydrogen production through dark fermentation is the low yield. Considerable research focusing on both process engineering and microbiology is needed to improve the overall hydrogen yield. With regard to process engineering, strategies such as improved reactor design, control of hydrogen partial pressure, and redirection of the metabolic pathway could be employed to obtain increased hydrogen yields. Microbiological research should focus on identifying, understanding the physiology of microbial populations, and developing genetically engineered microbes that could enhance hydrogen yield.

No economic analysis of hydrogen production by dark fermentation exits. A preliminary cost analysis conducted at National Renewable Energy Laboratory (NREL) revealed that in order to meet the hydrogen cost target of \$2.60/kg at the plant gate by 2010, a minimum yield of 4 mol of hydrogen must be produced per mole glucose, assuming glucose cost of 5 cents/lb (Maness, personal communication). This calculation is based on feedstock cost only without considering the additional cost of reactor construction, operation, and maintenance since feedstock cost is typically 75% of the cost of a chemical production process. It is therefore reasonable to assume that a hydrogen molar yield of 8 or higher is needed in order to develop a viable fermentation process.

Example 9.2

A food processing plant generates 2.7 ton/day (~3.0 ton/day) of organic waste. The plant plans to employ a two-stage process (hydrogen fermentation followed by methane production) for energy recovery from the waste. Based on laboratory-scale studies, 2.4 mol H_2/kg wet weight and 8.6 mol CH_4/kg wet weight were generated.

(a) How much energy could be generated if hydrogen is used in fuel cell?
(b) If methane gas is used for heating/steam generation, how much coal it could replace?

Solution

(a) *Energy generation from hydrogen fuel cell:*

Hydrogen production rate = (2.4 mol/kg wet wt) × (2.7 ton/day) × (1,000 kg/ton) = 6,480 mol/day

According to the ideal gas law, 1 mol of any gas at STP (1 atm and 0°C) occupies a volume of 22.4 L.

Thus, hydrogen production rate = (6,480 mol/day) × (22.4 L/mol) = 145,152 L/day or 145 m^3/day

1 m^3 of hydrogen generates 2.97 kWh of energy (or 1 kg H_2 produces 33.3 kWh of energy).

With a fuel cell efficiency of 50%, the total energy generation from hydrogen = (2.97 kWh) × (145 m^3/day) × (0.5) = 215 kWh

(b) *Energy generation through methane production*

Methane production rate = (8.6 mol/kg wet wt) × (2.7 ton/day) × (1,000 kg/ton) = 23,220 mol/day

According to the ideal gas law, 1 mol of any gas at STP (1 atm and 0°C) occupies a volume of 22.4 L.

Thus, methane production rate = (23,220 mol/day) × (22.4 L/mol) = 520,128 L/day or 520 m^3/day

1 m^3 of methane has an energy content of 35,310 Btu.

Total energy generation from methane = (35,310 Btu) × (520 m^3/day) = 18 MBtu/day

This energy is equivalent to 1 ton (2,000 lb) of coal per day.

1 m^3 of hydrogen generates 2.97 kWh of energy (or one kilogram H_2 produces 33.3 kWh of energy). With a fuel cell efficiency of 50%, the total energy generation from hydrogen = (2.97 kWh) x× (145 m^3/day) × (0.5) = 215 kWh

References

Angelent, L. T., Karim, K., Al-Dahhan, M. H., Wrenn, B. A., and Spinosa, R. D. 2004. Production of bioenergy and biochemicals from industrial and agricultural wastewater. *Trends Biotechnol.* 22(9):477–485.

Bahl, H., Andersch, W., Braun, K., and Gottschalk, G. 1982. Effect on pH and butyrate concentration on the production of acetone and butanol by *Clostridium acetobutylicum* growing in continuous culture. *Eur. J. Appl. Microbiol. Biotech.* 1:17–20.

Ballongue, J., Maison, E., Amine, J., Petitdemange, H., and Gay, R. 1987. Inhibitor effects of products of metabolism on growth of *Clostridium acetobutylicum. Appl. Microbiol. Biotechnol.* 26:568–573.

Barbosa, M. J., Rocha, J. M. S., Tramper, J., and Wijffels, R. H. 2001. Acetate as a carbon source for hydrogen production by photosynthetic bacteria. *J. Biotechnol.* 85:25–33.

Beatty, J. T., and Gest, H. 1981. Biosynthetic and bioenergetic functions of citric acid cycle reactions in *Rhodopseudomonas capsulata. J. Bacteriol.* 148:584–593.

Brock, T. D., Madigan, M. T., Martinko, J. M., and Parker, J. 1994. *Biology of Microorganisms.* Prentice-Hall, New York, USA.

Brosseau, J. D., and Zajic, J. E. 1982. Hydrogen gas production with *Citrobacter intermedius* and *Clostridium pasteurianum. J. Chem. Tech. Biotechnol.* 32:496–502.

Cai, M., Liu, J., and Wei, Y. 2004. Enhanced biological production from sewage sludge with alkaline pretreatment. *Environ. Sci. Technol.* 38:3195–3202.

Cato, E. P., and Stackebrandt, E. 1989. Taxonomy and phylogeny. In *Clostridia*, edited by N. P. Minton and D. J. Clarke, pp. 1–26. Plenum, New York, USA.

Chang, J. S., Lee, K. S., and Lin, P. J. 2002. Biohydrogen production with fixed-bed bioreactors. *Int. J. Hydrogen Energy* 27:1167–1174.

Chen, C. C., Lin, C. Y., and Chang, J. S. 2001. Kinetics of hydrogen production with continuous anaerobic cultures utilizing sucrose as the limiting substrate. *Appl. Microbiol. Biotechnol.* 57:56–64.

Chen, C. C., Lin, C. Y., and Lin, M. C. 2002. Acid-base enrichment enhances anaerobic hydrogen production process. *Appl. Microbiol. Biotechnol.* 8:224–228.

Chen, W. H., Chen, S. Y., Khanal, S. K., and Sung, S. 2006. Kinetic study of biological hydrogen production by anaerobic fermentation. *Int. J. Hydrogen Energy* 31:2170–2178.

Cheng, S. S., Lin, C. Y., Tseng, I. C., Lee, C. M., Lin, H. I., Lin, M. R., Chen, S. T., Chen, S. D., and Liu, P. W. 2003. *Biohydrogen Production Mechanisms and Processes Application on Multiple Substrates*, pp. 33–39. In Proceeding of 1st NRL International Workshop on Innovative Anaerobic Technology, Daejeon, Korea.

Classen, P. A. M., van Lier, J. B., Contreras, A. M. L., van Niel E. W. J., Sijtsma, L., Stams, A. J. M., de Vries, S. S., and Weusthuis, R. A. 1999. Utilisation of biomass for the supply of energy. *Appl. Microbiol. Biotechnol.* 52:741–745.

Dabrock, B., Bahl, H., and Gottschalk, G. 1992. Parameters affecting solvent production by *Clostridium pasteurianum. Appl. Environ. Microbiol.* 58:1233–1239.

Doremus, M. G., Linden, J. C., and Moreira, A. R. 1985. Agitation and pressure effects on acetone-butanol fermentation. *Biotechnol. Bioeng.* 27:852–860.

Doyle, E. M. 2002. *Clostridium perfrigens Growth During Cooling of Thermal Processed Meat Products.* FRI Briefings, in March 2002 Food Research Institute, University of Wisconsin-Madison, WI, USA.

Doyle, M. P. 1989. *Foodborne Bacterial Pathogens*, 10th edn. Marcel Dekker, Inc., New York, USA.

Duangmanee, T., Padmasiri, S., Simmons, J. J., Raskin, L., and S. Sung. 2002. *Hydrogen Production by Anaerobic Communities Exposed to Repeated Heat Treatment.* In CD-ROM Proceedings of Water Environment Federation 75th Annual Conference & Exposition. September 28–October 2, 2002, Chicago, IL, USA.

Fang, H. H. P., Li, C., and Zhang, T. 2006. Acidophilic biohydrogen production from rice slurry. *Int. J. Hydrogen Energy* 31:683–692.

Fang, H. H. P., and Liu, H. 2002. Effect of pH on hydrogen production from glucose by a mixed culture. *Bioresour. Technol.* 82(1):87–93.

Fang, H. H. P., Liu, H., and Zhang, T. 2002. Characterization of a hydrogen-producing granular sludge. *Biotechnol. Bioeng.* 78:44–52.

Ferchichi, M., Crabbe, E., Gil, G. H., Hintz, W., and Almadidy, A. 2005. Influence of initial pH on hydrogen production from cheese whey. *J. Biotechnol.* 120:402–409.

Gavala, H. N., Skiadas, I. V., and Ahring, B. K. 2006. Biological hydrogen production in suspended and attached growth anaerobic reactor systems. *Int. J. Hydrogen Energy* 31:1164–1175.

Hallenbeck, P. C. 2004. Fundamentals of the fermentative production of hydrogen. *Water Sci. Technol.* 52(1–2):21–29.

Hallenbeck, P. C., and Benemann, J. R. 2002. Biological hydrogen production, fundamentals and limiting processes. *Int. J. Hydrogen Energy* 27:1185–1193.

Han, S.-K, and Shin, H.-S. 2004. Performance of an innovative two-stage process converting food waste to hydrogen and methane. *J. Air Waste Manage. Assoc.* 54:242–249.

Hart, D. 1997. *Hydrogen Power: The Commercial Future of the Ultimate Fuel.* Financial Times Energy Publishing, London.

Hawkes, F. R., Dinsdale, R. Hawkes, D. L., and Hussy, I. 2002. Sustainable fermentative hydrogen production: Challenges for process optimization. *Int. J. Hydrogen Energy* 27(11–12):1339–1347.

Holt, R. A., Cairns, A. J., and Morris, G. J. 1988. Production of Butanol by *Clostridium puniceum* in batch and chemostat culture. *Appl. Microbiol. Biotechnol.* 27:319–324.

Horiuchi, J. I., Shimizu, T., Tada, K., Kanno, T., and Kobayashi, M. 2002. Selective production of organic acids in anaerobic acid reactor by pH control. *Bioresour. Technol.* 82:209–213.

Hui, Y. H., Gorham, J. R., Murrell, K. D., and Cliver, D. O. 1994. *Foodborne Disease Handbook: Diseases Caused by Bacteria,* Vol. 1, 10th edn, Marcel Dekker, Inc., New York, USA.

Hussy, I., Hawkes, F. R., Dinsdale, R., and Hawkes, D. L. 2003. Continuous fermentative hydrogen production from a wheat starch co-product by mixed microflora. *Biotechnol. Bioeng.* 84:619–626.

Hussy, I., Hawkes, F. R., Dinsdale, R., and Hawkes, D. L. 2005. Continuous fermentative hydrogen production from sucrose and sugar beet. *Int. J. Hydrogen Energy* 30:471–483.

Jones, D. T., and Woods, D. R. 1989. "*Solvent Production*" *Biotechnology Handbooks: Clostridia,* Vol. 3. Plenum Press, New York and London.

Kaneuchi, C., Watanabe, K., Terada, A., Benno, Y., and Mitsuoka, T. 1976. Taxonomic study of *Bacteroides clostridiiformis* subsp. *clostridiiformis* (Burri and Ankersmit) Holdeman and Moore and related organisms: Proposal of *Clostridium clostridiiforme* (Burri and Ankersmit) comb. nov. and *Clostridium symbiosum* (Stevens) comb. nov. *Int. J. Syst. Bacteriol.* 26:195–204.

Kataoka, N., Miya, K., and Kiriyama, K. 1997. Studies on hydrogen production by continuous culture system of hydrogen-producing anaerobic bacteria. *Water Sci. Techol.* 36(6–7):41–47.

Kawagoshi, Y., Hino, N., Fujimoto, A., Nakao, M., Fujita, Y., Sugimura, S., and Furukawa, K. 2005. Effect of inoculum conditioning on hydrogen fermentation and pH effect on bacterial community relevant to hydrogen production. *J. Biosci. Bioeng.* 100:524–530.

Khanal, S. K., Chen, W. H., Li, L., and Sung, S. 2004. Biological hydrogen production: Effects of pH and intermediate products. *Int. J. Hydrogen Energy* 29:1123–1131.

Khanal, S. K., Chen, W. H., Li, L., and Sung, S. 2006. Biohydrogen production in continuous flow reactor using mixed microbial culture. *Water Environ. Res.* 78(2):110–117.

Kumar, N., Ghosh, A., and Das, D. 2001. Redirection of biochemical pathways for the enhancement of H_2 production by *Enterobacter cloacae. Biotechnol. Lett.* 23:537–541.

Kumar, N., Monga, P. S., Biawas, A. K., and Das, D. 2000. Modeling and simulation of clean fuel production by *Enterbacter cloacae* IIT-BT 08. *Int. J. Hydrogen Energy* 25:945–952.

Lay, J. J. 2000. Modeling and optimization of anaerobic digested sludge converting starch to hydrogen. *Biotechnol. Bioeng.* 68(3):269–278.

Lay, J. J., Fan, K. S., Chang, J. I., and Ku, C. H. 2003. Influence of chemical nature of organic wastes on their conversion to hydrogen by heat-shock digested sludge. *Int. J. Hydrogen Energy* 28:1361–1367.

Lay, J. J., Lee, Y. J., and Noike, T. 1999. Feasibility of biological hydrogen production from organic fraction of municipal solid waste. *Water Res.* 33:2576–2586.

Lay, J. J., Tsai, C. J., Huang, C. C., Chang, J. J., Chou, C. H., Fan, K. S., Chang, J. I., and Hsu, P. C. 2005. Influence of pH and hydraulic retention time on anaerobes converting beer processing wastes. *Water Sci. Technol.* 52:123–129.

Lee, K. S., Lin, P. J., and Chang, J. S. 2006. Temperature effects on biohydrogen production in a granular sludge bed induced by activated carbon carriers. *Int. J. Hydrogen Energy* 31:465–472.

Lee, K. S., Wu, J. F., Lo, Y. S., Lin, P. J., and Chang, J. S. 2004. Anaerobic hydrogen production with an efficient carrier-induced granular sludge bed bioreactor. *Biotechnol. Bioeng.* 87:648–657.

Lee, Y. J., Miyahara, T., and Noike, T. 2001. Effect of iron concentration on hydrogen fermentation. *Bioresour. Technol.* 80(3):227–231.

Li, C., and Fang, H. H. P. 2007. Fermentative hydrogen production from wastewater and solid wastes by mixed cultures. *Crit. Rev. Environ. Sci. Technol.* 37(1):1–39.

Lin, C. Y., and Chang, R. C. 1999. Hydrogen production during the anaerobic acidogenic conversion of glucose. *J. Chem. Tehnol. Biotechnol.* 74(6):678–684.

Lin, C. Y., and Chang, R. C. 2004. Fermentative hydrogen production at ambient temperature. *Int. J. Hydrogen Energy* 29:715–720.

Lin, C. Y., and Jo, C. H. 2003. Hydrogen production from sucrose using an anaerobic sequencing batch reactor process. *J. Chem. Technol. Biotechnol.* 78:678–684.

Lin, C. Y., and Lay, C. H. 2004a. Carbon/nitrogen-ratio effect on fermentative hydrogen production by mixed microflora. *Int. J. Hydrogen Energy.* 29(1):41–45.

Lin, C. Y., and Lay, C. H. 2004b. Effect of carbonate and phosphate concentrations on hydrogen production using anaerobic sewage sludge microflora. *Int. J. Hydrogen Energy* 29(3):275–281.

Lin, C. Y., and Lay, C. H. 2005. A nutrient formulation for fermentative hydrogen production using anaerobic sewage sludge microflora. *Int. J. Hydrogen Energy* 30(3):285–292.

Liu, G. Z., and Shen, J. Q. 2004. Effect of culture and medium conditions on hydrogen production from starch using anaerobic bacteria. *J. Biosci. Bioeng.* 98(4):251–256.

Liu, H., and Fang, H. H. P. 2003. Hydrogen production from wastewater by acidogenic granular sludge. *Water Sci. Technol.* 47(1):153–158.

Liu, H., Grot, S. A., and Logan, B. E. 2005. Electrochemically assisted microbial production of hydrogen from acetate. *Environ. Sci. Technol.* 39(11):4317–4320.

Ljungdahl, L. G., Hugenholtz, J., and Wiegel, J. 1989. Clostridia. In *Biotechnology Handbooks III*, edited by N. P. Minton and D. J. Clarke. Plenum Press, New York, USA.

Logan, B. E., Oh, S., Kim, I. S., and van Ginkel, S. 2002. Biological hydrogen production measured in batch anaerobic respirometers. *Environ. Sci. Technol.* 36:2530–2535.

Mao, X. Y., Miyake, J., and Kawamura S. 1986. Screening photosynthetic bacteria for hydrogen production from organic acids. *J. Ferment. Technol.* 64:245–249.

Minton, N. P., and Clarke, D. J. 1989. *Clostridia-Biotechnology Handbook*, Vol. 3. Plenum Press, New York, USA.

Miyake, J. 1998. *Biohydrogen*, edited by O. R. Zaborsky. Plenum Press, New York, USA.

Miyake, J., Mao, X. Y., and Kawamura, S. 1984. Photoproduction of hydrogen by a co-culture of a photosynthetic bacterium and *Clostridium butyricum*. *J. Ferment. Technol.* 62:531–535.

Mizuno, O., Dinsdale, R., Hawkes, F. R., Hawkes, D. L., and Noike T. 2000. Enhancement of hydrogen production from glucose by nitrogen gas sparging. *Bioresour. Technol.* 73:59–65.

Morimoto, M., Atsuko, M., Atif, A. A. Y., Ngan, M. A., Fakhru'l-Razi, A., Iyuke, S. E., and Bakir, A. M. 2004. Biological hydrogen production from glucose using natural anaerobic microflora. *Int. J. Hydrogen Energy* 29(7):709–713.

Mu, Y., Zheng, X. J., Yu, H. Q., and Zhu, R. F. 2006. Biological hydrogen production by anaerobic sludge at various temperatures. *Int. J. Hydrogen Energy* 31:780–785.

Nath, K., and Das, D. 2004. Improvement of fermentative hydrogen production: Various approaches. *Appl. Microbiol. Biotechnol.* 65:520–529.

Nishio, N., and Nakashimada, Y. 2007. Recent development of anaerobic process for energy recovery from wastes. *J. Biosci. Bioeng.* 103(2):105–112.

Noike, T., and Mizuno, O. 2000. Hydrogen fermentation of organic municipal wastes. *Water Sci. Technol.* 42(12):155–162.

Oh, S. E., van Ginkel, S. W., and Logan, B. E. 2003. The relative effectiveness of pH control and heat treatment for enhancing biohydrogen gas production. *Environ. Sci. Technol.* 37(22):5186–5190.

Okamoto, M., Miyahara, T., Mizuno, O., and Noike, T. 2000. Biological hydrogen potential of materials characteristic of the organic fraction of municipal solid wastes. *Water Sci. Technol.* 41:25–32.

Rittmann, B. E., and McCarty, P. L. 2001. *Environmental Biotechnology: Principles and Application.* McGraw-Hill Companies Inc., New York, USA.

Roychowdhury, S. 2000. Process for production of hydrogen from anaerobically decomposed organic materials. U.S. Patent. US 006090266A.

Shin, H. S., and Youn, J. H. 2005. Conversion of food waste into hydrogen by thermophilic acidogenesis. *Biodegradation* 16:33–44.

Stadtman, E. R., Stadtman, T. C., Pastan, I., and Smith, L. D. 1972. *Clostridium barkeri* sp. n. *J. Bacteriol.* 110(2):758–760.

Thauer, R. K. 1976. Limitation of biological hydrogen formation via fermentation. In *Microbial Energy Conversion,* edited by H. G. Schlegel and J. Barnea, pp. 201–294. Erich Goltze, Gttingen.

Ueno, Y., Kawai, T., Sato, S., Otsuka, S., and Morimoto, M. 1995. Biological production of hydrogen from cellulose by nature anaerobic microflora. *J. Ferment. Bioeng.* 79:395–397.

Ueno, Y., Otsuka, S., and Morimoto, M. 1996. Hydrogen production from industrial wastewater by anaerobic microflora in chemostat culture. *J. Ferment. Bioeng.* 82:194–197.

Ueno, Y., Otsuki, S., Ishii, M., and Igarashi, Y. 2001. Microbial community in anaerobic hydrogen-producing microflora enriched from sludge compost. *Appl. Microbiol. Biotechnol.* 57(4):555–562.

van Ginkel, S., Sung, S., and J. J. Lay. 2001. Biohydrogen production as a function of pH and substrate concentration. *Environ. Sci. Technol.* 35(24):4726–4730.

van Groenestijn, J. W., Hazewinkel, J. H. O, Nienoord, M., and Bussmann, P. J. T. 2002. Energy aspects of biological hydrogen production in high rate bioreactors operated in the thermophilic temperature range. *Int. J. Hydrogen Energy* 27(11–12):1141–1147.

van Haandel, A. C., and Lettinga, G. 1994. *Anaerobic Sewage Treatment: A Practical Guide for Regions with a Hot Climate.* John Wiley & Sons, Chichester, England.

van Niel Ed, W. J., Claasen Pieternel, A. M., and Stams Alfons, J. M. 2003. Substrate and product inhibition of hydrogen production by the extreme thermophile, *Caldicellulosiruptor saccharolyticus. Biotechnol. Bioeng.* 81:255–262.

Wang, C. C., Chang, C. W., Chu, C. P., Lee, D. J., Chang, B. V., and Liao, C. S. 2003. Producing hydrogen from wastewater sludge by *Clostridium bifermentans. J. Biotechnol.* 102:83–92.

Wang, G., Mu, Y., and Yu, H. Q. 2005. Response surface analysis to evaluate the influence of pH, temperature, and substrate concentration on the acidogenesis of sucrose-rich wastewater. *Biochem. Eng. J.* 23:175–184.

Wood, W. A. 1961. Fermentation of carbohydrates and related compounds. *The Bacteria.* II. Academic Press, New York, USA.

Wu, S. Y., Lin, C. N., and Chang, J. S. 2003. Hydrogen production with immobilized sewage sludge in three-phase fluidized-bed bioreactors. *Biotechnol. Prog.* 19:828–832.

Wu, S. Y., Lin, C. N., Chang, J. S., and Chang, J. S. 2005. Biohydrogen production with anaerobic sludge immobilized by ethylene-vinyl acetate copolymer. *Int. J. Hydrogen Energy* 30:1375–1381.

Yokoyama, H., Waki, M., Moriya, N., Yasuda, T., Tanaka, Y., and Haga, K. 2007. Effect of fermentation temperature on hydrogen production from cow waste slurry by using anaerobic microflora within the slurry. *Appl. Microbiol. Biotechnol.* 74(2):474–483.

Yu, H., Hu, Z., and Hong, T. 2003. Hydrogen production from rice winery wastewater by using a continuously-stirred reactor. *J. Chem. Eng. Japan* 36:1147–1151.

Yu, H., Zhu, Z., Hu, W., and Zhang, H. 2002. Hydrogen production from rice winery wastewater in an upflow anaerobic reactor by using mixed anaerobic cultures. *Int. J. Hydrogen Energy* 27:1359–1365.

Zhang, T., Liu, H., and Fang, H. H. P. 2003. Biohydrogen production from starch in wastewater under thermophilic condition. *J. Environ. Manage.* 69:149–156.

Zhang, Y. F, and Shen, J. Q. 2006. Effect of temperature and iron concentration on the growth and hydrogen production of mixed bacteria. *Int. J. Hydrogen Energy* 31:441–446.

CHAPTER

10

Microbial Fuel Cell: Novel Anaerobic Biotechnology for Energy Generation from Wastewater

Hong Liu

10.1 Background

World energy demand is expected to rise from 421 quadrillion British Thermal Units (BTUs) in 2003 to 563 quadrillion BTUs in 2015 and 722 quadrillion BTUs in 2030 (USDOE 2006). Energy production from renewable feedstocks holds great potential to meet these needs in a sustainable and environmentally sound manner, and to reduce dependence on fossil fuels.

At the same time, impairment of water resources requires increased investment in water and wastewater treatment infrastructure. Some estimates of the needed investment in the United States approach $1 trillion over the next 20 years (Water Infrastructure Network 2007). Energy generation from "negative-value" waste streams can simultaneously help meet the world's energy needs, reduce pollution, and reduce costs associated with water and wastewater treatment.

Anaerobic digestion has been used for methane recovery from solid and liquid waste streams for over a century. Methane fermentation has several intrinsic advantages over aerobic treatment processes. These include renewable energy (methane) generation, reduced energy costs through elimination of aeration, and reduced sludge treatment and disposal expenses. Anaerobic technology has been successfully commercialized for the treatment of waste, and several full-scale anaerobic treatment plants are in operation worldwide (Gallert et al. 2003).

In recent years, biohydrogen production from waste and wastewater through dark fermentation has also drawn considerable attention due to interest in clean energy production using hydrogen fuel cells. Despite a stoichiometric potential of 12 mol H_2/mol glucose, current fermentation techniques can unfortunately produce a maximum of only 2–3 mol H_2/mol glucose, because most organic matter remains mired as volatile fatty acids and alcohols. The process is limited to feedstocks with suitable fermentation substrates, that is, those rich in carbohydrates, such as glucose (Li and Fang 2007; Liu 2002; Logan 2004).

Microbial fuel cell (MFC) technology, which uses microorganisms to catalyze the direct generation of electricity from organic matter, provides a completely new approach for energy generation from wastes (Cheng et al. 2006a; He et al. 2006; Liu et al. 2004, 2005a; Rabaey et al. 2003). It has been known for nearly a century that bacteria can generate electricity (Potter 1911). But MFCs with sufficient power output and capable of utilizing a variety of biodegradable organic materials, and therefore practical for wastewater treatment, have been developed only recently.

MFC technology has great potential for both renewable energy generation and waste remediation for a number of reasons:

1. Clean energy can be generated that helps offset costs.
2. A variety of organic materials can be used for electricity generation, including carbohydrates (Liu and Logan 2004; Rabaey et al. 2003), fatty acids (Liu et al. 2005b), proteins (Heilmann and Logan 2006), and wastewater from various sources (Liu et al. 2004; Min et al. 2005b).
3. No aeration is needed if a passively aerated air cathode is used (Liu and Logan 2004).
4. Compared to other biogas (e.g., methane and hydrogen) production processes, an MFC directly generates electrical energy, thus eliminating the post–gas purification process.
5. Significantly less sludge is produced by the anaerobic microbes in an MFC system (Rabaey and Verstraete 2005).
6. An MFC can denitrify wastewater while generating electricity using biocathode (Clauwaert et al. 2007).
7. Ammonia can be removed during the treatment of animal manure (Min et al. 2005b).

The remainder of this chapter focuses on the principles, stoichiometry, energetics, microbiology, design, and operation of MFCs, with particular attention to wastewater treatment applications.

10.2 How Does a Microbial Fuel Cell Work?

A microbial fuel cell is a device that directly converts chemical energy stored in organic matter into electricity through the catalytic activities of microorganisms. Figure 10.1 illustrates the essential components of an MFC: anode, cathode, electrochemically active microorganisms, and electrolyte (water). Microbes catalyze the

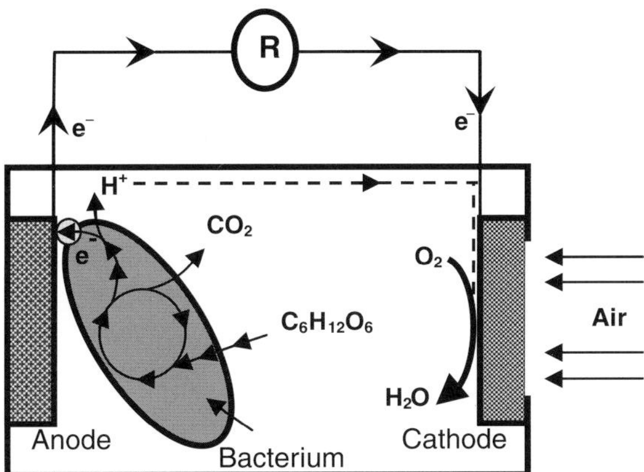

FIG. 10.1. Schematic of single-chamber MFC, showing the anode, where bacteria form a biofilm on the surface, and a cathode that is exposed to air.

oxidation of organic materials, such as glucose, which results in the production of electrons, protons, and CO_2. The protons travel through the electrolyte solution by direct diffusion and/or via other proton carrier ions, such as phosphates or carbonates (Fan et al. 2007b). The electrons flow through the external load (R) to the cathode, where they combine with protons and oxygen to form water and create electrical current. The potential difference between the anode and the cathode drives the electron flow.

10.3 Stoichiometry and Energetics

With a glucose substrate and oxygen serving as the electron acceptor, the electrode reactions can be writen as follows:
Anode

$$C_6H_{12}O_6 + 6H_2O \rightarrow 6CO_2 + 24H^+ + 24e^- \tag{10.1}$$

Cathode

$$O_2 + 4H^+ + 4e^- \rightarrow 2H_2O \tag{10.2}$$

The overall reaction in an MFC is as follows:

$$C_6H_{12}O_6 + 6O_2 \rightarrow 6CO_2 + 6H_2O \tag{10.3}$$

If there is no energy loss in the MFC, or the reaction is reversible, the work (W (J)) produced by the cell (nFE_{emf}) equals the Gibb's free energy of the electrochemical reaction:

$$W = nFE_{emf} = -\Delta G_r \tag{10.4}$$

where n (mol) is the number of electrons transferred in the reaction, F (96,485 C/mol) is Faraday's constant, and E_{emf} (V) is the cell electromotive force, which is the potential difference between the cathode and anode. G_r (J) is the Gibb's free energy of the reaction, which can be calculated by the following equation:

$$\Delta G_r = \Delta G_r^0 + RT \ln Q \tag{10.5}$$

where $G_r{}^0$ (J) is the Gibb's free energy under standard temperature and pressure (STP) (298.15 K, 1 atm, 1 mol/L), R (8.31447 J/mol/K) is the universal gas constant, T (K) is the absolute temperature, and Q (unitless) is the reaction quotient, which can be calculated by dividing the product of the activities of the products by the product of the activities of the reactants.

Combining Eqs (10.4) and (10.5),

$$E_{emf} = E_{emf}^0 - \frac{RT}{nF} \ln Q \tag{10.6}$$

where E_{emf}^0 (V) is the cell electromotive force under standard conditions, or $Q = 1$.

For the anode reaction (Eq. (10.1)), the potential can be written as follows:

$$E_{anode} = E_{anode}^0 - \frac{RT}{24F} \ln \left(\frac{[C_6H_{12}O_6]}{[CO_2]^6[H_{anode}^+]^{24}} \right) \tag{10.7}$$

For the cathode reaction (Eq. (10.2)), the electrode potential can be written as follows:

$$E_{cathode} = E_{cathode}^0 - \frac{RT}{4F} \ln \left(\frac{1}{[O_2][H_{cathode}^+]^4} \right) \tag{10.8}$$

The cell electomotive force (E_{emf}) equals the cathode potential minus anode potential, or

$$E_{emf} = E^0 - \left\{ \frac{RT}{4F} \ln \left(\frac{1}{[O_2][H_{cathode}^+]^4} \right) - \frac{RT}{24F} \ln \left(\frac{[C_6H_{12}O_6]}{[CO_2]^6[H_{anode}^+]^{24}} \right) \right\} \tag{10.9}$$

where $E^0 = E_{cathode}^0 - E_{anode}^0$ is the electromotive force under standard conditions. Equation (10.9) indicates that the electromotive force is independent of cell pH if the anodic pH equals the cathodic pH. In an MFC, however, cathodic pH is usually greater than the anodic pH due to the production of protons at the anode and the consumption of protons at the cathode. The equation also illustrates that the variation in anodic and cathodic pH can affect cell potential more severely than the concentration of oxygen, CO_2, glucose, et al.

Example 10.1

Calculate the cathode potentials of MFCs at pH 4.0 and pH 10.0 ($E_0 = 1.229$ V; O_2 partial pressure = 0.21 atm; $T = 303$ K).

Solution

Using Eq. (10.8), the cathode potential at pH 4 is:

$$E_{pH4} = E_0 - RT/4F\ \ln(1/(0.21 \times 10^{-4})) = 0.978\ \text{V}$$

Similarly, the cathode potential at pH 10.0 equals to 0.618 V.

Therefore, the difference between the cathode potentials at pH 4.0 and pH 10.0 is 0.36 V.

10.4 Electrochemically Active Microbes and Electron Transfer Mechanisms

10.4.1 Electron Transfer Mechanisms

Based on the studies of electrochemically active microbes and possible extracellular electron transfer mechanisms, a new area of microbial ecology is emerging. Studies of metal-reducing bacteria such as *Geobacter* and *Shewanella* sp. provided an initial understanding of electron transfer from bacteria to electrodes. In a natural environment, these microorganisms can use insoluble metal oxides such as Fe(III) as electron acceptors to oxidize organic matter. When the metal oxide is substituted with an anode connected in circuit to a cathode, and with oxygen serving as the electron acceptor, electricity can be generated (Bond and Lovley 2003; Kim et al. 2002; Magnuson et al. 2001; Myers and Myers 2001). Other genera of bacteria, such as *Escherichia* (Park et al. 2001) and *Pseudomonas* (Rabaey et al. 2004), have also been identified that are capable of producing electricity in MFCs.

There are several mechanisms involved in the electron transfer from the bacteria to the anode, illustrated as follows:

1. *Direct electron transfer via the cell's outer-membrane proteins* (Fig. 10.2a). A bacterial cell membrane is not usually conducive to electron transfer. Recent biochemical and genetic characterization studies indicated that outer-membrane cytochromes, the enzymes on the bacterial respiratory chain, might be involved

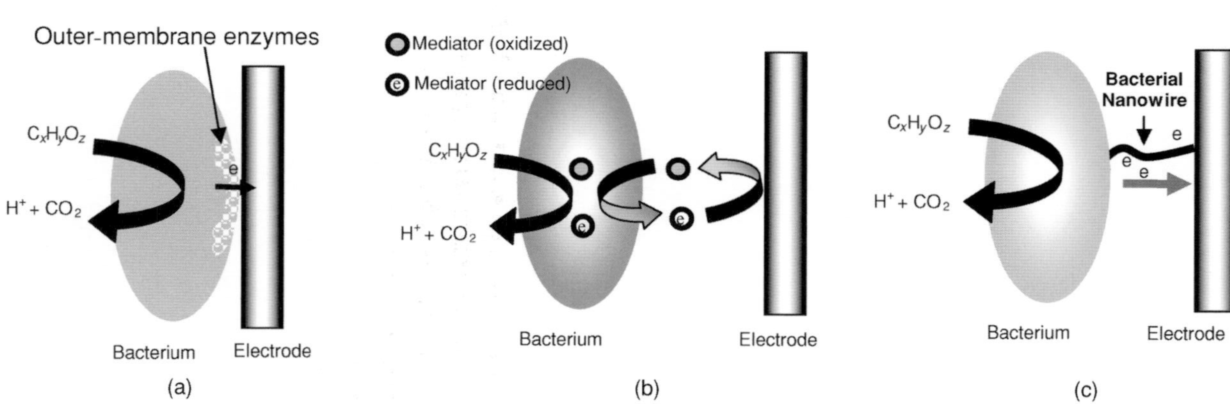

FIG. 10.2. Three anodic electron transfer mechanisms: (a) direct electron transfer via bacterial outer-membrane enzymes; (b) electron transfer via mediators; (c) electron transfer via pilus-like nanowires.

in the electron transfer from *Geobacter sulfurreducens* and *Shewanella putrefaciens* to electrodes (Magnuson et al. 2001; Myers and Myers 2001). Direct contact of cytochromes with the electrode is needed for this electron transfer mechanism.

2. *Mediated electron transfer* (Fig. 10.2b).
   Efficient electron transfer can be achieved by adding artificial mediators, such as neutral red and methylene blue, which are capable of crossing cell membranes, accepting electrons from intracellular electron carriers, exiting the cell in the reduced form, and then releasing the electron onto the electrode surface. This eliminates the need for direct contact between the cell and electron acceptor. Electron transfer from *Escherichia coli* to the electrode can be enhanced with the addition of neutral red to MFCs (Park and Zeikus 2003). Some bacteria produce their own electron mediators that can also be used by other species. For example, *Pseudomonas aeruginosa* can produce phenazine to stimulate electron transfer for several bacterial strains (Rabaey et al. 2004). The microbes that transfer electrons via this mechanism are not suitable, however, for wastewater treatment due to the cost and toxicity of many artificial mediators. More importantly, there is a rapid loss of mediators in a continuous-flow system.
3. *Electron transfer via bacterial nanowires* (Fig. 10.2c).
   The recent discovery of bacterial nanowires (Gorby et al. 2006; Reguera et al. 2005) indicated that the conductive, piluslike structures grown on the cell membrane might be directly involved in extracellular electron transfer and allow the direct reduction of a distant electron acceptor. These nanowires have been identified in *G. sulfurreducens* PCA, *Shewanella oneidensis* MR-1, a phototrophic cyanobacterium *Synechocystis* PCC6803, and the thermophilic fermentative bacteria *Pelotomaculum thermopropionicum* (Gorby et al. 2006: Reguera et al. 2005).

10.4.2 Mixed Culture

For waste and wastewater treatments, engineers prefer mixed cultures rather than pure cultures. A mixed culture can be easily adapted to utilize complex organic materials in waste streams. Processes using mixed cultures are simpler to operate and easier to control.

Mixed cultures enriched from domestic wastewater (Liu et al. 2004), animal wastes (Min et al. 2005b) and anaerobic sewage sludge (Kim et al. 2005), and ocean sediments (Reimers et al. 2001) have been used to generate electricity in MFCs. Mixed-culture bacterial communities are very diverse. δ-Proteobacteria predominate in marine sediment MFCs, and α-, β-, γ-, or δ-Proteobacteria, Firmicutes, and uncharacterized clones populate other types of MFCs (Logan and Regan 2006). Although MFC tests were conducted with mixed-culture isolates, a higher power density was generated from the mixed culture (Rabaey et al. 2004).

The exact mechanisms for this observation are still unclear. More research is needed to better understand the microbiology of MFC biofilms and the electrochemical capabilities of bacterial species and consortia.

10.5 Evaluation of MFC Performance

10.5.1 Circuit Voltage

The most direct way to examine electricity generation in MFCs is through the measurement of the circuit voltage using a multimeter. Circuit voltage (V) can be used to calculate the circuit current (I) through an external resistance (R) based on Ohm's law as follows:

$$I = \frac{V}{R} \tag{10.10}$$

The circuit voltage (V) can also be described as follows:

$$V = \text{OCV} - \eta \tag{10.11}$$

$$\eta = I R_{\text{int}} \tag{10.12}$$

where OCV (V) is the open circuit voltage, which can be measured in the absence of current and reflects the maximum voltage that can be obtained in the tested system. R_{int} is the overall internal resistance, and $\eta(V)$ is the current-related overpotentials, which include the activation losses (V_{act}) caused by the slow reactions taking place on the electrode surface, the ohmic losses (V_{ohm}) caused by the resistance of the electrolyte and the electrode, and concentration polarization or mass transfer losses (V_{tran}) associated with the loss of reactant on the electrode surface (Larminie and Dicks 2000):

$$\eta = \Delta V_{\text{act}} + \Delta V_{\text{ohm}} + \Delta V_{\text{tran}} \tag{10.13}$$

A polarization curve is commonly used to study MFC performance and the various current-related losses. Figure 10.3 is simply a plot of steady-state voltage versus current. The reader can identify three distinct regions of overpotentials. The cell voltage drops rapidly and nonlinearly in the activation region, followed by a slow and near-linear drop in the ohmic region. The continuous current increase may reflect concentration polarization or a condition where mass transport of reactants becomes limiting. In this region, the voltage drops rapidly and nonlinearly with current increase.

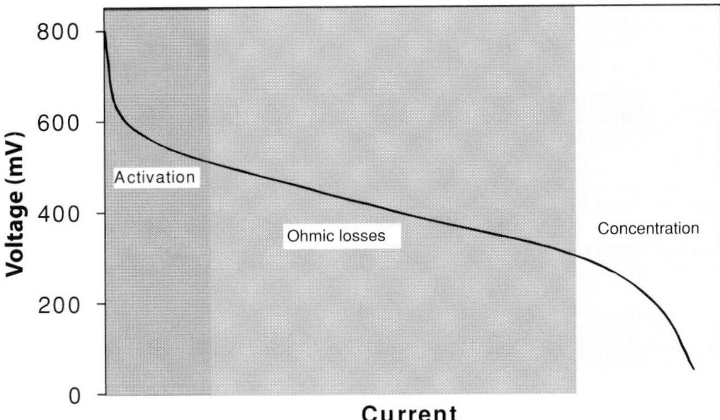

FIG. 10.3. The polarization curve of an MFC shows the three regions of polarization overpotentials.

The open circuit voltage (OCV) of MFC systems ranges from 0.5 to 0.8 V. The reader should note that this voltage is over 0.4 V lower than the E_{emf} (1.2 V at pH 7.0) calculated from Eq. (10.9). The difference between theoretical E_{emf} and measured OCV is known as the current irrelative overpotential, which is caused by the internal current at open circuit (Larminie and Dicks 2000).

10.5.2 Power Density and Current Density

Power density is one of the most important parameters for evaluating MFC performance. The power output (P) of an MFC can be calculated according to $P = IV$. Power output is often normalized to projected electrode surface area, defined as surface power density, to make it possible to compare different MFC systems. However, compared to the simple flat plate type of a conventional chemical fuel cell, the diversity and complexity of MFCs makes it difficult to compare the power density directly. For example, the surface area of a two-chamber MFC using a graphite granule as anode material cannot be directly compared with that of an MFC with carbon cloth anode. Volumetric power density, which is normalized to the reactor volume, is more suitable for the evaluation of overall performance of MFCs.

Maximum current density is seldom reported in MFC studies, probably because the primary goal of these investigations is to maximize the power output, while the power density of MFCs at maximum current density is not the highest. In addition, MFCs operated at maximum current density are often unstable. The maximum current density, however, is a very important parameter when evaluating the bacterial electrochemical capabilities and the electrode materials. This is because it reflects the maximum reactant conversion rate that is directly related to the organic removal rate in wastewater treatment.

10.5.3 Coulombic Efficiency and Energy Recovery

Coulombic efficiency (CE) is used to evaluate the electron recovery as current from the organic matter. CE can be calculated as a ratio of total recovered coulombs (C_p) obtained by integrating the current overtime to the theoretical coulombs (C_t) that can be produced from the substrate.

For MFCs operated in a batch mode, CE can be calculated as follows:

$$\mathrm{CE} = \frac{\int_0^t I dt}{Fbv\Delta C} \tag{10.14}$$

where F is Faraday's constant (96,485 C/mol), b is the number of moles of electrons produced per mole of substrate ($b = 24$ for glucose), v is the volume of anolyte, and C is the concentration difference of the substrate from time 0 to time t (mol/L).

For MFCs operated in a continuous-flow mode and under a steady-state condition, CE is given by:

$$\mathrm{CE} = \frac{I}{Fbq\Delta C} \tag{10.15}$$

where q is the anolyte flow rate (L/s) and C is the substrate concentration difference between the influent and the effluent (mol/L).

Energy recovery (ER) or energy efficiency is very useful for comparing the performance of MFCs with traditional energy conversion methods, such as combustion. The ER is the ratio of power produced by the cell to the theoretical heat energy of the degraded organic substrates, as given by:

$$\mathrm{ER} = \frac{\int_0^t VI dt}{\Delta H \Delta C v} \tag{10.16}$$

where ΔH is the heat of combustion of the substrate (J/mol), ΔC is the substrate concentration (mol/L) change for batch operation or difference between influent and effluent for continuous-flow operation, and v is the liquid volume (L) of the anode chamber for batch operation, or the total anolyte input from time 0 to time t for continuous-flow operation.

Example 10.2

An MFC is operated in a continuous-flow mode (flow rate 3.6 L/h) using glucose as a substrate. The concentration of glucose in the influent is 5 and 0.5 mmol/L in the effluent. The hydraulic retention time is 2 h. The average current output of the MFC is 8 A at 0.4 V. Please calculate (1) the volumetric power density, (2) the Coulombic efficiency, and (3) the energy recovery of the MFC (data: standard heat of combustion of glucose, $\Delta H = -2{,}805$ kJ/mol).

Solution

The volumetric power density, p = power/volume
$= 8\text{ A} \times 0.4\text{ V}/(3.6\text{ L/h} \times 2\text{ h}) = 0.44\text{ W/L}$
The flow rate, $q = 3.6\text{ L/h} = 1\text{ mL/s}$.
$\Delta C_{glucose} = 4.5\text{ mmol/L}$
Using Eq. (10.15),
$CE = 8/(96{,}485 \times 24 \times 0.001 \times 0.0045) = 76.8\%$
Using Eq. (10.16),
$ER = 0.4 \times 8/(2{,}805 \times 10^3 \times 0.0045 \times 0.001) = 25.4\%$

10.5.4 COD/BOD Removal

The organic matter in wastewater (expressed collectively as chemical oxygen demand/biochemical oxygen demand (COD/BOD)) can be removed in an MFC through conversion to electrical current, biomass, and/or through sulfate and nitrate reduction, or aerobic oxidation. Oxygen sources include that originally contained in the influent and/or that diffused from air to the MFC chamber (anode) through the air cathode and/or through the membrane. The COD/BOD removal rate is normally higher than the CE, which represents only a portion of the organic matter converted to electrical current. For readily biodegradable substrates such as volatile fatty acids and simple sugars, the COD/BOD removal efficiency is over 90% (Table 10.1). For complex substrates like domestic wastewater, the COD/BOD removal efficiencies range from 40 to 95% (Table 10.2).

10.6 MFC Designs and Electrode Materials

10.6.1 Two-Chamber System

MFCs are typically designed as a two-chamber system with the bacteria in the anode chamber separated from the cathode chamber by a membrane as shown in Fig. 10.4. Some MFCs use aqueous cathodes where air is bubbled through water to

Table 10.1. MFC processes and performance using defined substrate.

MFC Type[a]	Culture	Substrates	Anode	Cathode	pH	HRT (h)	Temperature (°C)	Power Density W/m^3	Power Density mW/m^2	CE (%)	COD (%)	References
SC-continuous	Mixed	Acetate	Carbon cloth	Carbon cloth/Pt	9.0	0.1	30	1,550	2,770	—	—	Fan et al. (2007b)
SC-batch	Mixed	Acetate	Carbon cloth	Carbon cloth/Pt	7.0		30	627	1,010	—	—	Fan et al. (2007a)
SC-batch	Mixed	Acetate	NH_3-treated carbon cloth	Carbon cloth/CoTMPP	7.0	—	30	73	2,400	60	—	Logan et al. (2007)
SC-batch	Mixed	Acetate	NH_3-treated carbon cloth	Carbon cloth/Pt	7.0	—	30	115	1,970	30–60	—	Cheng and Logan (2007)
TC-continuous tubular	Mixed	Acetate	Granular graphite	Graphite mat/ferricyanide	—	7.0	—	90	—	75	—	Rabaey et al. (2005b)
TC-continuous stack	Mixed	Acetate	Granular graphite	Granular graphite ferricyanide	7.0	2.9–8.9	22	258	—	71–76	—	Aelterman et al. (2006)
SC-batch	Mixed	Acetate	Carbon paper	Carbon cloth/Pt	7.0	—	30	13	506	10–31	99	Liu et al. (2005c)
SC-batch	Mixed	Butyrate	Carbon paper	Carbon cloth/Pt	7.0	—	30	8	305	8–15	98	Liu et al. (2005c)
TC-continuous	*Shewanella oneidensis*	Lactate	Graphite felt	Graphite felt	7.4	0.8–3.3 min	—	96	16	—	—	Biffinger et al. (2007)
TC-continuous	*Shewanella oneidensis*	Lactate	Graphite felt	Graphite/ferricyanide	7.4	0.06–2.0 min	19	500	3,000	8	—	Ringeisen et al. (2006)
SC-batch	Mixed	Ethanol	Carbon paper	Carbon cloth/Pt	7.0	—	30	12	488	10	—	Kim et al. (2007)
SC-batch	Mixed	Glucose	Carbon paper	Carbon cloth/Pt	7.0	—	30	13	494	9–12	98	Liu et al. (2004)
TC-continuous tubular	Mixed	Glucose	Granular graphite	Graphite mat/ferricyanide	—	7	—	66	—	59	—	Rabaey et al. (2005b)
SC-continuous	Mixed	Glucose	Carbon cloth	Carbon cloth/Pt	7.0	4–16 h	30	51	1,540	60	89–93	Cheng et al. (2006a)
Two-chamber	Mixed	Glucose	Graphite	Graphite/ferricyanide	7.1	18–42	28	216	3,600	89	—	Rabaey et al. (2003)
TC-continuous	Mixed	Glucosa and glutamate	Graphite felt	Graphite felt with Pt	—	0.6	35	102	560	—	97	Moon et al. (2006)
TC-continuous tubular		Sucrose	GAC[b]	GAC/ferricyanide	5.8–6.3	6.0	35	29.2	—	51	>90	He et al. (2006)
SC-batch	Mixed	Corn stover	Carbon paper	Carbon cloth/Pt	7.0	—	30	24	971	20–30	60–70	Zuo et al. (2006)
SC-batch	Mixed	BSA[c]	Carbon paper	Carbon cloth/Pt	7.0	—	30	8.8	354	21	86	Heilmann and Logan (2006)
TC-continuous	*Paracoccus denitrificans*	Sulfide	Granular graphite	Granular graphite ferricyanide	7.0	—	35	101	47	15–30	—	Rabaey et al. (2006)

[a] SC, single chamber; TC, two chambers.
[b] Granular activated carbon.
[c] Bovine serum albumin.

Table 10.2. MFC processes and performance using wastewater.

MFC Type[a]	Substrate	Anode	Cathode	pH	HRT (h)	Temperature (°C)	Power Density W/m³	Power Density mW/m²	CE (%)	COD (%)	References
TC-continuous tubular	Hosipital wastewater	Granular graphite	Graphite mat/ferricyanide	—	7.0	—	8.0	—	—	23	Rabaey et al. (2005b)
TC-continuous tubular	Domestic wastewater	Granular graphite	graphite mat/ferricyanide	—	7.0	—	5.0	—	20	—	Rabaey et al. (2005b)
SC-continuous	Domestic wastewater	Graphite rod	Carbon cloth/Pt	7.3–7.6	3–33	30	—	26	20	40–80	Liu and Logan (2004)
SC-continuous	Domestic wastewater	Carbon cloth	Carbon cloth/Pt	7.5	3.4–4.6	30	15.5	464	27	40–50	Cheng et al. (2006a)
SC-batch	Domestic wastewater	Carbon paper	Carbon cloth/Pt	7.3–7.6	—	30	3.7	147	20	75	Liu et al. (2004)
TC-batch	Food-processing wastewater	Carbon paper	Carbon cloth/Pt	7.0	—	30	—	81	27–41	95	Oh and Logan (2005)
SC-batch	Fermentation effluent	Carbon paper	Carbon cloth/Pt	7.0	—	30	9.0	371	—	—	Oh and Logan (2005)
SC-batch	Meat-packing wastewater	Carbon paper	Carbon cloth/Pt	—	—	30	2.0	80	5–12	87	Heilmann and Logan (2006)
SC-batch	Swine wastewater	Carbon paper	Carbon cloth/Pt	—	—	30	6.5	261	8	86	Min et al. (2005a)

[a]SC, single chamber; TC, two chambers.

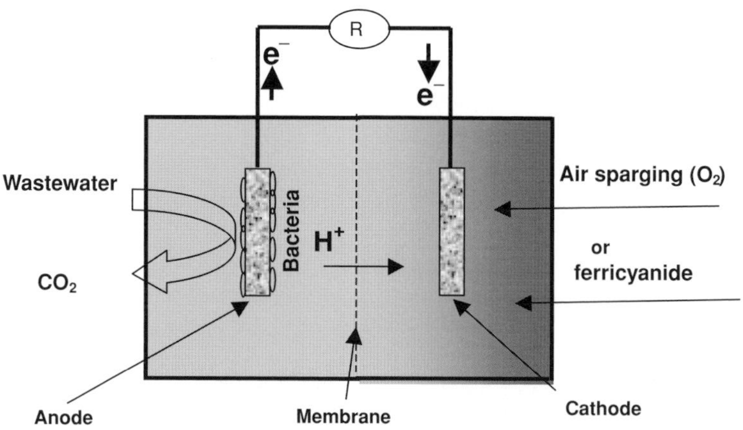

FIG. 10.4. Schematic of a two-chamber MFC system.

provide dissolved oxygen to the electrode. This approach not only consumes a fair amount of energy but also is less efficient. Using ferricyanide as the catholyte can greatly improve the MFC performance compared to the use of air (Oh and Logan 2006). The power generated from this type of MFC, however, is not sustainable because the ferricyanide is consumed in the cathodic reaction and needs to be replenished.

A basic requirement in the design of an MFC system is to minimize the internal resistance. H-shaped MFCs, consisting of two bottles connected by a tube containing a proton exchange membrane (PEM) membrane or a salt bridge, produce low power density due to the high internal resistance (Logan et al. 2006). Flat-plate design, in which a membrane sheet is sandwiched between the anode and cathode chamber, can effectively reduce the internal resistance (Fig. 10.5). The advantage of flat-plate MFCs is that they can be stacked together (Fig. 10.5d) to achieve higher voltage (in series) and/or current (in parallel).

Another design is a two-chamber tubular-type MFC with graphite granules serving as anode material and ferricyanide used as the catholyte (Fig. 10.6). Compared to an upflow MFC with the cathode column on the top of the anode column (He et al. 2005), the internal resistance of the MFC can be greatly reduced when the cathode is placed directly inside (He et al. 2006) or outside (Rabaey et al. 2005b) of the anode column, thereby resulting in much higher power density. However, the power density of such tubular MFCs is still lower than that of flat-plate ones using similar electrode material (Aelterman et al. 2006). This could probably be due to the relatively larger average electrode spacing.

A cation exchange membrane is required for a two-chamber MFC system using ferricyanide as an electron acceptor to avoid the diffusion of toxic ferricyanide into the anode chamber while allowing the transfer of protons or other cations

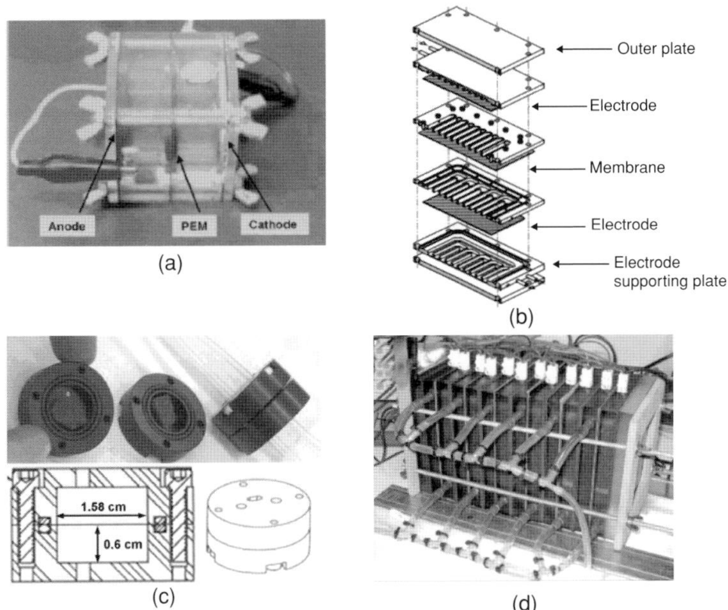

FIG. 10.5. Flat-plate two-chamber MFCs: (a) a two-chamber air cathode MFC with PEM (Kim et al. 2007); (b) a flat-plate MFC with biopolar membrane and ferric ion cathode (Heijne et al. 2006); (c) a miniature MFC using Nafion membrane and ferricyanide as catholyte (Ringeisen et al. 2006); and (d) stacked MFC using Ultrex membrane and ferricyanide as catholyte (Aelterman et al. 2006). Reprinted with Permission.

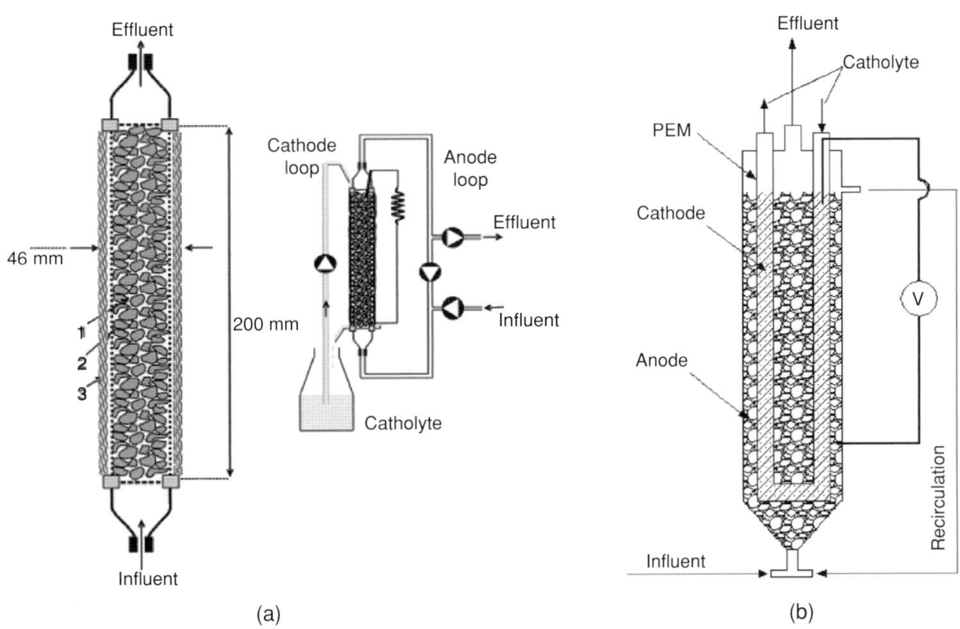

FIG. 10.6. Tubular MFCs: (a) with an outer cathode (1, granular anode; 2, membrane; 3, cathode electrode) (Rabaey et al. 2005b) and (b) with an interior cathode (He et al. 2006). Reprinted with permission.

to the cathode chamber. The Nafion membrane or alternatives (such as Ultrex CMI-7000) are most commonly used in a two-chamber MFC system (Bond and Lovley 2003; Gil et al. 2003; Kim et al. 2007; Oh and Logan 2006; Rozendal et al. 2006; Schröder et al. 2003), possibly due to their popularity in chemical fuel cells for excellent proton conductivity and thermal and mechanical stability. However, the transfer of alkali cations from the anodic to the cathodic chamber can cause pH reductions in the anolyte (Gil et al. 2003; Kim et al. 2007; Rozendal et al. 2006), which may negatively affect the performance of the MFC. For MFCs other than those using ferricyanide, anion (Kim et al. 2007) or bipolar (Heijne et al. 2006) exchange membranes, nanoporous polymer filters (Biffinger et al. 2007), ultrafiltration membranes (Kim et al. 2007), and even J cloth (Fan et al. 2007a) could be better choices because they result in a relatively stable anolyte pH, which is critical for biological activity.

10.6.2 Single-Chamber System

Two-chamber MFCs may be unsuitable for wastewater remediation because of low power density (dissolved oxygen as electron acceptor) or because catholyte such as ferricyanide cannot be sustained. A single-chamber system, however, may have great potential in a similar situation. By hot pressing PEM and carbon cloth to form the cathode, single-chamber MFCs (SCMFCs) can achieve much better performance than a two-chamber air cathode system due to the higher mass transfer coefficient of oxygen in air compared to water. A cylindrical-type SCMFC (Fig. 10.7a) consists of a single cylindrical chamber with graphite rods (anode) placed in a concentric arrangement and a single air-porous cathode with a carbon cloth/platinum catalyst/Nafion membrane fused together. A plate-type SCMFC, illustrated in Fig. 10.7b, is configured with a carbon cloth or carbon paper anode, PEM, and carbon cloth cathode arranged in parallel.

In single-chamber air cathode MFCs, the main function of the PEM membrane is to block oxygen diffusion. This expensive and complicated membrane system can be eliminated with the development of a biofilm on the cathode surface (Liu and Logan 2004). A biofilm populated with aerobic bacteria can function as a membrane to minimize oxygen diffusion into the anode chamber. A higher power density can be achieved in the membrane-free system due to decrease in internal resistance (Liu and Logan 2004). One of the challenges of a membrane-free system is the fast air diffusion through the cathode, which results in low CE due to substrate utilization by aerobic bacteria (Liu and Logan 2004). The diffusion of oxygen also limits the minimal distance between anode and cathode to about 1–2 cm (Cheng et al. 2006a), which in turn limits the maximum volumetric power density.

In a recent study, a cloth layer was applied on the water-facing side of an SCMFC air cathode (Fig. 10.8a). The CE improved by twofold compared to cathodes

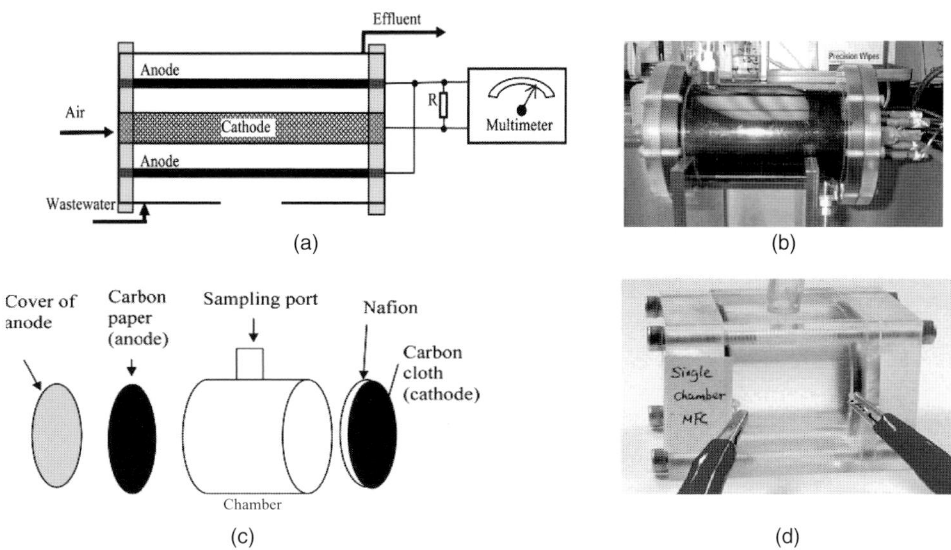

Fig. 10.7. (a) Schematic and (b) laboratory-scale prototype of the cylindrical-type single-chamber MFC (Liu and Logan 2004); (c) schematic and (d) laboratory-scale prototype of flat-plate single-chamber MFC (Liu et al. 2004). Reprinted with permission.

without the cloth layer (71% vs 35%) at the same current density of 0.6 mA/cm^2 (Fan et al. 2007a). This was due to the decrease in oxygen diffusion as a result of the cloth. The CE was comparable to that of two-chamber anoxic MFCs using ferricyanide as catholyte (Table 10.1). The cloth enables the electrode spacing to be reduced to less than 1 mm for an air cathode MFC system with a cloth electrode assembly (CEA). Significant improvement in volumetric power density

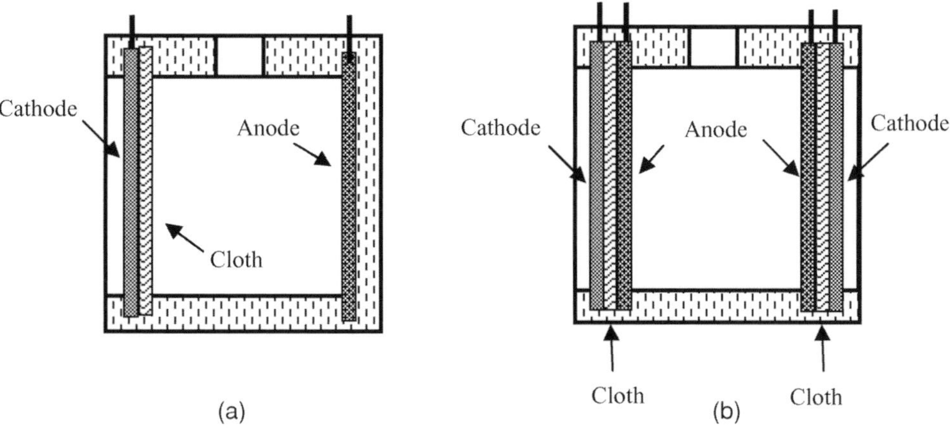

Fig. 10.8. Schematic of MFCs with (a) cloth layer and (b) double-cloth electrode assemblies (Fan et al. 2007a). Reprinted with permission.

(627 W/m^3) has been achieved using MFCs with double CEAs operated in a batch-fed mode (Fig. 10.8b). This could be due to an increased surface/volume ratio and reduced internal resistance. More recently, the power density was further increased to 1,550 W/m^3 (2,770 mW/m^2) using the same MFC operated in a continuous mode with bicarbonate as buffer (Fan et al. 2007b).

10.6.3 Electrode Materials

10.6.3.1 Anode

Carbon-based materials are commonly used as an anode for MFCs due to their superior conductivity, chemical stability, structural strength, favorable surface properties for biofilm development, and versatility for creating a large surface area. Various carbon materials, including graphite rods (Liu et al. 2004), plain graphite (Chaudhuri and Lovley 2003; Rabaey et al. 2003), graphite foam (Chaudhuri and Lovley 2003), woven graphite (Park and Zeikus 2003), graphite felt (Biffinger et al. 2007), graphite granules (Rabaey et al. 2005b), reticulated vitreous carbon (He et al. 2005), granular-activated carbon (He et al. 2006), carbon paper (Liu and Logan 2004), carbon cloth (Liu et al. 2005a), and a graphite fiber brush (Logan et al. 2007) have been examined as anode electrodes for MFCs. Among these materials, carbon fiber showed the best performance, probably, due to its large surface area, relatively high conductivity, and favorable surface properties for biofilm development. In a recent study, a power density of 7,200 mW/m^2 was achieved using microfiber carbon cloth in the author's group, which is the highest MFC power density reported up to the time this document was produced.

Chemical modifications of the anodes can further improve MFC performance. Electron transfer improvements were reported when a carbon electrode surface was coated with self-assembled monolayers (SAMs) (Crittenden et al. 2006). Modifications of the electrode surface with Mn(IV), Fe(III), neutral red (Park and Zeikus 2003), Ni^{2+} and Fe_3O_4 (Lowy et al. 2006), and ammonia (Cheng and Logan 2007) have also been reported.

10.6.3.2 Cathode

Development of high-performance, low-cost cathode material is critical for the successful application of MFC technology in wastewater remediation. Carbon material such as graphite plate (Tender et al. 2002) and graphite felt (Biffinger et al. 2007) can be directly used as a cathode. The power density is, however, much lower than that with a catalyst. Noble metals such as platinum can be used in laboratory MFC systems as a cathode catalyst to catalyze oxygen reduction in the cathode. The high cost of platinum reduces the appeal of this approach.

The recently developed high-performance, noble-metal-free catalysts such as pyrolyzed iron (II) phthalocyanine or cobalt tetramethoxyphenylporphyrin (Cheng et al. 2006b; Zhao et al. 2006) for MFC cathodes showed promise for reducing the cost of cathode materials.

While a two-chamber system using ferricyanide is not practical for wastewater remediation, air cathode MFCs also face technical challenges, such as cathode flooding due to water accumulation on the air side and water leakage through the fibrous material.

10.7 Operational Factors Affecting MFC Performance

10.7.1 Substrates

Electricity can be generated from various defined substrates, including glucose (Liu and Logan 2004; Rabaey et al. 2005a), starch (Min and Logan 2004), cellulose (Niessen et al. 2005), acetate, butyrate (Liu et al. 2005b), lactate (Ringeisen et al. 2006), ethanol (Kim et al. 2005), cysteine (Logan et al. 2005), peptone, and bovine serum albumin (Heilmann and Logan 2006). Electricity generation from pentose sugars, such as xylose, arabinose, ribose, and their pentitols, such as xylitol and arabitol, has also been demonstrated in a recent study by the author's group (Catal et al. 2008). Varying MFC configurations, inocula, solution chemistry, and operational conditions from study to study make power generation comparisons difficult among these substrate species. Variations from substrate to substrate have been observed when the experiments were performed in a similar single-chamber air cathode MFC using the same inocula and similar solution chemistry, except the carbon sources. Similar power densities were observed for glucose (494 mW/m^2; Liu and Logan 2004), acetate (509 mW/m^2; Liu et al. 2005b), and ethanol (488 mW/m^2; Kim et al. 2007); a second, lower tier of similar power densities was observed for butyrate (305 mW/m^2; Liu et al. 2005b), peptone (269 mW/m^2), and bovine serum albumin (354 mW/m^2) (Heilmann and Logan 2006).

Electricity generation from real waste streams has also been studied. Table 10.2 shows the performance of MFCs using wastewaters, for example, from domestic sources (146 mW/m^2), swine (264 mW/m^2; Min et al. 2005b), food processing (81 mW/m^2; Oh and Logan 2005), and meat packing (80 mW/m^2; Heilmann and Logan 2006). Although the CE of MFCs obtained from using real wastewater (5–47%; Table 10.2) was generally lower than that for defined substrates (8–89%; Table 10.1), the COD removal efficiencies were comparable (40–95% for wastewaters; 60–99% for defined substrates). Hospital wastewater was an exception, with a COD removal efficiency of only 23%. This was likely due to the low biodegradability of this wastewater.

Further investigations on the effect of substrates on microbial activity and power generation need to be conducted either in similar MFC systems with a tested anode process as the limiting factor or using a potentiostat, which can characterize the anode potential at fixed current and eliminate the limitations resulting from the cathode and/or internal resistance. Research efforts should also be directed toward the optimization of electrochemically active microbial communities that could result in increased electron transfer efficiency and substrate degradation.

Inorganic substrates such as hydrogen sulfide have also been evaluated for electricity generation in an MFC with the purpose of removing sulfide produced anaerobically (Rabaey et al. 2006). Ammonia was also found to be removed from animal wastewater in an MFC (Min et al. 2005b), but the use of ammonia for electricity generation in an MFC has yet to be demonstrated.

10.7.2 Solution Chemistry

10.7.2.1 pH

pH is critical for all microbial-based processes. In MFCs, pH not only affects bacterial metabolism and growth, but also affects the proton transfer, the cathode reaction, and thus the MFC performance. Most MFCs are operated at near-neutral pH to maintain optimal growth conditions for microbial communities involved in electricity generation (Table 10.1). Gil et al. (2003) reported maximum power generation at pH 7.0 in a two-chamber MFCs using a mixed culture enriched from activated sludge. The low concentration of protons at this pH, however, makes the internal resistance of the cell relatively high compared to chemical fuel cells that use acidic or alkaline electrolytes. The use of pH buffers can help facilitate proton transfer, thereby reducing the internal resistance and enhancing power generation (Fan et al. 2007b). In addition, buffering also helps maintain a favorable pH for electricity-generating bacteria. Phosphate buffer is commonly used in MFCs because it has a high buffering capacity at neutral pH, and more importantly its ability to facilitate proton transfer. MFCs using 200 mM phosphate buffer produced significantly higher power density than those using 50 mM phosphate buffer (Cheng and Logan 2007; Logan et al. 2007).

Bicarbonate can also buffer an MFC system. Although the buffering capacity of bicarbonate is limited at pH < 8, bicarbonate buffer has shown excellent performance in lowering the internal resistance and enhancing the power generation of MFCs at pH 8.0–9.0. At pH 9.0, 200 mM bicarbonate buffer produced a power density of 1,550 W/m^3 (2,770 mW/m^2) in a recent study (Fan et al. 2007b), far higher than any previously reported value.

Example 10.3

A single-chamber microbial fuel cell (Fig. 10.7c) was operated in a batch mode without any mechanical motion of the electrolyte. The anodic and cathodic pHs were 5.0 and 9.0, respectively, when a stable current of 10 mA was generated. The cross-sectional area of electrodes was 7 cm^2 and the electrode spacing was 1 cm. Calculate (1) the diffusion rate of protons and (2) the total proton transfer rate, using simplified Fick's equation:

$$W = \frac{-DA\Delta C}{\delta} \tag{10.17}$$

where W (mol/s) is diffusion rate, D (cm^2/s) is the diffusion coefficient (for protons in water, $D = 9.3 \times 10^{-5} cm^2/s$), δ (cm) is diffusion distance, A (cm^2) is cross-sectional area, and ΔC (mol/L) is concentration difference.

Solution

1. $W = 6.5 \times 10^{-10}$ mmol/s $= 2.3 \times 10^{-6}$ mol/h
2. $I = 10$ mA $= 10 \times 10^{-3}$ C/s
3. Total proton transfer rate $= 10 \times 10^{-3}$ C/(98,485 C/mol)/s
$= 1.1 \times 10^{-4}$ mmol/s $= 0.76$ mol/h

Therefore, the majority of protons were not transferred by the diffusion of free protons.

10.7.2.2 Ionic Strength

Ionic strength affects the conductivity of the solution in the MFC chamber and thus the internal resistance, which thereby affects the MFC performance. Liu et al. (2005c) reported that power increased up to 85% with the addition of NaCl (300 mM) to the solution in the anode chamber, due to a decrease in internal resistance. The reader should take note that it may not be practical to enhance MFC performance in this manner. MFCs may be highly effective, however, in generating electricity from saline industrial wastewaters and municipal wastewater in cities such as Hong Kong where seawater is used for toilet flushing (Li et al. 2002).

10.7.3 Temperature and Hydraulic Retention Time

Bacterial kinetics, mass transfer rate of protons through the electrolyte, and oxygen reaction rates in the cathode govern the MFC performance and are all temperature dependent. Typically, biochemical reaction rate constants double for every 10°C increase in temperature until the optimal temperature is reached. Most MFC studies have been conducted at a temperature range of 28–35°C (Table 10.1). Liu et al. (2005c) investigated the effect of temperature on the performance of a single-chamber membrane-free MFC and found that there was only a slight reduction in power density (~9%) when the temperature was reduced from 32 to 20°C. This apparently suggests that MFC systems can be operated at lower temperatures without significantly sacrificing the electricity generation capability, potentially reducing operating costs. A high volumetric power density (500 W/m^3) that has been reported was obtained from an MFC at 19°C (Ringeisen et al. 2006). The reader should bear in mind that other factors, such as reactor configuration and microbial species, may also affect the power output. Additional research is needed to thoroughly investigate MFC operation at different temperature ranges and to determine how temperature fluctuations affect MFC performance using different bacterial consortia, MFC configuration, and solution chemistry.

Hydraulic retention time (HRT) is another important variable in wastewater treatment. It affects both COD/BOD removal and power generation in an MFC (Tables 10.1 and 10.2). Liu et al. (2004) observed maximum power density at an HRT of 12 h and COD removal rate of 78% using a single-chamber tubular system and domestic wastewater. An MFC operated in a flow-through mode using glucose as substrate saw power generation increase by 60% with an HRT increase from 4.2 to 15.6 h. The COD removal efficiency remained higher than 89% at all HRTs. At lower HRTs of 3.4–4.6 h, only 40–50% of COD removal was achieved with domestic wastewater as a substrate (Cheng et al. 2006a).

10.8 Opportunities and Challenges for MFCs in Wastewater Treatment

As an emerging technology, MFC shows great potential for the simultaneous generation of electricity and the treatment of wastewater. The surface power density of MFCs has been dramatically improved in recent years. Liu reported an increase from 26 mW/m^2 (2004) to 506 mW/m^2 (2005), and 2,770 mW/m^2 in a recent study (Fan et al. 2007b). Cheng et al. (2006a) and Logan et al. (2007) have also reported impressive power densities in recently published research. A volumetric power density of 1,010 W/m^3 was achieved recently using MFCs with double-cloth electrode assemblies operated in a continuous-flow mode (Fan et al. 2007a). A much higher power density of 1,550 W/m^3 was achieved using the same MFC with bicarbonate buffer (Fan et al. 2007b). These power densities exceed 1,000 W/m^3, a critical value that would make a 10-year payback possible for wastewater treatment (Logan et al. 2006; Rabaey and Verstraete 2005). Based on a recent power density of 7,200 mW/m^{2}anode surface achieved in the author's laboratory, a power density of 3–10 kW/m^3 seems achievable if cathode limitations can be solved. For comparison, it is estimated that combustion of biogas produced by an upflow anaerobic sludge blanket reactor can generate a power output of 0.5–1 kW/m^3, while aeration energy cost for treating domestic wastewater is about 0.5 kWh/m^3 (Rabaey and Verstraete 2005).

The high power densities reported in these studies were all achieved using simple, defined substrates, such as glucose and acetate, with the solution buffered. Power densities generated from actual wastewater using the same systems are expected to be lower than those achieved with well-defined substrates. Enhancing power generation while at the same time lowering material costs is the greatest challenge to making this technology feasible for practical wastewater treatment. The need for sustainability and sound environmental solutions will quickly drive developments in MFC technology, even though the field is still in its infancy.

References

Aelterman, P., Rabaey, K., Pham, T. H., Boon, N., and Verstraete. W. 2006. Continuous electricity generation at high voltages and currents using stacked microbial fuel cells. *Environ. Sci. Technol.* 40(10):3388–3394.

Biffinger, J. C., Ray, R., Little, B., and Ringeisen, B. R. 2007. Diversifying biological fuel cell designs by use of nanoporous filters. *Environ. Sci. Technol.* 41:1444–1449.

Bond, D. R., and Lovley, D. R. 2003. Electricity production by *Geobacter sulfurreducens* attached to electrodes. *Appl. Environ. Microbiol.* 69:1548–1555.

Catal, T., Li, K., Bermek, H., and Liu, H. 2008. Electricity production from twelve monosaccharides using microbial fuel cells. *J. Power Sources.* 175:196–200.

Chaudhuri, S. K., and Lovley, D. R. 2003. Electricity generation by direct oxidation of glucose in mediatorless microbial fuel cells. *Nat. Biotechnol.* 21:1229–1232.

Cheng, S., Liu, H., and Logan, B. E. 2006a. Increased power generation in a continuous flow MFC with advective flow through the porous anode and reduced electrode spacing. *Environ. Sci. Technol.* 40(7):2426–2432.

Cheng, S., Liu, H., and Logan, B. E. 2006b. Power densities using different cathode catalysts (Pt and CoTMPP) and polymer binders (Nafion and PTFE) in single chamber microbial fuel cells. *Environ. Sci. Technol.* 40(1):364–369.

Cheng, S., and Logan. B. E. 2007. Ammonia treatment of carbon cloth anodes to enhance power generation of microbial fuel cells. *Elec. Comm.* 9:492–496.

Clauwaert, P., Rabaey, K., Aelterman, P., Schamphelaire, L. D., Pham, T. H., Boeckx, P., Boon, N., and Verstraete, W. 2007. Biological denitrification in microbial fuel cells. *Environ. Sci. Technol.* 41(9):3354–3360.

Crittenden, S. R., Sund, C. J., and Sumner, J. J. 2006. Mediating electron transfer from bacteria to a gold electrode via a self-assembled monolayer. *Langmuir* 22(23):9473–9476.

Fan, Y., Hu, H., and Liu, H. 2007a. Enhanced columbic efficiency and power density of air-cathode microbial fuel cells with an improved cell configuration. *J. Power Sources* 171(2):348–354.

Fan, Y., Hu, H., and Liu, H. 2007b. Sustainable power generation in microbial fuel cells using bicarbonate buffer and proton transfer mechanisms. *Environ. Sci. Technol.* 41(23):8154–8158.

Gallert, C., Henning, A., and Winter, J. 2003. Scale-up of anaerobic digestion of the biowaste fraction from domestic wastes. *Water Res.* 37(6):1433–1441.

Gil, G. C., Chang, I. S., Kim, B. H., Kim, M., Jang, J. K., Park, H. S., and Kim, H. J. 2003. Operational parameters affecting the performance of a mediator-less microbial fuel cell. *Biosens. Bioelectron.* 18:327–334.

Gorby, Y. A., Yanina, S., McLean, J. S., Rosso, K. M., Moyles, D., Dohnalkova, A., Beveridge, T. J., Chang, I. S., Kim, B. H., Kim, K. S., Culley, D. E., Reed, S. B., Romine, M. F., Saffarini, D. A., Hill, E. A., Shi, L., Elias, D. A., Kennedy, D. W., Pinchuk, G., Watanabe, K., Ishii, S., Logan, B. E., Nealson, K. H., and Fredrickson, J. K. 2006. Electrically conductive bacterial nanowires produced by *Shewanella oneidensis* strain MR-1 and other microorganisms. *Proc. Natl. Acad. Sci. U. S. A.* 103:11358–11363.

He, Z., Minteer, S. D., and Angenent, L. T. 2005. Electricity generation from artificial wastewater using an upflow microbial fuel cell. *Environ. Sci. Technol.* 39:5262–5267.

He, Z., Wagner, N., Minteer, S. D., and Angenent, L. T. 2006. An upflow microbial fuel cell with an interior cathode: Assessment of the internal resistance by impedance spectroscopy. *Environ. Sci. Technol.* 40:5212–5217.

Heijne, A. T., Hamelers, H. V. M., Wilde, V. D., Rozendal, R. A., and Buisman, C. J. N. 2006. A bipolar membrane combined with ferric iron reduction as an efficient cathode system in microbial fuel cells. *Environ. Sci. Technol.* 40:5200–5205.

Heilmann, J., and Logan, B. E. 2006. Production of electricity from proteins using a single chamber microbial fuel cell. *Water Environ. Res.* 78(5):1716–1721.

Kim, H. J., Park, H. S., Hyun, M. S., Chang, I. S., Kim, M., and Kim B. H. 2002. A mediator-less microbial fuel cell using a metal reducing bacterium, *Shewanella putrefaciens. Enzyme Microb. Technol.* 30:145–152.

Kim, J. R., Min, B., and Logan, B. E. 2005. Evaluation of procedures to acclimate a microbial fuel cell for electricity production. *Appl. Microbiol. Biotechnol.* 68(1):23–30.

Kim, J. R., Oh, S. E., Cheng, S., and Logan, B. E. 2007. Power generation using different cation, anion and ultrafiltration membranes in microbial fuel cells. *Environ. Sci. Technol.* 41(3):1004–1009.

Larminie, J., and Dicks, A. 2000. *Fuel Cell Systems Explained.* John Wiley & Sons, Chichester, West Sussex, UK.

Li, C., and Fang, H. H. P. 2007. Fermentative hydrogen production from wastewater and solid wastes by mixed cultures. *Environ. Sci. Technol.* 37:1–39.

Li, X. Y., Ding, F., Lo, P. S. Y., and Sin, S. H. P. 2002. Electrochemical disinfection of saline wastewater effluent. *J. Environ. Eng.* 128(8):697–704.

Liu, H. 2002. *Bio-Hydrogen Production from Carbohydrate-Containing Wastewater.* PhD 1025 thesis, University of Hong Kong, Hong Kong.

Liu, H., and Logan, B. E. 2004. Electricity generation using an air-cathode single chamber microbial fuel cell in the presence and absence of a proton exchange membrane. *Environ. Sci. Technol.* 38:4040–4046.

Liu, H., Cheng, S., and Logan, B. E. 2005a. Power generation in fed-batch microbial fuel cells as a function of ionic strength, temperature, and reactor configuration. *Environ. Sci. Technol.* 39:5488–5493.

Liu, H., Cheng, S., Logan, B. E. 2005b. Production of electricity from acetate or butyrate in a single chamber microbial fuel cell. *Environ. Sci. Technol.* 39:658–662.

Liu, H., Grot, S., and Logan, B. E. 2005c. Electrochemically assisted microbial production of hydrogen from acetate. *Environ. Sci. Technol.* 39(11):4317–4320.

Liu, H., Ramnarayanan, R., and Logan, B. E. 2004. Production of electricity during wastewater treatment using a single chamber microbial fuel cell. *Environ. Sci. Technol.* 38:2281–2285.

Logan, B. E. 2004. Biologically extracting energy from wastewater: Biohydrogen production and microbial fuel cells. *Environ. Sci. Technol.* 38:160A–167A.

Logan, B. E., Aelterman, P., Hamelers, B., Rozendal, R., Schröeder, U., Keller, J., Freguiac, S., Verstraete, W., and Rabaey, K. 2006. Microbial fuel cells: Methodology and technology. *Environ. Sci. Technol.* 40(17):5181–5192.

Logan, B. E., Cheng, S., Watson, V., and Estadt, G. 2007. Graphite fiber brush anodes for increased power production in air-cathode microbial fuel cells. *Environ. Sci. Technol.* 41(9):3341–3346.

Logan, B. E., Murano, C., Scott, K., Gray N. D., and Head, I. M. 2005. Electricity generation from cysteine in a microbial fuel cell. *Water Res.* 39(5):942–952.

Logan, B. E., and Regan, J. M. 2006. Electricity-producing bacterial communities in microbial fuel cells. *Trends Microbiol.* 14(12):512–518.

Lowy, D. A., Tender, L. M., Zeikus, J. G., Park, D. H., and Lovley, D. R. 2006. Harvesting energy from the marine sediment-water interface II. Kinetic activity of anode materials. *Biosens. Bioelectron.* 21:2058–2063.

Magnuson, T. S., Isoyama, N., Hodges-Myerson, A. L., Davidson, G., Maroney, M. J., Geesey, G. G., and Lovley, D. R. 2001. Isolation, characterization and gene sequence analysis of a membrane-associated 89 kDa Fe(III) reducing cytochrome c from *Geobacter sulfurreducens. Biochem. J.* 359:147–152.

Min, B., Cheng, S., and Logan, B. E. 2005a. Electricity generation using membrane and salt bridge microbial fuel cells. *Water Res.* 39(5):942–952.

Min, B., Kim, J. R., Oh, S. E., Regan, J. M., and Logan, B. E. 2005b. Electricity generation from animal wastewater using microbial fuel cells. *Water Res.* 39(20):4961–4968.

Min, B., and Logan, B. E. 2004. Continuous electricity generation from domestic wastewater and organic substrates in a flat plate microbial fuel cell. *Environ. Sci. Technol.* 38(18):4900–4904.

Moon, H., Chang, I. S., and Kim, B. H. 2006. Continuous electricity production from artificial wastewater using a mediator-less microbial fuel cell. *Bioresour. Technol.* 97:621–627.

Myers, J. M., and Myers, C. R. 2001. Role for outer membrane cytochromes OmcA and OmcB of *Shewanella putrefacians* MR-1 in reduction of manganese dioxide. *Appl. Environ. Microbiol.* 67:260–269.

Niessen, J., Schröder, U., Harnisch, F., and Scholz, F. 2005. Gaining electricity from in situ oxidation of hydrogen produced by fermentative cellulose degradation. *Lett. Appl. Microbiol.* 41:286–290.

Oh, S. E., and Logan, B. E. 2005. Hydrogen and electricity production from a food processing wastewater using fermentation and microbial fuel cell technologies. *Water Res.* 39(19):4673–4682.

Oh, S. E., and Logan, B. E. 2006. Proton exchange membrane and electrode surface areas as factors that affect power generation in microbial fuel cells. *Appl. Microbiol. Biotechnol.* 70:162–169.

Park, D. H., and Zeikus, J. G. 2003. Improved fuel cell and electrode designs for producing electricity from microbial degradation. *Biotechnol. Bioeng.* 81:348–355.

Park, H. S., Kim, B. H., Kim, H. S., Kim, G. T., Kim, M., Chang, I. S., Park, Y. K., and Chang, H. I. 2001. A novel electro-chemically active and Fe(III)-reducing bacterium phylogenetically related to *Clostridium butyricum* isolated from a microbial fuel cell. *Anaerobe* 7:297–306.

Potter, M. C. 1911. Electrical effects accompanying the decomposition of organic compounds. *Proc. R. Soc. Lond. B. Biol. Sci.* 84:260–276.

Rabaey, K., Boon, N., Höfte, M., and Verstraete, W. 2005a. Microbial phenazine production enhances electron transfer in biofuel cells. *Environ. Sci. Technol.* 39:3401–3408.

Rabaey, K., Boon, N., Siciliano, S. D., Verhaege, M., and Verstraete, W. 2004. Biofuel cells select for microbial consortia that self-mediate electron transfer. *Appl. Environ. Microbiol.* 70:5373–5382.

Rabaey, K., Clauwaert, P., Aelterman, P., and Verstraete, W. 2005b. Tubular microbial fuel cells for efficient energy generation. *Environ. Sci. Technol.* 39(20):8077–8082.

Rabaey, K., Lissens, G., Siciliano, S. D., and Verstraete, W. A. 2003. Microbial fuel cell capable of converting glucose to electricity at high rate and efficiency. *Biotechnol. Lett.* 25:1531–1535.

Rabaey, K., Van de Sompel, K., Maignien, L., Boon, N., Aelterman, P., Clauwaert, P., De Schamphelaire, L., Pham, H. T., Vermeulen, J., Verhaege, M., Lens, P., and Verstraete, W. 2006. Microbial fuel cells for sulfide removal. *Environ. Sci. Technol.* 40:5218–5224.

Rabaey, K., and Verstraete, W. 2005. Microbial fuel cells: Novel biotechnology for energy generation. *Trends Biotechnol.* 23:291–298.

Reguera, G., McCarthy, K. D., Mehta, T., Nicoll, J. S., Tuominen, M. T., and Lovley, D. R. 2005. Extracellular electron transfer via microbial nanowires. *Nature* 435:1098–1101.

Reimers, C. E., Tender, L. M., Fertig, S., and Wang, W. 2001. Harvesting energy from the marine sediment-water interface. *Environ. Sci. Technol.* 35:192–195.

Ringeisen, B. R., Henderson, E., Wu, P. K., Pietron, J., Ray, R., Little, B., Biffinger, J. C., and Jones-Meehan, J. M. 2006. High power density from a miniature microbial fuel cell using *Shewanella oneidensis* DSP10. *Environ. Sci. Technol.* 40:2629–2634.

Rozendal, R. A., Hamelers, H. V. M., and Buisman, C. J. N. 2006. Effects of membrane cation transport on pH and microbial fuel cell performance. *Environ. Sci. Technol.* 40:5206–5211.

Schröder, U., Niessen, J., Scholz, F. 2003. A generation of microbial fuel cells with current outputs boosted by more than one order of magnitude. *Angew. Chem. Int. Ed.* 42:2880–2883.

Tender, L. M., Reimers, C. E., Stecher III, H. A., Holmes, D. E., Bond, D. R., Lowy, D. A., Pilobello, K., Fertig, S. J., and Lovley, D. R. 2002. Harnessing microbially generated power on the seafloor. *Nature Biotechnol.* 20(8):821–825.

USDOE, 2006. *International Energy Outlook 2006.* Energy Information Administration, U.S. Department of Energy.

Water Infrastructure Network. 2007. Clean safe water for the 21st century. Available at: http://www.win-water.org/reports/winreport2000.pdf (accessed April 8, 2008).

Zhao, F., Harnisch, F., Schröder, U., Scholz, F., Bogdanoff, P., and Herrmann, I. 2006. Constraints and challenges of using oxygen cathodes in microbial fuel cells. *Environ. Sci. Technol.* 40(17):5191–5199.

Zuo, Y., Maness, P. C., and Logan, B. E. 2006. Electricity production from steam exploded corn stover biomass. *Energy Fuels* 20(4):1716–1721.

CHAPTER

11

Pretreatment of High-Solids Wastes/Residues to Enhance Bioenergy Recovery

Santha Harikishan

11.1 Background

Most wastes with a high-solids content, for example, municipal sludge (primary solids and waste-activated sludge (WAS)), animal manure, food wastes, and agricultural residues contain a significant amount of biodegradable organic carbon. These organic-rich wastes and residues are ideal feedstocks for renewable energy (methane, hydrogen, or butanol) generation through nonoxidative metabolism (anaerobic fermentation). One major challenge to these feedstocks is their slow digestibility due to a rate-limiting hydrolysis step. Pretreatment of these feedstocks is essential to enhance their digestibility and bioenergy generation potential. Various pretreatments such as mechanical, thermal, chemical, or biological help solubilize particulate matter, accelerating digestion.

Pretreatment has mainly been confined to municipal sludge, particularly WAS for mass reduction and stabilization. Other high-solids wastes and residues are often generated in remote locations with unlimited land availability, resulting in little incentive for bioenergy recovery. Interest on bioenergy generation is growing, however, with increasing emphasis toward maximizing bioenergy recovery from all available renewable feedstocks. Pretreatment could become a standard practice for all high-solids organic wastes and residues in the coming years.

Municipal sludge, particularly WAS, is more difficult to digest than primary solids and other high-solids residues. This is because the cell wall and membrane of prokaryotic organisms in WAS are composed of complex organic materials such as peptidoglycan, teichoic acids, and complex polysaccharides, which tenaciously resist biodegradation. Thus, the readers should bear in mind that WAS is not typical of all wastes and that the information presented here mainly derives from WAS pretreatment. The same concept should, therefore, apply at least as effectively to other high-solids waste streams. Several innovative pretreatment technologies for promoting cell disintegration and optimizing anaerobic digestion have been developed. Only a few of these technologies, however, have shown promise on a large scale. This chapter focuses on three established pretreatment technologies: ultrasound, MicroSludge® (chemical treatment followed by application of high pressures), and thermal hydrolysis. Information on other disintegration technologies can be found elsewhere (Weemaes and Verstraete 1998).

11.2 Efficiency of Sludge Pretreatment

The parameters commonly used to determine the effectiveness of sludge disintegration can be classified into the three categories discussed as follows (Khanal et al. 2007):

11.2.1 Physical Characteristics

Particle size distribution and microscopic examination have been widely used as a qualitative measure of sludge disintegration. Researchers have assessed disintegration by changes in particle size distribution and turbidity (Tiehm et al. 2001). Disintegration reduces the size of the sludge particles and flocs, which subsequently increases turbidity. Light and electron microscopy examinations reveal structural changes that occur in the cells and flocs.

11.2.2 Chemical Evaluation

Cell disintegration is measured by the increase in released soluble chemical oxygen demand (SCOD). However, pretreatment also disintegrates extracellular materials, including organic debris and extracellular polymers, which become part of the SCOD. A parameter known as "degree of disintegration" (DD) has often been used to quantify the efficiency of sludge disintegration (Schmitz et al. 2000). The degree of disintegration can be evaluated by determining COD using Eq. (11.1):

$$DD = \left[\frac{COD_{treatment} - COD_{original}}{COD_{NaOH_{22h}} - COD_{NaOH_0}}\right]\left[\frac{COD_{NaOH^*}}{COD_{homogenization}}\right] \times 100(\%) \tag{11.1}$$

where $COD_{treatment}$ is supernatant COD of treated sample (mg/L), $COD_{original}$ is supernatant COD of untreated sample (mg/L), $COD_{NaOH_{22h}}$ is supernatant COD at 22 h after addition of 1 M NaOH (mg/L), COD_{NaOH_0} is supernatant COD at time zero after addition of 1 M NaOH (mg/L), COD_{NaOH^*} is sample COD immediately after addition of 1 M NaOH (mg/L), and $COD_{homogenization}$ is COD of original sample after homogenization (mg/L).

In Eq. (11.1), the term ($COD_{treatment} - COD_{original}$) represents the SCOD released following pretreatment, and the term ($COD_{NaOH_{22h}} - COD_{NaOH_0}$) represents SCOD released by chemical disintegration. It is assumed that COD released by NaOH addition constitutes complete disintegration of sludge and is taken as the reference COD. The ratio of COD_{NaOH^*} and $COD_{homogenization}$ represents the COD of the sample before and after addition of 1 M NaOH in the ratio 1:3.5 at 20°C.

A modified version of Eq. (11.1) was proposed by Müller to determine the degree of disintegration (Schmitz et al. 2000).

$$DD_M = \left[\frac{COD_{treatment} - COD_{original}}{COD_{NaOH} - COD_{original_0}}\right] \times 100(\%) \tag{11.2}$$

where $COD_{treatment}$ is supernatant COD of treated sample (mg/L), $COD_{original}$ is supernatant COD of untreated sample (mg/L), and COD_{NaOH} is maximum COD release in the supernatant after NaOH digestion.

The NaOH digestion is conducted by treating the sludge samples with 1 M NaOH in the ratio 1:2 for 10 min at 90°C. The supernatant is obtained by centrifugation for 10 min at 30,000 g and a temperature of 4°C.

Depending on sludge characteristics and solids concentration, the chemical dosage required for complete disintegration of solids can vary. Therefore, it may be necessary to conduct some investigative tests to determine the actual concentration and dosage of NaOH required and the reaction time and temperature for the reaction to proceed to completion, with the objective being maximum release of organics from the sludge.

11.2.3 Biological Evaluation

Since WAS is composed primarily of microbial cells, a measure of their survival rate following pretreatment will furnish data on the efficiency of the pretreatment. The parameter typically used to determine survival is the oxygen uptake rate (OUR). OUR is the decrease in oxygen concentration over time (Eq. (11.3)) directly correlated to the bacterial cells surviving pretreatment. Rai et al. (2004) quantified the drop in OUR of the cells using the term "degree of inactivation" DD_{OUR}, which is calculated using Eq. (11.4):

$$OUR = -\frac{d[O_2]}{dt} \tag{11.3}$$

$$DD_{OUR} = \left[1 - \frac{OUR_{treated}}{OUR_{original}}\right] \times 100(\%) \tag{11.4}$$

where $OUR_{treated}$ is oxygen uptake rate of the treated sample and $OUR_{original}$ is oxygen uptake rate of the original sample.

11.3 Ultrasound Pretreatment

Ultrasound is sound frequency exceeding normal hearing range of humans (>15–20 kHz). Ultrasound produces cavitation that generates powerful hydromechanical shear forces in the liquid phase. The hydrodynamic shear force ruptures the cell walls and membranes of microbial cells. In addition, sonochemical reactions that result in the formation of highly reactive radicals (e.g., $OH^{\bullet}$, $HO_2^{\bullet}$, $H^{\bullet}$) and hydrogen peroxide have been reported to contribute to sludge disintegration (Tiehm et al. 2001). Cavitation results in localized "hot spots" in the liquid phase with temperatures up to 5,000 K and pressures up to 7,250 psi (Suslick 1990), which also aids in cell disintegration.

Ultrasonic waves are commonly produced by a transducer containing a piezoelectric substance that converts high-frequency electric current into vibrating ultrasonic waves. A typical ultrasound system consists of three main components: (1) a transducer that converts electrical energy to ultrasonic waves, (2) a booster that is a mechanical amplifier for increasing the wave amplitude, and (3) a sonotrode or horn that delivers the ultrasonic waves to the sludge. Figure 11.1 shows the arrangement of the three components in an ultrasound system.

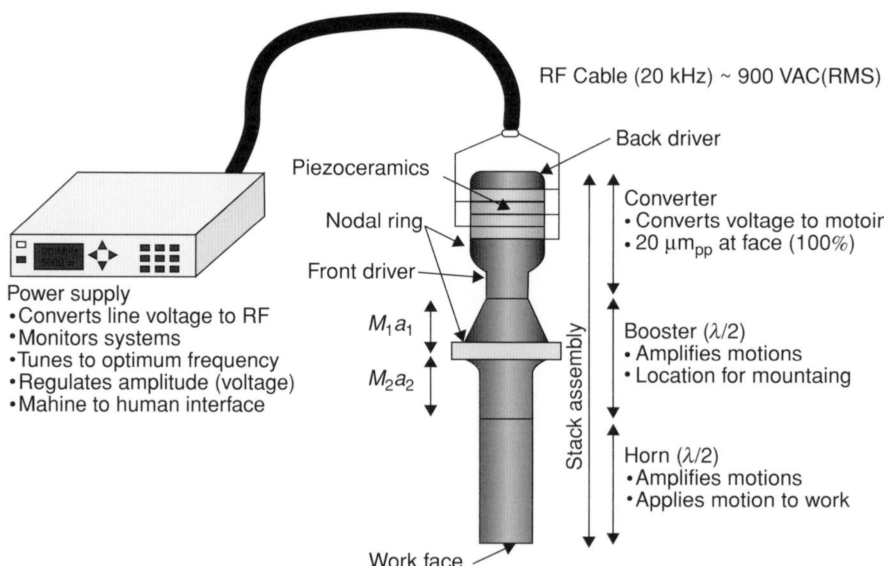

FIG. 11.1. Components of an ultrasound system.
Source: Khanal et al. (2007).

11.3.1 Ultrasound System Operating Variables

Cavitation bubble size depends on the operating frequency of the ultrasound system. Lower frequency results in higher cavitation intensities that produce larger bubbles, which on collapsing exert strong shear forces on the liquid.

Ultrasound system efficiency is determined by the power or energy input required to achieve the desired degree of disintegration of the solids. The specific energy input or the energy consumed per unit of solids is proportional to exposure time and inversely proportional to the total solids concentration of the sludge. A higher sample volume results in a reduced specific energy input at the same ultrasonic power input. The specific energy input can be calculated using Eq. (11.5) as follows:

$$E_{\text{spec}} = \frac{P \times t}{\text{TS} \times V} \tag{11.5}$$

where E_{spec} is specific energy input in kW s/kg of solids (kJ/kg of solids), P is ultrasonic power input (kW), t is duration of exposure to ultrasonic waves (s), V is volume of sonicated sludge in (L), and TS is total solids concentration of the sludge (kg/L).

The other parameters used to express energy input to the sludge include: (1) ultrasonic density that refers to the power supplied per unit volume of sludge (W/L or kW/L) and (2) ultrasonic intensity that is defined as the power supplied to the sludge per unit of transducer area (W/cm^2). Higher ultrasonic intensities result in better sludge disintegration.

The intensity of cavitation is also affected by factors other than equipment design, such as feed solids concentrations, pH, feed substrate viscosity, dissolved gases, and line pressures (Roxburgh et al. 2005). The soluble COD release increases with an increase in the total solids concentration of the feed (Grönroos et al. 2005; Wang et al. 2005). Researchers have also found the SCOD release to increase at higher feed pH and temperatures. It may be possible that the higher pH is weakening the bacterial cell wall aiding disintegration (Wang et al. 2005).

Example 11.1

A student conducted a series of ultrasound tests to optimize the power requirement for effective disintegration of high-solids feedstock. At a TS concentration of 5%, the specific energy was 10 kW s/kg TS with a power input of 900 W. Calculate the sonication time in seconds if the flow rate was 1 gal/min and the retention time in the sonication chamber was 2 min.

Solution

The sludge volume = flow rate × retention time

$= 1\ (\text{gal/min}) \times 3.785\ (\text{L/gal}) \times 2\ (\text{min}) = 7.57\ \text{L}$

$\text{TS} = 5\% = 50{,}000\ \text{mg/L} = 0.05\ \text{kg/L}$

From Eq. (11.5), specific energy input $= E_{\text{spec}} = \frac{P \times t}{\text{TS} \times V}$

$10\ \text{kW s/kg TS} = (0.9\ \text{kW}) \times (t)/(0.05\ \text{kg/L}) \times 7.57\ \text{L})$

$T = 4.2\ \text{s}$

11.3.2 Application in Sludge Pretreatment

Ultrasound technology provides an easy retrofit option for wastewater treatment plants with existing anaerobic digestion facilities. With separate primary solids and WAS streams, the ultrasound unit is typically installed on the WAS stream. The primary solids are readily biodegradable and digestion is not positively affected by disintegration. However, if the wastewater treatment plant has cosettled primary solids and WAS, it may be necessary to install the ultrasound system on the digester feed line. The other possible location for the ultrasound unit is along the sludge recirculation line from the digesters. Figure 11.2 illustrates potential locations for the ultrasound system in a wastewater treatment plant.

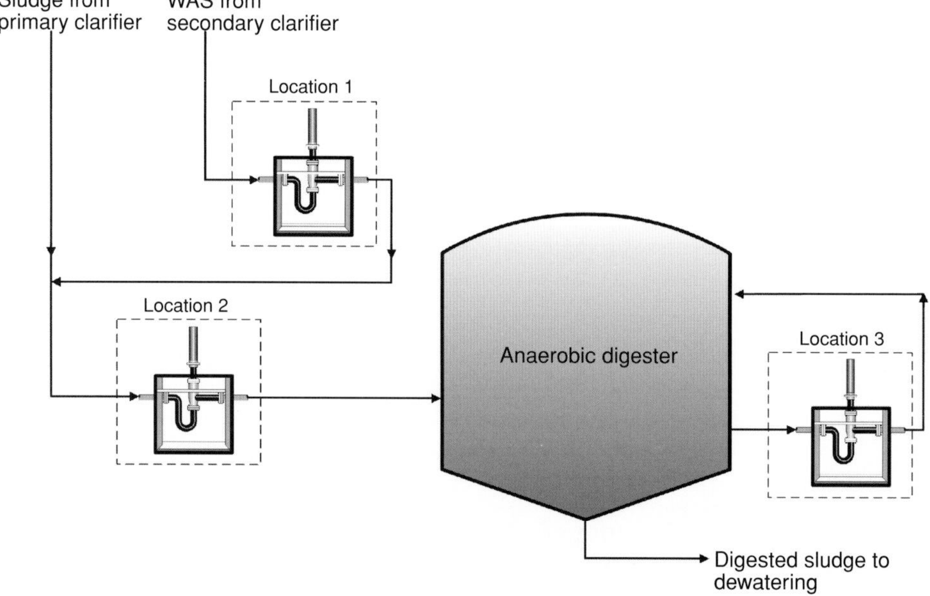

FIG. 11.2. Potential locations of ultrasound system in a wastewater treatment plant (WWTP).

There are two main hypotheses on how ultrasonic pretreatment should be implemented for effective cell disintegration: (1) full-stream treatment and (2) partial-stream treatment. The full-stream concept is based on sonicating all of the WAS flow to the anaerobic digesters to maximize the release of intracellular material and make it amenable for digestion. The assumption here is that biological activity is directly proportional to the liquid-phase substrate concentration. The partial-stream approach sonicates approximately 30–50% of the WAS flow. It is based on the hypothesis that microbial activity is influenced by substrate concentration only up to a certain critical level, above which there is no effect. Commercially available ultrasound systems employ one of the two concepts.

Full-scale ultrasound systems consist of a series of ultrasonic horns configured in a flow-through vessel. As WAS flows through the vessel, it directly contacts the horn that transmits the ultrasonic waves to the solids, creating cavitation. The ultrasound units are packaged in modules that can be incorporated into the existing piping with minor modifications. Most systems have the flexibility to shut down individual horns as solids flow is reduced. The three main ultrasound vendors are (1) Sonix by Sonico, a joint venture between WS Atkins plc and Purac Limited; (2) Ultrawaves, marketed in North America by Dorr Oliver Eimco; and (3) IWE Tec system by Dirk Holdings.

The Sonix ultrasound system uses doughnut-shaped, radial horns, mounted in series in a V-shaped configuration positioned either horizontally or vertically along the pipeline with the liquid flowing through the horn openings. A Sonix system contains between three and five horns. Sonico's standard equipment uses 6-kW ultrasonic horns, and the system is designed to operate at 65% of the maximum rated power. The maximum hydraulic capacity of currently available Sonix systems is 7.95–9.08 m^3/h (35–40 gal/min), and the retention time in the system is approximately 30 s.

The Ultrawaves system uses a baffled chamber reactor operating in an upflow mode. Each unit houses five 2-kW sonotrodes, with a total operating power draw of 5 kW. The largest commercially available unit has a rated capacity of 1.36–1.59 m^3/h (6–7 gal/min), and the retention time in the system ranges from 83 to 90 s. Figure 11.3 shows an Ultrawaves ultrasound system.

The IWE Tec ultrasound system employs "cascade" probes, each installed within a cylindrical reactor. The commercially available units are commonly equipped with either 2- or 4-kW probes and are designed to operate at between 60 and 70% of the maximum rated power. The retention times in the system range from 30 to 60 s.

11.3.3 Effect of Chemical Pretreatment on Ultrasound Disintegration

Chemical pretreatment of WAS using acid, base, or enzyme may enhance the sonication effect. Such chemicals weaken the cell wall, effectively disrupting the

FIG. 11.3. Ultrawaves ultrasound system.
Source: Courtesy of Eimco.

biological cells with less energy input. Bases such as NaOH and KOH are widely used to improve hydrolysis. Literature describing the effect of such pretreatment on ultrasound disintegration is not currently available. Batch-test studies conducted by the authors showed significant improvements in SCOD release when sonication was preceded by chemical treatment. These results are presented in Table 11.1.

Table 11.1. Comparison of SCOD released with ultrasound only and ultrasound preceded by chemical pretreatment at a power input of 190 W.

	SCOD (mg/L)				
		Ultrasound Preceded by Chemical Treatment			
Sonication Times (s)	Ultrasound Only	10 mg/g TS	15 mg/g TS	20 mg/g TS	25 mg/g TS
0	458	698	698	698	698
30	2,007	3,939	4,827	5,738	6,249
60	5,498	7,621	7,560	9,017	9,183
120	7,418	10,178	11,567	12,478	12,118

11.3.4 Effects of Ultrasound Pretreatment on Biogas Production

Ultrasound conditioning of thickened WAS (TWAS) prior to anaerobic digestion has been implemented in Europe and North America as a technique to condition WAS for more complete digestion and to enhance volatile solids reduction (VSR). Increased VSR translates into increased biogas (and bioenergy) generation from sludge.

Wang et al. (1999) reported approximately 15, 38, 68, and 75% improvement in cumulative methane yield for WAS sonicated for 10, 20, 30, and 40 min, respectively, in comparison to control during 11 days of anaerobic digestion. The authors used floatation-thickened WAS at a TS content of 3.3–4.0% for their studies. The sludge was sonicated using a 200-W ultrasonic unit at a frequency of 9 kHz. Based on serum bottle tests, Tiehm et al. (1997) observed nearly 28% higher biogas yield for sonicated sludge in comparison to untreated sludge during 28 days of digestion. Interestingly, in a continuous study at a solids retention time (SRT) of 22 days, the cumulative biogas production did not improve for sonicated sludge in comparison to unsonicated sludge during 100 days of digester operation. This could be due to the long SRT, which provided sufficient time even for the unsonicated sludge to achieve hydrolysis of particulate matter comparable to that for the sonicated sludge. The test was conducted using municipal sludge that consisted of 53% primary sludge and 47% WAS (dry weight). Sonication was conducted using a 3.6-kW ultrasound unit at a frequency of 31 kHz for 64 s.

In another study, Tiehm et al. (2001) reported cumulative biogas generation of 2.93, 2.79, 3.39, 3.38, and 4.15 L, respectively, from five continous stirred tank reactors (CSTRs) fed with WAS sonicated for 0 (control), 7.5, 30, 60, and 150 min. The CSTRs were operated at an SRT of 8 days. Sonication was conducted at a frequency of 41 kHz using a disk transducer. It is important to point out that even the shortest sonication time of 7.5 min is relatively long for full-scale applications.

Ultrasonic design, particularly that of the horn and converter, has improved significantly in recent years. These improvements have made it possible to achieve high amplitudes, with the result being more power delivered to the sludge in a short time.

Navaratnam (2007) examined the effect of full- and partial-stream sonication on biogas production using WAS with 3% TS content (Fig. 11.4). The sludge was sonicated for 45 s at a specific energy input of 2.85 kJ/g TS. The biogas production rate for full- and partial-stream digesters increased by 102 and 91%, respectively, with respect to control at an SRT of 20 days. At a 15-day SRT, the respective increases were only 81 and 57% with respect to control for full- and partial-stream digesters. From an energy balance perspective, partial-stream sonication was more energy efficient.

Pilot-scale demonstration trials using V-shaped sonication chambers with doughnut horn were conducted at the Avonmouth Wastewater Treatment Plant

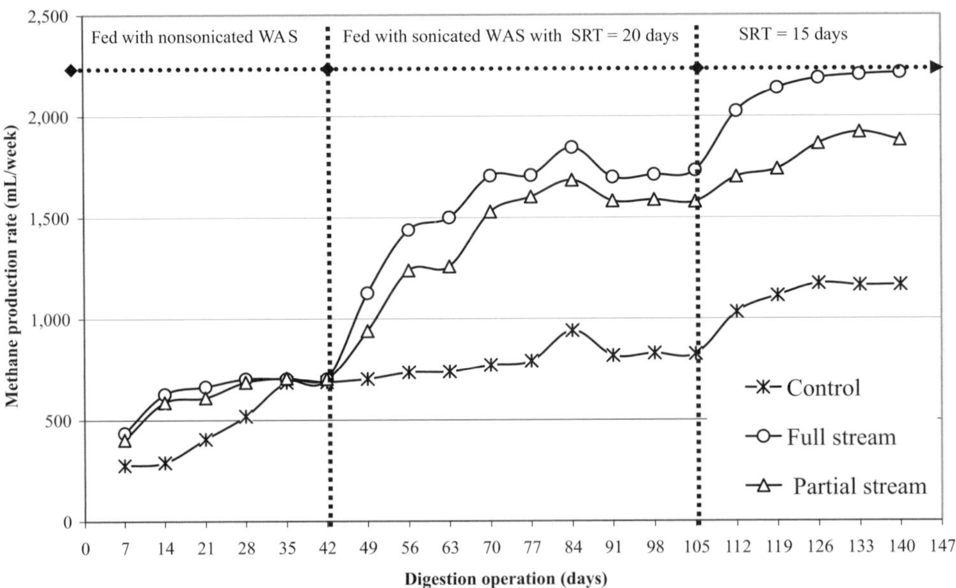

FIG. 11.4. Methane production rate under different conditions.

(UK; Hogan et al. 2004). Thickened municipal sludge (70% TWAS by weight) was sonicated at a frequency of 20 kHz. The authors reported up to 100% more biogas production with sonicated sludge than with unsonicated sludge (also with 70% TWAS). The authors also tested the volatile solids (VS) removal efficiency of a mesophilic anaerobic digester fed both unsonicated and sonicated TWAS (100%) at Severn Trent Water (UK). Sonication also produced positive results in this study related to VS destruction in sonicated sludge, with a 40% increased in biogas yield after full digester acclimation. Hogan et al. (2004) conducted a demonstration trial to examine the effect of sonication on biogas generation from waste at the Orange County Sanitation District (CA, USA). Biogas production from the sonicated sludge was 50–55% higher in comparison to control.

The biogas generation from WAS sonicated at different specific energy inputs was evaluated in a series of anaerobic digestion-batch tests during 16 days of incubation (Bougrier et al. 2004). The WAS (2% TS) was sonicated using an ultrasonic unit with a power supply of 225 W at a frequency of 20 kHz and various specific energy inputs. The authors found that the biogas yields were 1.48, 1.75, 1.88, and 1.84 times higher for the sonicated WAS in comparison to control (unsonicated WAS) at specific energy inputs of 1,355, 2,707, 6,951, and 14,547 kJ/kg TS, respectively. The biogas yield clearly showed improvement as specific energy inputs were increased to 6,951 kJ/kg TS. Further energy input increases to 14,547 kJ/kg TS did not improve biogas yields, even though increased amounts of

SCOD were released. Why biogas yields did not continue to increase is unknown. No data on VS destruction were presented.

11.3.5 Ultrasound in Lignocellulose Biomass Pretreatment

Lignocellulosic biomass consists mainly of cellulose, hemicellulose, and lignin. Due to its heterogeneity and crystallinity, microbes use this material extremely slowly. This makes pretreatment important so that the underlying monomeric sugars needed for the economic production of ethanol and other valuable chemicals can be released. Enzymatic hydrolysis is the most cost-effective method to obtain monomeric sugars. Pretreatment is required to make the cellulose/hemicellulose more accessible to enzymes. One of the most important factors affecting enzymatic hydrolysis is availability of a cleavage site for attack. Particle size reduction increases the available surface area of the substrate, opening up the cleavage sites and making the substrate susceptible to enzyme action. High-power ultrasound has the potential to increase the pore volume of cellulosic biomass, thereby improving enzymatic activities. Thus, employing ultrasonics prior to enzymatic hydrolysis could increase sugar yield at low enzyme dosing. The effect of ultrasound pretreatment on the microstructure of alkaline-pretreated corn stover is shown in Fig. 11.5.

11.4 Chemical and Physical Pretreatment

MicroSludge is a patented process developed by Paradigm Environmental Technologies, Inc., Vancouver, Canada, which uses an alkaline solution to raise the pH of biological sludge to about 9–10. The high pH weakens the bacterial cell walls and lowers the viscosity of the solids stream. The treated sludge is sheared in a grinder pump to reduce particle size and is then screened to remove large debris that could damage downstream equipment. The screened solids are introduced into a homogenizer that is a valve with a narrow passage <1 mm through which the solids are pumped at around 82,700 kPa (12,000 psi) pressure. The high pressure accelerates the solids to 305 m/s (speed of sound) in about 2 μs. The solids then impinge on an impact ring, causing a sudden pressure drop, which lyses the cells. The homogenized sludge is then fed to an anaerobic digester with or without primary solids (Stephenson and Dhaliwal 2000). Figure 11.6 illustrates the MicroSludge process.

11.4.1 MicroSludge System Operating Variables

The two key process variables affecting disintegration are homogenization pressure and chemical dosage used for weakening the biological cell walls.

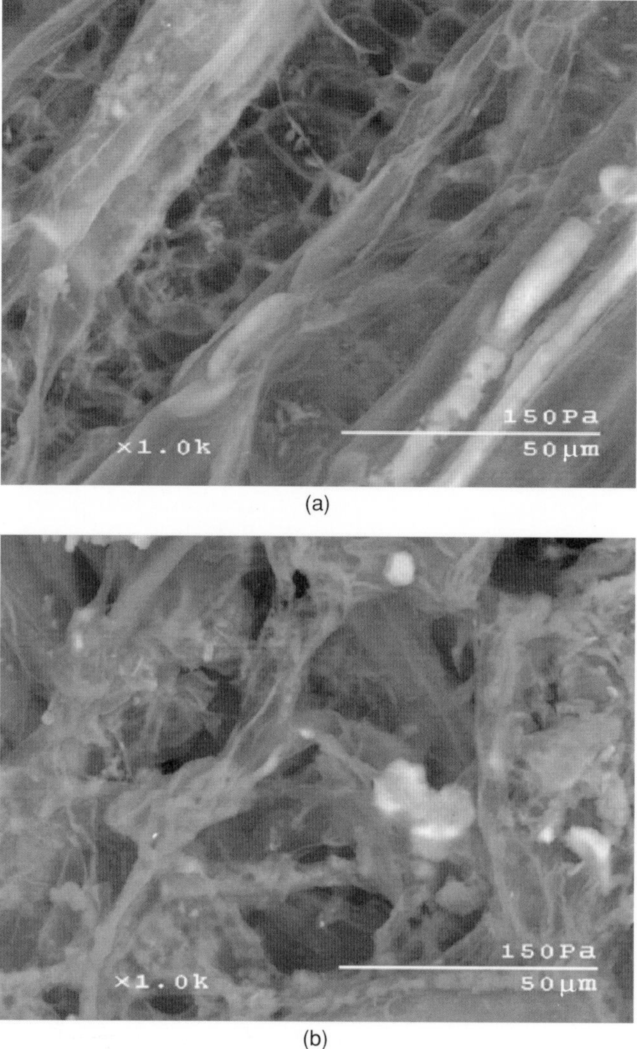

FIG. 11.5. Scanning electron micrographs of alkaline-pretreated corn stover: (a) control (unsonicated) and (b) sonicated for 40 s.

Dhaliwal (1996) studied the effect of homogenization pressures on the VSR achieved during digestion. Higher pressures ranging from 82,700 to 124,050 kPa (12,000–18,000 psi) produced only minor increases in VSR in the anaerobic digesters. In contrast, lower pressures of 68,920–82,700 psi (10,000–12,000 psi) lowered the VSR by 40% in the anaerobic digesters. Based on these findings, the system was designed to operate at 82,700 kPa (12,000 psi) so as to optimize digestion.

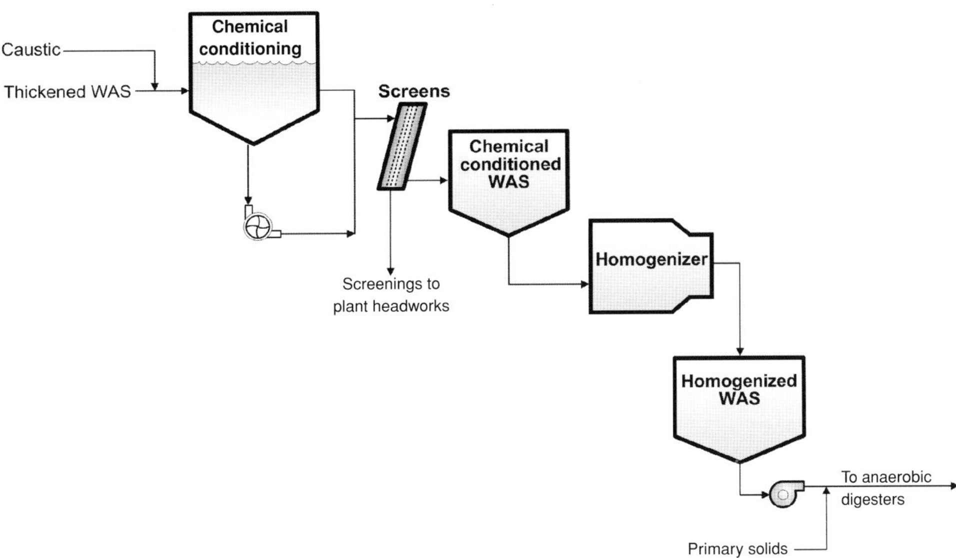

FIG. 11.6. Schematic of the MicroSludge process.
Source: Courtesy of MicroSludge.

Sodium hydroxide is used to condition the WAS prior to homogenization. The caustic dosage must be optimized to reduce viscosity for cell disruption in the homogenizer without adversely affecting the equipment. The caustic dosages used should also not have any detrimental effects on the anaerobic digestion process. The MicroSludge process is typically set to achieve a target Na^+ concentration below 500 mg/L in the anaerobic digester. Studies have found that Na^+ between 100 and 200 mg/L stimulates digestion (McCarty 1964). Concentrations between 3,500 and 5,500 mg/L are moderately inhibitory and those above 8,000 mg/L are strongly inhibitory.

11.4.2 Application in Sludge Pretreatment

The MicroSludge system can also be easily incorporated into an existing treatment scheme. Since the MicroSludge process focuses on the biodegradability of the bacterial cells, it is typically only used on the WAS stream.

The MicroSludge system is available in modules that include the caustic system, a grinder pump for reducing particle size, a screen for eliminating large debris, and a high-pressure homogenizer (Fig. 11.7). The maximum hydraulic capacity of currently available homogenizers ranges from 8 m^3/h (2,110 gal/h) to 24 m^3/h (6,340 gal/h). Multiple units can be configured in parallel to increase flow capacity.

Fig. 11.7. Full-scale MicroSludge system.
Source: Courtesy of MicroSludge.

11.4.3 Effects of MicroSludge on Biogas Production

The MicroSludge process was studied at laboratory- and pilot-scale conditions between 2000 and 2003 with support from Canada's National Research Council. The first full-scale MicroSludge plant was installed at the Chilliwack WWTP near Vancouver, Canada, in January 2004 as a demonstration project. The technology was also tested at the Los Angeles County (USA) Sanitation Districts' (LACSD) Joint Water Pollution Control Plant in October 2005.

As part of the demonstration project at Chilliwack, all of the WAS generated at the WWTP was thickened and pretreated in a single 8 m^3/h (2,110 gal/h) homogenizer. The MicroSludge-processed WAS was mixed with gravity-thickened primary solids prior to the anaerobic digesters. The digesters were operated at a retention time of 13 days (Rabinowitz and Stephenson 2005). At LACSD, two 4 m^3/h (1,060 gal/h) MicroSludge system modules were used to process thickened WAS containing 5–6% TS. The processed solids were blended with primary solids (32–68% (by mass)) and fed to one of the digesters. Digester performance was compared to another full-scale digester operated at identical conditions, but with pretreatment. The digesters at LACSD were operated at an SRT of 19 days (Stephenson et al. 2007).

At the Chilliwack trials, the VSR in the anaerobic digesters following pretreatment averaged 40–60%, which was higher than the 40–50% achieved without the MicroSludge process. At LACSD, VSR in the digester receiving pretreated WAS was approximately 58%, compared to 50% achieved without MicroSludge pretreatment. Since the overall VSR was assumed to be dampened by the larger portion of primary solids fed to the digesters at both locations, Stephenson et al. (2007) attempted to quantify the VSR increase and digester gas production resulting from pretreatment. The VSR of the WAS stream was estimated using Eq. (11.6) as follows:

$$\mathrm{VSR_T} = (X_{\mathrm{PS}} \times \mathrm{VSR_{PS}}) + (X_{\mathrm{WAS}} \times \mathrm{VSR_{WAS}}) \tag{11.6}$$

where $\mathrm{VSR_T}$ is total VSR in the digester (%), X_{PS} is mass fraction of primary solids in the digester feed (%), $\mathrm{VSR_{PS}}$ is VSR of primary solids in the digester (%), X_{WAS} is mass fraction of waste-activated solids in the digester feed (%), and $\mathrm{VSR_{PS}}$ is VSR of waste-activated solids in the digester (%).

Assuming a VSR of 60% for the primary solids, the VSR of WAS without MicroSludge pretreatment at LACSD was calculated as follows:

$$\text{Without MicroSludge: } 50\% = (68\% \times 60\%) + (32\% \times \mathrm{VSR_{WAS}}),$$
$$\mathrm{VSR_{WAS}} = 29\%$$
$$\text{With MicroSludge: } 58\% = (68\% \times 60\%) + (32\% \times \mathrm{VSR_{WAS}}),$$
$$\mathrm{VSR_{WAS}} = 54\%$$

Similar analysis at Chilliwack indicated a 30% increase in the VSR without pretreatment to approximately 90% with pretreatment. Based on these findings, Stephenson et al. (2007) concluded that the VSR of WAS could be increased considerably with MicroSludge pretreatment.

Gas production from the anaerobic digestion process was not monitored at Chilliwack due to faulty gas meters. At LACSD, biogas production from the digester receiving pretreated WAS increased in accordance with the higher VSR achieved in the digester.

11.5 Thermal Hydrolysis

Thermal pretreatment is also an "add-on" to conventional anaerobic digestion. Raw solids are briefly heated that enhances digestion and dewatering. Unlike the ultrasound and MicroSludge processes, thermal hydrolysis targets both primary solids and WAS.

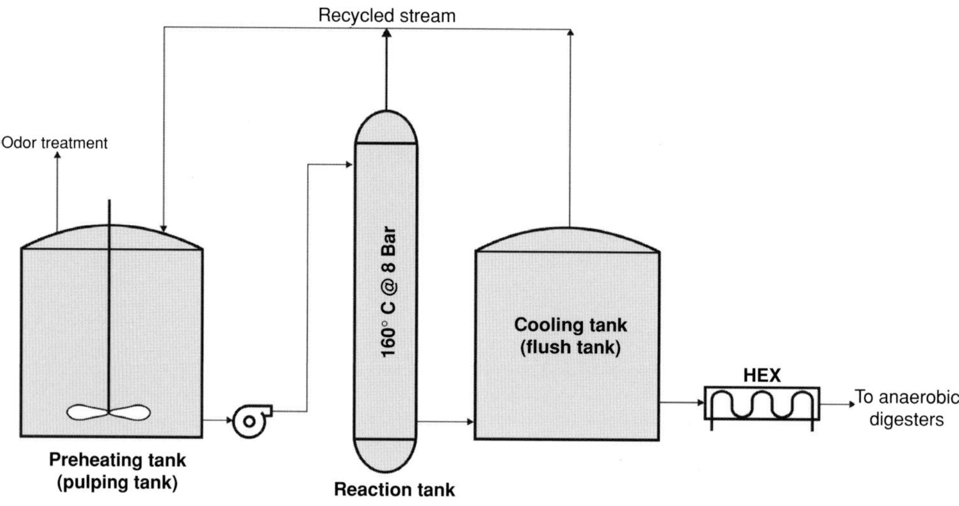

FIG. 11.8. Zimpro wet air oxidation process.
Source: Courtesy of Zimpro.

With the increased emphasis on sludge biodegradability and biogas production, researchers studied the process at lower temperatures to achieve the best compromise between dewaterability and digestion (Haug et al. 1978). Stuckey and McCarty (1984) studied the effect of thermal pretreatment on digestibility of WAS between 150 and 275°C (302–527°F). Optimum methane production was observed at 175°C (347°F). This finding was supported by Pinekamp (1989), who also observed optimum digestion and dewatering. At higher temperatures, a drop in digestibility was observed, which was attributed to formation of toxic refractory compounds. Li and Noike (1992) reported pretreatment at 170°C (338°F) for 60 min as the optimum conditions for improving COD removal and digester gas production during mesophilic digestion.

Of the several proprietary systems available from manufacturers for thermal pretreatment, Zimpro and CAMBI are the most widely used technologies.

11.5.1 Zimpro Process

One of the first thermal hydrolysis processes developed in the 1950s was the Zimpro process. It was originally designed as a wet oxidation process conducted at 250°C (482°F) to improve solids dewaterability. The operating temperatures were later lowered below 200°C (392°F), as research indicated the formation of toxic refractory compounds at temperatures in excess of 175°C (347°F). A schematic of the Zimpro system is shown in Fig. 11.8.

At present, there are a few Zimpro installations operating in the United States and China. However, the technology is gradually being phased out in the United

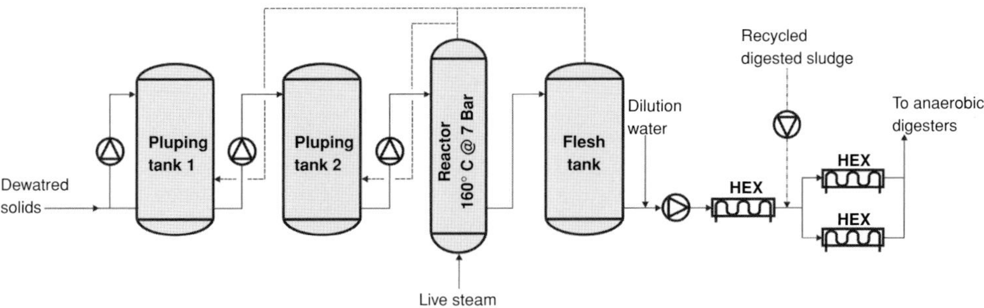

FIG. 11.9. CAMBI thermal hydrolysis process.
Source: Courtesy of CAMBI.

States, with several Zimpro units decommissioned in recent years due to system complexity, maintenance requirements, and sidestream odors.

11.5.2 CAMBI Thermal Hydrolysis

CAMBI is also a low-pressure, wet air oxidation process in which the solids are dewatered to 10–15% TS prior to thermal conditioning. The dewatered solids are continuously fed to a preheating tank. The preheated solids are then batch fed to a second tank, where pressurized steam is added to achieve temperatures of 150–160°C (300–320°F) and pressures of about 8–9 bar (115–130 psi). After approximately 30 min of reaction time, the pressure is released and the steam is recirculated to the first tank for preheating of the incoming raw solids. The hydrolyzed solids are cooled to 35°C (95°F), diluted, and fed to anaerobic digesters. Figure 11.9 illustrates the CAMBI process.

The CAMBI thermal hydrolysis process is more prevalent in Europe, where there are around ten full-scale installations operating (Weisz and Solheim 1999).

11.5.3 Effects of Thermal Hydrolysis on Anaerobic Digestion

The operating data from full-scale installations in Europe suggest that the digesters achieve high VSR even at high-solids loading rates. The Naestved WWTP (Denmark) implemented the thermal hydrolysis process upstream of existing anaerobic digesters to reduce biosolids for recycling. VSR in the anaerobic digesters increased from approximately 30% to between 54 and 60% with thermal pretreatment (Fjordside 2002). The installation at HIAS, Norway, has also been achieving VSR from 60 to 65% (Panter 2001). A comparison of digestion performance between conventional and thermally treated solids at the Ringsend STW shows VSR improvements between 42 and 62% following thermal hydrolysis (Pickworth et al. 2005). Even though the feed composition to the digesters remained unchanged before and after implementation of thermal hydrolysis, information is lacking on the exact ratios of primary solids and WAS fed to the digesters.

Benefits notwithstanding, the thermal hydrolysis process is complex for plant operators. The process operates at high pressures, requiring steam at around 10 bar (145 psi) pressure during the react phase. Any plant considering the process in the United States will need to employ a full-time, certified boiler operator, which will have a significantly increase in operating costs.

11.6 Impact of Improved Digestibility on Overall Process Economics

Solids pretreatment is still an emerging field and process comparisons are difficult with limited data. The focus of bench-scale or pilot-scale studies for most of the emerging pretreatment technologies is often narrow, and the reported conclusions must be synthesized from different sources to obtain a complete and balanced perspective. Key information is also lacking on factors related to the operating parameters and costs involved to quantify the economic benefits from solids pretreatment. One of the key considerations in pretreatment options would be net energy gain from the improved bioenergy production.

References

Bougrier, C., Carrère, H., Battimelli, A., and Delgenès, J. P. 2004. *Effects of Various Pre-treatments on Waste Activated Sludge in Order to Improve Matter Solubilization and Anaerobic Digestion*, Vol. 2, pp. 998–103. In Proceedings of 10th World Congress on Anaerobic Digestion. Montréal, Canada.

Dhaliwal, H. 1996. *Biodegradability of Processed Waste Activated Sludge.* Graduate thesis, University of British Columbia.

Fjordside, C. 2002. *Full Scale Experience of Retrofitting Thermal Hydrolysis to an Existing Anaerobic Digester for the Digestion of Waste Activated Sludge.* Proceedings of the 16th Annual Residuals and Biosolids Management Conference. March 3–6, 2002, Austin, TX, USA.

Grönroos, A., Kyllönen, H., Korpijärvi, K., Pirkonen, P., Paavola, T., Jokela, J., and Rintala, J. 2005. Ultrasound assisted method to increase soluble chemical oxygen demand (SCOD) of sewage sludge for digestion. *Ultrason. Sonochem.* 12:115–120.

Haug, R. T., Stucky, D. C., Gossett, I. M., and McCarty, P. L. 1978. Effect of thermal treatment on digestibility and dewaterability of organic sludges. *J. Water Poll. Control Fed.* 50(1):73–85.

Hogan, F., Mormede, S., Clark, P., and Crane, M. 2004. Ultrasonic sludge treatment for enhanced anaerobic digestion. *Water Sci. Technol.* 50(9):25–32.

Khanal, S. K., Grewell, D., Sung, S., and van Leeuwen, J. 2007. Ultrasound application in wastewater sludge pretreatment: A review. *Crit. Rev. Environ. Sci. Technol.* 37:277–313.

Li, Y. Y., and Noike, T. 1992. Upgrading of anaerobic digestion of waste activated sludge by thermal pre-treatment. *Water Sci. Technol.* 26(3–4):857–866.

McCarty, P. L. 1964. Anaerobic waste treatment fundamentals. Part three: Toxic material and their control. *Public Works* 95:91–94.

Navaratnam, N. 2007. *Anaerobic Digestion of Ultrasound Pretreated Thickened Waste Activated Sludge (TWAS) to Enhance Bioenergy Production.* M.S. thesis, Asian Institute of Technology, Bangkok, Thailand.

Panter, K. 2001. *Getting the Bugs Out of Digestion and Dewatering with Thermal Hydrolysis.* Proceedings of the WEF/AWWA/CWEA Joint Residuals and Biosolids Management Conference. February 21–24, 2001, San Diego, CA, USA.

Pickworth, B., Adams, J., Panter, K., and Solheim, O. E. 2005. *Maximizing Biogas in Sludge Digestion by Using Engine Waste Heat for Thermal Hydrolysis Pre-Treatment of Anaerobic Digestion at Dublin's Ringsend STW.* Proceedings of IWA Sustainable Management of Residues Emanating from Water and Wastewater Treatment Conference. August 2005, Johannesburg, South Africa.

Pinekamp, J. 1989. Effects of thermal pre-treatment of sewage sludge on anaerobic digestion, *Water Sci. Technol.* 21(4–5):97–108.

Rabinowitz, B. and Stephenson, R. 2005. *Improving Anaerobic Digester Efficiency by Homogenization of Waste Activated Sludge.* Proceedings of WEFTEC, 79th Annual Technical Exhibition and Conference. November 1, 2005, Washington, DC, USA.

Rai, C. L., Struenkmann, G., Mueller, J., and Rao, P. G. 2004. Influence of ultrasonic disintegration on sludge growth and its estimation by respirometry. *Environ. Sci. Technol.* 38:5779–5785.

Roxburgh, R., Soroushian, F., Sieger, R., and Burrowes, P. 2005. *Ultrasound for Improved Solids Management—How, Who, When and Why.* Proceedings of the Joint Residuals and Biosolids Management Conference. April 17–20, 2005, Nashville, TN, USA.

Schmitz, U., Berger, C. R., and Orth, H. 2000. Protein analysis as a simple method for the quantitative assessment of sewage sludge disintegration. *Water Res.* 34(14):3682–3685.

Stephenson, R., and Dhaliwal, H. 2000. Method of liquefying microorganisms derived from wastewater treatment processes, U.S. Patent No. 6,013,183.

Stephenson, R., Laliberte, S., Hoy, P., Drew, A., and Britch, D. 2007. *Full Scale and Laboratory Scale Results from the Trial of Microsludge at the Joint Water Pollution Control Plant at Los Angeles County.* Proceedings of the WEF/AWWA Joint Residuals and Biosolids Management Conference. April 14–17, 2007, Denver, CO, USA.

Stuckey, D. C., and McCarty, P. L. 1984. The effect of thermal pre-treatment on the anaerobic biodegradability and toxicity of waste activated sludge. *Water Res.* 18(11):1343–1353.

Suslick, K. 1990. Sonochemistry. *Science* 247:1439.

Tiehm, A., Nickel, K., and Neis, U. 1997. The use of ultrasound to accelerate the anaerobic digestion of sewage sludge. *Water Sci. Technol.* 36(11):121–128.

Tiehm, A., Nickel, K., Zellhorn, M., and Neis, U. 2001. Ultrasound waste activated sludge disintegration for improving anaerobic stabilization. *Water Res.* 35(8):2003–2009.

Wang, F., Wang, Y., and Ji, M. 2005. Mechanisms and kinetic models for ultrasonic waste activated sludge disintegration. *J. Hazard. Mat.* B123:145–150.

Wang, Q., Kuninobu, M., Kakimoto, K., Ogawa, H. I., and Kato, Y. 1999. Upgrading of anaerobic digestion of waste activated sludge by ultrasonic pretreatment. *Biores. Technol.* 68:309–313.

Weemaes, M. P. J., and Verstraete, W. H. 1998. Evaluation of current wet sludge disintegration techniques. *J. Chem. Technol. Biotechnol.* 73:83–92.

Weisz, N., and Solheim, O. E. 1999. *International Applications of CAMBI Thermal Hydrolysis for Sludge and Biowaste Treatment.* Proceedings of the 4th European Biosolids and Organic Residuals Conference. November 1999, Wakefield, West Yorkshire, UK.

CHAPTER

12

Biogas Processing and Utilization as an Energy Source

Santha Harikishan

12.1 Background

Dwindling supplies of conventional energy sources and the drive to increase the share of renewable energy in our total energy consumption have increased the significance of biogas as a renewable fuel.

Biogas, produced during anaerobic digestion of biodegradable organic solids, typically contains about 60–65% methane, which is a valuable resource and can be used to offset part of the energy requirements for in-plant use. With rising electricity and natural gas costs, there is an incentive for utilization of biogas as an energy source. Biogas collection and utilization technologies have steadily improved over the years, and energy recovery from biogas is developing into a successful "waste/residues-to-bioenergy" technology.

The most common use of biogas in the United States is as fuel for boilers to generate steam or hot water for heating. Increasingly, biogas is being used in combined heat and power applications, where biogas is converted to electricity using on-site power generation equipment. Heat is recovered from the power generation units in the form of hot water or steam for heating applications. Overall efficiency of biogas use can approach 80% if all the recovered heat is used. Other developing avenues are the use in fuel cell and synthetic gas (syngas) production. This chapter provides an overview of digester gas production from various feedstocks along with a discussion of the cleaning requirements and the utilization options as an energy source.

12.2 Biogas Production

The quantity and quality of gas produced during anaerobic digestion depends on the feed characteristics. Several methods are available to estimate methane

Table 12.1. Typical digester gas production rates.

	Specific Gas Production per Unit Mass Destroyed	
Substrate	m^3/kg	ft^3/lb
Fats	1.2–1.6	19.2–25.6
Scum	0.9–1.0	14.4–16.0
Grease	1.1	17.6
Crude fibers	0.8	12.8
Protein	0.7	11.2
Carbohydrates	0.7	11.2

Source: Adapted from *WEF Manual of Practice 8* (1998).

generation from a waste stream during anaerobic digestion. Knowing the chemical composition of the waste stream, the methane production can be estimated using Bushwell equation as discussed in Chapter 2. Methane production can also be estimated from chemical oxygen demand (COD) or ultimate biochemical oxygen demand (BOD_L) stabilization based on the fact that 1 kg COD destroyed produces 0.35 m^3 CH_4 (5.62 ft^3/lb COD destroyed) at standard temperature and pressure (STP) (see Example 12.1). The typical gas production rates for different substrates are shown in Table 12.1

Example 12.1

An anaerobic digester receiving 7,600 m^3/day (2.0 million gal/day) of food waste generates 2,500 m^3 of biogas per day at 37°C. If the biogas contains 65% (by volume) methane, calculate organic feeding rate (as COD). Make all valid assumptions.

Solution

1. Methane generation rate from the biogas at mesophilic conditions:

   2,500 m^3/day of biogas × 65% methane = 1,625 m^3/day of CH_4

2. Methane generation rate at STP:

   1,625 m^3/day × (273 K/310 K) = 1,431 m^3/day

3. Every kilogram of COD destroyed in the digester generates 0.35 m^3 of CH_4 at STP:

   COD destroyed = (1,431 m^3 CH_4/day)/(0.35 m^3/kg COD destroyed)

   = 4,106 kg/day

4. Organic (COD) feeding rate to the digester
   Feed COD per day = 4,106 kg/0.6 = 6,843 kg

Table 12.2. Growth constants and endogenous respiration rates.

Waste Stream	Growth Constant, *a*	Endogenous Respiration Rate, *b*
Fatty acids	0.054	0.038
Carbohydrates	0.240	0.033
Proteins	0.014	0.014

Source: McCarty (1964).

Methane production can also be estimated from the waste strength using Eq. (12.1) (McCarty 1964) as follows:

$$C = 5.62(eF - 1.42A) \tag{12.1}$$

where C is methane produced per day at STP (ft^3); e is efficiency of waste utilization (%), typically ranging from 0.80 to 0.95 under satisfactory operating conditions; F is BOD_L added per day (lb); A is volatile biological solids produced per day (lb); 5.62 is constant, theoretical methane production from 1 lb of BOD_L; 1.42 is constant, factor for conversion of pounds of volatile biological solids to BOD_L.

The portion of the organic waste stream converted to biological cells (A) during the anaerobic digestion process will not be stabilized by conversion to methane. The growth of microorganisms can be approximated using Eq. (12.2).

$$A = \frac{aF}{1 + b(\mathrm{SRT})} \tag{12.2}$$

where SRT is solids retention time (days), a is growth constant, and b is endogenous respiration rate.

The growth constants and endogenous respiration rates for various waste streams are listed in Table 12.2.

12.3 Factors Affecting Digester Gas Production

Typical biogas yield values for an anaerobic digester treating primary and waste-activated solids range from 0.8 to 1.0 m^3/kg (12.8–16.0 f^3/lb) of volatile solids (VS) destroyed at mesophilic conditions. For thin stillage the yield varies from 1.1 to 1.3 m^3/kg (17.6–20.8 f^3/lb) of VS destroyed under thermophilic conditions. The amount of gas produced, however, depends on several factors such as temperature, pH and alkalinity, hydraulic and organic loading rates, toxic compounds, substrate type, and total solids (TS)/volatile solids (VS) content. The detailed discussion is presented in Chapter 2.

Table 12.3. Theoretical methane percent for different waste streams.

Waste Type	Reaction	CH_4 in Biogas (%)
Carbohydrate	$(C_6H_{10}O_5)n + nH_2O \rightarrow 3nCH_4 + 3nCO_2$	50
Protein	$4C_{11}H_{24}O_5N_4 + 58H_2O \rightarrow 33CH_4 + 15CO_2 + 19NH_4^+ + 16HCO_3^-$	69
Lipids	$4C_{15}H_{90}O_6 + 98H_2O \rightarrow 139CH_4 + 61CO_2$	70
Primary sludge	$C_{10}H_{19}O_3N + 4.5H_2O \rightarrow 6.25CH_4 + 3.75CO_2 + NH_3$	62.5
Waste-activated sludge	$C_5H_7O_2N + 3H_2O \rightarrow 2.5CH_4 + 2.5CO_2 + NH_3$	50

12.4 Biogas Composition

Biogas is composed primarily of methane and carbon dioxide. The methane contained in biogas imparts fuel value to the gas. The methane concentration in biogas can vary depending on the type of waste stream as given in Table 12.3.

Table 12.4 summarizes the characteristics and biogas yields from some of the agricultural feedstocks.

Based on a typical fuel value of 22,400 kJ/normal m^3 (600 Btu/ft^3), the gross energy available from biogas can be calculated using Eq. (12.3). The net energy available can vary depending on the efficiency of the gas utilization equipment:

$$E = G \times 22{,}400\,\text{kJ/m}^3 \tag{12.3}$$

where E is gross energy from digester gas (kJ/day) and G is digester gas production at STP (m^3/day).

Table 12.4. Biogas yields and methane content from agricultural feedstocks.

Feedstocks	Total Solids (% dissolved solids (DS))	Volatile Solids (% DS)	Retention Time (days)	Biogas Yield[a] (m^3/kg VS)	CH_4 Content (%)
Pig slurry	3–81[b]	70–80	20–40	0.25–0.50	70–80
Cow slurry	5–12[b]	75–85	20–30	0.20–0.30	55–75
Chicken slurry	10–30[b]	70–80	>30	0.35–0.60	60–80
Whey	1–5	80–95	3–10	0.80–0.95	60–80[d]
Leaves	80	90	8–20	0.10–0.30[c]	NA
Straw	70	90	10–50[c]	0.35–0.45[c]	NA
Garden wastes	60–70	90	8–30	0.20–0.50	NA
Grass silage	15–25	90	10	0.56	NA
Fruit wastes	15–20	75	8–20	0.25–0.50	NA
Food remains	10	80	10–20	0.50–0.60	70–80

Source: Adapted from Steffen et al. (1998).
[a] Depends on retention time.
[b] Depends on dilution.
[c] Depends on drying rate.
[d] Depends on particle size.
NA, not available.

Example 12.2

A farm anaerobic digester receives 350 m^3/day (87,180 gal/day) of cow slurry with 6% TS and specific gravity of 1.04. The biogas generated from the digester is used in an engine generator to generate electricity for the farm operations.

(a) Determine the size of the engine generator required to use all the biogas effectively.

(b) If the total electricity demand for the farm operations is 10 MW h, what fraction of the total electricity demand can be met by biogas?

Assume a VS fraction of 75% and a volatile solids reduction (VSR) of 40%.

Solution

1. Determine biogas production from the farm digester
   VSR in the digester

$$\begin{aligned} \text{VSR} &= (350\,\text{m}^3/\text{day} \times 1{,}000\,\text{kg/m}^3 \times 1.04 \times 0.06) \times 0.75 \times 0.40 \\ &= 6{,}552\,\text{kg/day} \end{aligned}$$

Assuming 0.25 m^3 biogas/kg VSR, biogas production can be estimated as:

$$G = 6{,}552\,\text{kg VSR} \times 0.25\,\text{m}^3/\text{kg} = 1{,}638\,\text{m}^3/\text{day}$$

2. Estimate maximum electricity generation potential using engine generators
   Gross heat available from digester gas:

$$E = 1{,}638\,\text{m}^3/\text{day} \times 22{,}400\,\text{kJ/m}^3 = 36.7 \times 10^6\,\text{kJ/day}$$

Electricity generation potential, assuming 35% efficiency for engine generators:

$$P = 36.7 \times 10^6\,\text{kJ/day} \times 0.35 \times 0.000278 \times \text{kW h/kJ} = 3{,}571 \times \text{kW h/day}$$

(i) Provide seven 500-kW engine generators

Total electricity demand for farm operations = 10 MW h

(ii) Fraction of the total electricity from biogas

% of total from biogas = 3.57 MW h/10 MW h = 36%

Caution has to be exercised while handling biogas because of high combustibility of methane. Methane forms an explosive mixture with air and can also asphyxiate at higher concentrations. The initial generation of biogas from an anaerobic system has to be bled off as it may be mixed with significant quantities of air. The physical and chemical characteristics of methane are summarized in Table 12.5.

Table 12.5. Physical and chemical properties of methane.

Physical characteristics	Colorless, odorless
Specific gravity	0.55 at 21°C (70°F)
Density	0.042 kg/m^{3}at 21°C (70°F)
Hazard	Extremely combustible
Flammability limits in air	Forms an explosive mixture with air (5–15% volume). Avoid naked flames or spark-producing tools when there is unburnt gas in the air
Toxicity	Asphyxiant at high concentrations (cause insufficient intake of oxygen)
Typical heating value	37,750 kJ/m^3 (1,016 Btu/ft^3) Biogas has a lower heating value of 22,400 kJ/m^3 (600 Btu/ft^3) since it typically contains only 60–65% methane

12.5 Biogas Impurities

Biogas also contains some impurities that include particulates or dust, hydrogen sulfide, and siloxanes. The biogas will also be saturated with moisture at the operating temperature of the digesters.

12.5.1 Carbon Dioxide

The carbon dioxide present in digester gas is not necessarily a contaminant, but it dilutes the energy content of the biogas, lowering its calorific value. Removal of carbon dioxide from the gas mixture helps in enriching the fuel value of digester gas. Carbon dioxide removal technologies are expensive and may be economically feasible only if the gas is to be upgraded to natural gas quality and sold commercially. The power generation equipment designed specifically for alternative fuels are capable of handling 30–50% (by volume) of carbon dioxide in the gas. Therefore, use of biogas for power generation does not call for carbon dioxide removal.

12.5.2 Moisture

Biogas from anaerobic reactor is typically saturated with moisture at the operating temperature. If the moisture is not removed, it will condense as it comes in contact with the cooler piping and cause increased corrosion by dissolving any hydrogen sulfide present in the biogas. Moisture can also accelerate the deterioration of the membranes or diaphragms in check valves, relief valves, gas meters, and regulators. The condensed water typically accumulates in the lower sections of piping, hindering gas flow and causing pressure losses in the piping and pressure built up in the digesters. Any blockage in the gas line due to condensation can be detected by observing the manometer pressure readings at various points in the biogas system. The manometer will show a zero pressure downstream of the blockage. Moisture removal is accomplished by cooling the gas and removing the condensate.

FIG. 12.1. Damage caused by hydrogen sulfide.
Source: Applied Filter Technology. Reprinted with permission.

12.5.3 Hydrogen Sulfide

H_2S is an extremely reactive biogas constituent that forms sulfuric acid in presence of moisture. The acid formed can corrode pipelines, gas storage tanks, and gas utilization equipment. The H_2S present in digester gas is the primary contributory factor to shortened usable life of many of the components in the digester biogas handling system. Figure 12.1 shows the damage caused by hydrogen sulfide to the gas handling equipment.

In addition, H_2S can be lethal at concentrations above 700 ppmv. H_2S has the smell of rotten eggs, but after a short exposure, the nose becomes numb to odor, which can lead to the dangerously mistaken conclusion that the hazard has diminished or disappeared. Therefore, it is prudent to rely on a gas detector at the slightest indication of H_2S gas. The gas detection instruments must be regularly calibrated to ensure accurate readings in the field. Table 12.6 shows the effects and standards of hydrogen sulfide at different concentrations.

Typical hydrogen sulfide removal strategies include absorption using iron sponge, water scrubbing (simultaneous removal of CO_2), dosing of iron salts to the feed, and biological oxidation.

12.5.4 Siloxanes

Siloxanes are a contaminant of growing concern for facilities contemplating the use of biogas as an energy source. Siloxanes are organic silicon polymers that are used in many forms in a wide range of commercial, personal care, industrial,

Table 12.6. Effects of hydrogen sulfide at different concentrations.

Concentration (mg/m^3)	Effects/Standards
0.011	Odor threshold
2.8	Bronchial constriction in asthmatic individuals
5.0	Increased eye complaints
7 or 14	Increased blood lactate concentration, decreased skeletal muscle citrate synthase activity, decreased oxygen uptake
5–29	Eye irritation
28	Fatigue, loss of appetite, headache, irritability, poor memory, dizziness
>140	Olfactory paralysis
>560	Respiratory distress
≥ 700	Death

Source: Adapted from World Health Organization Report (2003).

medical, and even food products. Personal care products such as shampoos, hair conditioners, cosmetics, deodorants, detergents, and antiperspirants are thought to be the main sources of siloxanes found in digester gas. These compounds are volatile and are released as gas during the anaerobic digestion process of municipal sludge or municipal solid waste. Oxidation of these compounds in gas utilization equipment produces an abrasive solid, similar to fine sand, which accumulates on moving parts or heat exchange surfaces, resulting in accelerated wear and loss of heat transfer efficiency. Figure 12.2 shows the accumulation of fine sand-like material in boilers and the damage caused to an engine piston by siloxanes.

Since siloxanes are highly volatile at relatively low temperatures, the possibility of a relationship between digester operating temperature and the concentrations of siloxanes in the digester gas may be implied. There are reasons to believe that additional siloxanes may be found in the biogas where digesters are heated to higher temperatures than for mesophilic digestion. Typically siloxanes are removed by using activated carbon or graphite media filters or refrigerant dryers.

12.6 Biogas Cleaning for Effective Utilization

The primary challenge associated with the use of biogas as a fuel is the need for gas cleaning to ensure that the gas meets the quality requirements for the utilization equipment. Biogas cleaning is a capital-intensive, multistage operation that can also carry high maintenance costs due to media replacements and/or power costs. However, if the impurities in the gas are left untreated, they can increase the maintenance requirements of the equipment fueled by the gas and can reduce equipment life. Therefore, gas cleaning to reduce condensation, lower H_2S levels, and removal siloxanes is a prerequisite for effective gas utilization.

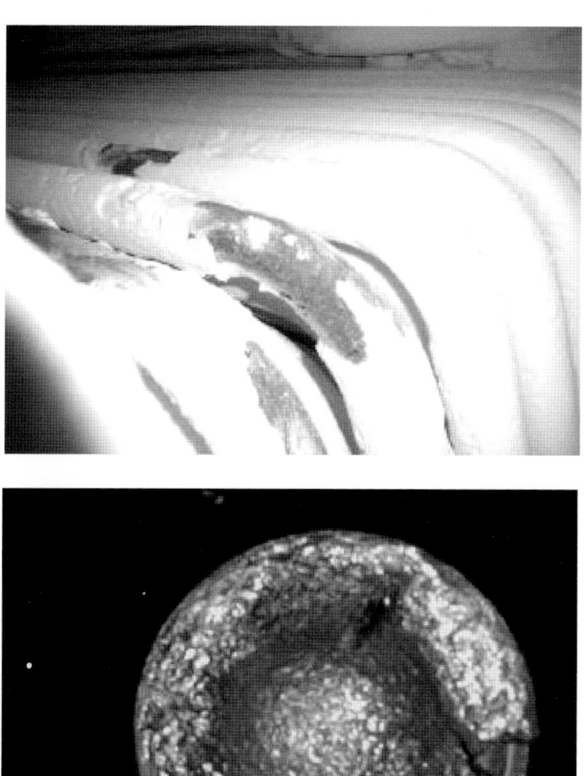

Fig. 12.2. SiO_2 depositions on boiler tubes and damage to an engine piston.
Source: Applied Filter Technology. Reprinted with permission.

12.6.1 Foam and Sediment Removal

Any foam and sediments entrained in the gas stream are separated using a foam separator in the digester gas piping. The foam separator is a large vessel separated down the middle by a baffle wall. The roof of the vessel is fitted with water nozzles to provide a continuous spray wash. The foam- and sediment-laden gas enters the vessel near the top. The gas travels down through the spray wash under the baffle wall and backs up through a second spray wash to the discharge nozzle. The gas exiting the foam separator will mostly be free of any foam or sediment. Figure 12.3 shows the cross section of a foam separator (Santha et al. 2007).

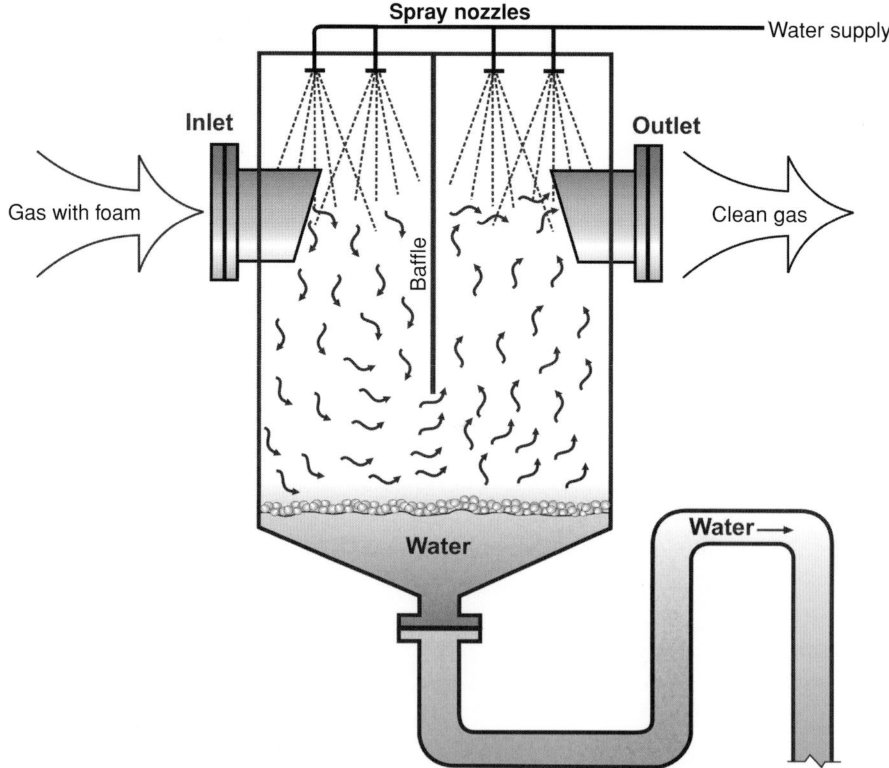

FIG. 12.3. Foam separator by Varec Biogas. Reprinted with permission.

12.6.2 H_2S Removal Options

There are several methods available for scrubbing H_2S from biogas. The most commonly used methods include the use of iron sponge or chemical scrubbers and the addition of ferric (Fe^{3+}) salts to the feed. The biological sulfide removal is discussed in greater detail in Section 7.7.3.

12.6.2.1 Iron Sponge

The digester gas is passed through a permeable bed of iron sponge (hydrated ferric oxide). The hydrated ferric oxide media used is in the form of iron-dipped wood chips, soaked in water. As H_2S passes through the iron media, an exothermic reaction takes place, converting H_2S to ferric sulfide (black solid) and water as illustrated by Eq. (12.4):

$$Fe_2O_3 \cdot H_2O + 3H_2S \rightarrow Fe_2S_3 + 4H_2O \qquad (12.4)$$

The iron sponge can be regenerated and reused multiple times before replacement. However, the process of regenerating iron sponge is a highly exothermic reaction that uses water and air to release sulfur from the iron and reform the hydrated ferric hydroxide. If not submerged in flowing water bath and controlled properly, the iron chips can overheat and combust spontaneously. This can be even more dangerous if all the digester gas has not been removed from the vessel. The exhausted media is usually a combination of ferric sulfide and wood chips. This is not a hazardous waste and can be taken to any landfill for disposal.

12.6.2.2 Proprietary Scrubber Systems

In proprietary scrubber systems, the H_2S-laden digester gas is passed through a bed of media that selectively reacts with H_2S in the gas. The proprietary systems typically use a free-flowing granular media that will not harden like the iron sponge media, but the media cannot be regenerated. These systems have multiple vessels arranged in a lead/lag setup. The sulfide-laden biogas passes through the lead vessel first where most of the sulfide reacts with the media. Subsequently, the gas flows through the lag vessel, which renders the biogas nearly free of H_2S. Once the media in the lead vessel is used up, the vessel is taken off the line and the media is replaced. While the lead vessel is off-line, the lag vessel becomes the primary scrubber. After media replacement, the original vessel is returned to service in backup capacity.

12.6.2.3 Ferric Chloride Injection

Ferric chloride injection does not remove H_2S in the digester gas stream like the other methods, but it reduces the amount of H_2S generated in the digesters. Ferric chloride when injected into a sludge stream oxidizes H_2S and forms insoluble ferrous sulfide that falls out of solution and is taken out during dewatering (Eq. (12.5)).

$$3H_2S + 2FeCl_3 \rightarrow S + 2FeS + 6HCl \tag{12.5}$$

There are two added benefits to using ferric chloride. First, since the ferric salts added react with the H_2S before it is introduced into the gas stream, the odors caused by the presence of H_2S will be reduced at the digesters. Second, ferric chloride has been shown to reduce the formation of struvite (magnesium ammonium phosphate, $MgNH_4PO_4 \cdot 6H_2O$) on process mixers, pump impellers, and in heat exchangers downstream of the anaerobic digesters. Any ferric chloride added to the digesters will preferentially react with hydrogen sulfide. If struvite control is also desired, then sufficient quantities of iron must be added to account for both sulfide and phosphorus reactions.

12.6.3 Moisture Reduction

Moisture removal is typically accomplished by providing a wide spot in the gas collection line to reduce flow velocity. To successfully remove the moisture, the gas flow rate should not exceed 3.7 m/s (12 ft/s) through the pipe. At this reduced velocity, moisture and sediment settle out of the flow and can be removed by providing drip taps at all low spots along the gas collection system. It is recommended that the gas piping be sloped a minimum 1% (10.4 mm/m (0.125 in/ft)) toward the collection point (*WEF Manual of Practice 8* 1998).

Additional moisture reduction can also be achieved by cooling the gas to around 4°C (40°F), using a refrigerant-type dryer. The dryer is typically made of stainless steel or other corrosion-resistant materials to minimize corrosion from H_2S. Corrosion from condensing acid can be minimized by removing H_2S from the gas prior to drying.

12.6.4 Siloxane Removal

The two common types of siloxane removal systems include (1) low-temperature drying systems and (2) graphite molecular sieve scrubbers.

12.6.4.1 Gas Drying

A significant fraction of the siloxanes can be removed along with the moisture when the gas is dried since the siloxanes, being relatively heavy, tend to adhere to water vapor in the biogas stream. Low-temperature gas-drying systems consist of a refrigerant-type dryer that cools the digester gas to −23°C (−10°F) as it passes through the dryer, causing the siloxanes to drop out with condensed moisture. Dryer manufacturers have reported 90–95% removal of siloxanes by low-temperature drying. While gas-drying systems are relatively simple in concept, operating at temperatures of −23°C (−10°F) or lower has the disadvantage of being subject to significant amounts of ice formation. If the ice is not removed, it can cause damage to the refrigeration system.

12.6.4.2 Activated Carbon/Graphite Media Scrubbers

The activated carbon or graphite media scrubbers used to remove siloxanes from biogas operate on the same principle as the carbon scrubbers used for odor control. The biogas is passed through a vessel filled with the media, which absorb siloxanes, H_2S, and other volatile organic compounds (VOC) in the digester gas. Typically, the medium used for siloxane removal is not selective for siloxanes. Consequently, H_2S and other VOCs in the biogas will also be adsorbed on to the media if their

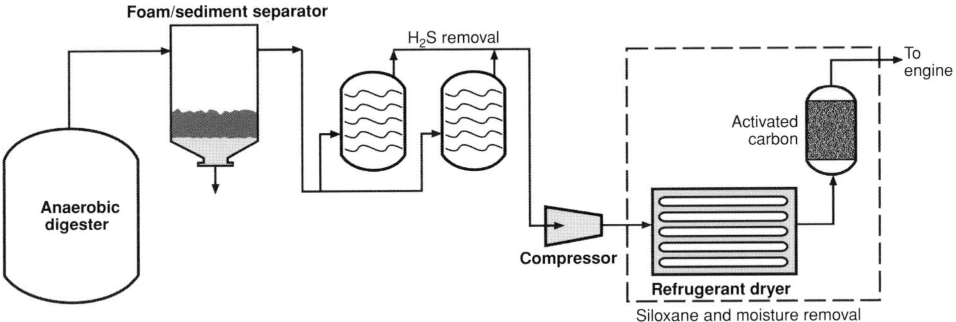

FIG. 12.4. Digester gas treatment module.

concentrations are not reduced prior to the siloxane scrubbers, requiring more frequent media replacement and increased operating costs.

If used in conjunction with a gas dryer, activated carbon or graphite media scrubbers offer a cost-effective means of removing siloxanes from digester gas. Chilling the gas to 4°C (40°F) with a gas dryer would remove some of the water and approximately 30–40% of the siloxanes. The remaining siloxanes can be removed using a scrubber system.

Figure 12.4 illustrates a complete digester gas treatment module (Santha et al. 2007). It includes a foam separator for removing any entrained foam and sediment from the gas, a H_2S removal system to reduce H_2S concentrations in the gas upstream of the siloxane scrubber, a gas dryer to lower the gas temperature to 4°C (40°F) for removing water, heavy organics, and approximately 30–40% of the siloxanes, and finally, a siloxane scrubber. Depending on the gas pressures available from the digesters or the storage equipment, a compressor may also be required to provide the pressure needed to convey the digester gas through the dryer and the siloxane scrubber.

The gas-cleaning requirements can vary depending on the application and the concentrations of the various contaminants in the gas.

12.7 Biogas Utilization

With biogas collection and utilization technologies improving over the years and energy recovery from biogas becoming one of the more mature and successful waste/residues-to-energy technologies, the question arises as to what is the best use for the biogas? This depends on a number of factors, including amount of biogas produced, energy cost, plant's energy demand, and other incentives. Traditionally, biogas has been used as fuel for boilers to provide the heat needed to maintain the process temperatures in the anaerobic digesters. At most plants, the digestion process generates more gas than is needed to support the process and the excess gas is flared. This excess biogas represents a potential energy source for other plant processes and functions.

The most common alternative use for biogas has been combustion in an engine generator or combustion turbines to provide electrical power for plant operations or for sale or credit to the local power utility. For this use, all biogas is used by the engine generator and waste heat is recovered from the engine generator to heat the digesters. With rising energy costs, this method of digester gas utilization is expected to become more cost effective for utilities and other close-by industries.

In developing countries, another possible avenue for digester gas use is as fuel for cooking and lighting. Cheap and improved biogas stoves and lamps are being made available to people who have biogas resources that could be used for cooking or generating electricity.

The potential gas use scenarios are illustrated in Fig. 12.5.

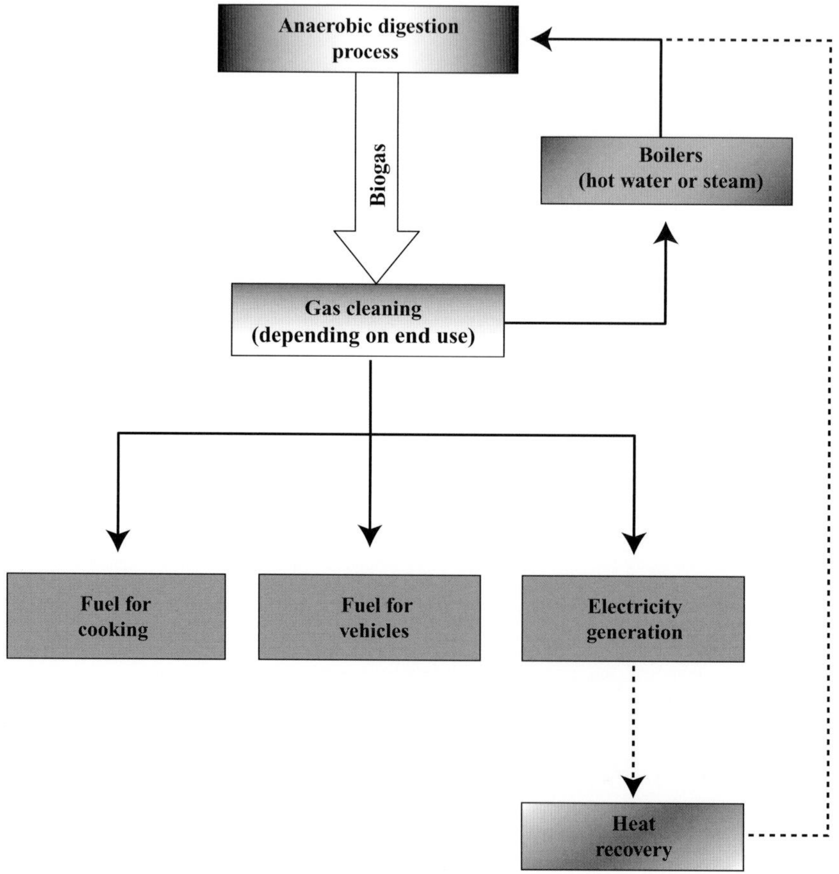

FIG. 12.5. Avenues for biogas use.

12.7.1 Digester Gas for Heating Feedstock

The digester heating requirements consist of two components: (1) heating of raw solids to the required process temperature and (2) replacing heat lost through the digester tank walls and roof, typically referred to as transmission losses. The sludge heating requirements are based on raw sludge feed rate, and the feed and digester operating temperatures. Transmission heating losses are based on the insulation value of the tank walls and roof construction and ambient temperatures. Assuming the digesters are well insulated, transmission losses constitute only a small fraction of the total heating loads.

The energy requirement for raising the temperature of the feed sludge to the required digestion process temperature can be calculated using Eq. (12.6) as follows:

$$H = Q \times \rho \times C \times (T_1 - T_2) \tag{12.6}$$

where H is energy required (kJ/h), Q is sludge feed rate to the digesters (m^3/h), ρ is density of feed sludge (assumed equal to water) (kg/m^3), C is specific heat capacity of feed sludge (assumed equal to water) (kJ/kg °C), T_1 is digestion process temperature (°C), and T_2 is feed sludge temperature (°C).

The digester heating requirements will depend on the operating temperature of the digesters. Heating loads for thermophilic operation (55°C (131°F)) are approximately double compared to mesophilic operation (35°C (95°F)). The heating requirements are also proportional to the volume of the feed solids. The generally accepted practice in the United States has been to thicken the feed solids to at least 4% solids. Figure 12.6 illustrates the reduction in heating loads with increasing feed solids concentrations to the anaerobic digestion process.

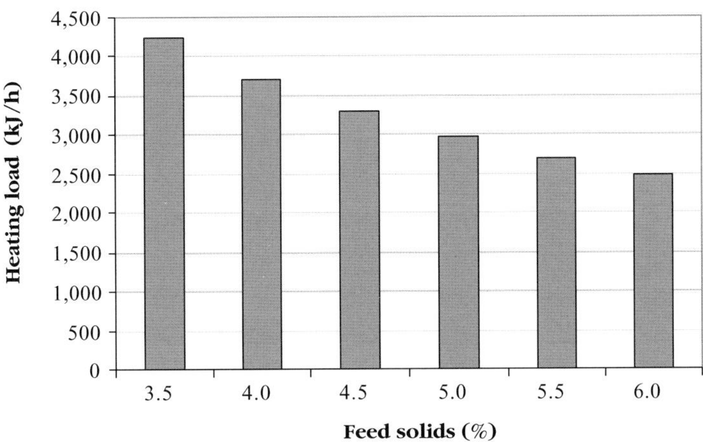

FIG. 12.6. Variation of digester heating loads with feed solids concentration.

Increasing the feed solids concentration from 3.5 to 6.0% can lower the incoming sludge heating requirements by approximately 40%. The currently available mechanical thickening methods, such as centrifuge, gravity belt thickener, and rotary drum thickener, can readily achieve the desired 5–6% feed solids concentration. Thickening of the feed solids takes on added importance for digestion processes operating at higher temperatures.

Knowing the thermal efficiency of the boiler used to generate hot water or steam, the energy requirement for process heating can be estimated. The typical boiler efficiencies range from 70 to 80%.

Example 12.3

The sludge feed to a digester is to be heated from 10°C (50°F) to a thermophilic temperature of 55°C (131°F). The sludge flow rate is 250 m^3/day (66,000 gal/day). Assuming digester transmission losses are approximately 10% of the total sludge heating loads:

(i) Determine the required heat exchanger capacity.
(ii) Determine the biogas requirement to maintain process temperatures.

Solution

1. Energy requirement for raising the temperature of the feed sludge to the required digestion process temperature:

$$H = 250\,\text{m}^3/\text{day} \times 1{,}000\,\text{kg/m}^3 \times 4.187\,\text{kJ/kg}^\circ\text{C} \times (55 - 10)^\circ\text{C}$$

$$= 47{,}103{,}750\,\text{kJ/day}$$

Digester transmission losses = 10% of sludge heating loads

$$= 47{,}103{,}750\ \text{kJ/day} \times 10\% = 4{,}710{,}375\ \text{kJ/day}$$

Total energy requirement = 47,103,750 kJ/day + 4,710,375 kJ/day

$$= 51{,}814{,}125\,\text{kJ/day}$$

2. Biogas required for meeting the energy requirement

$$= (51{,}814{,}125\ \text{kJ/day})/(22{,}400\ \text{kJ/m}^3)$$

$$= 2{,}313\,\text{m}^3/\text{day}$$

12.7.2 Biogas for Electricity Generation

The most common alternative use for biogas has been for power generation. The electrical power generated can be used for plant operations or for sale or credit to the local power utility.

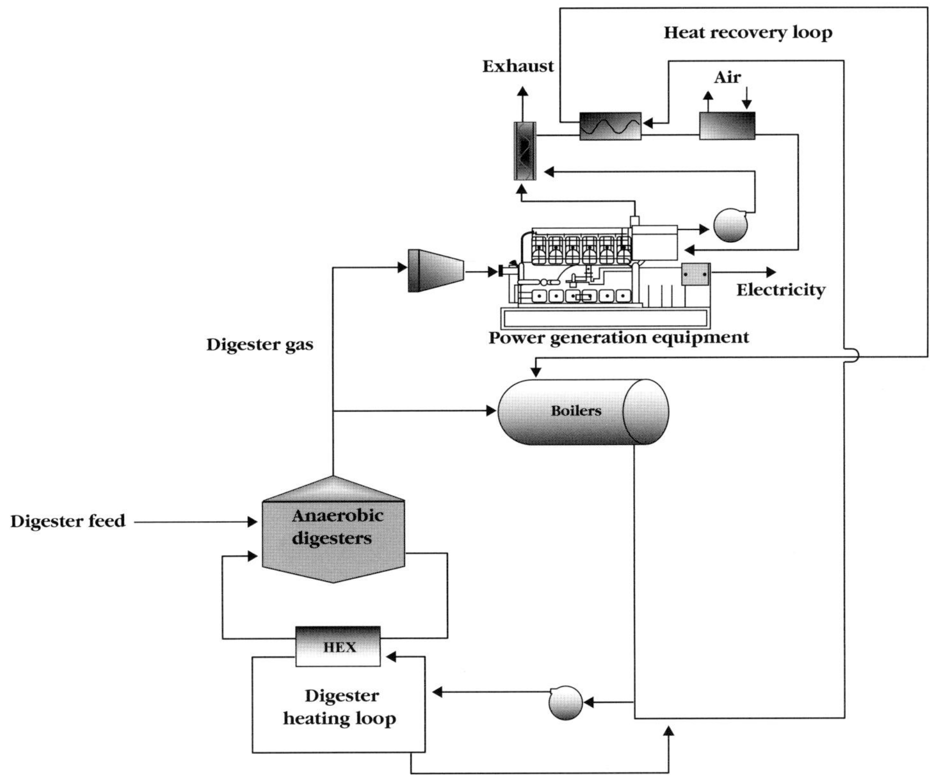

FIG. 12.7. Biogas for electricity generation.

Power production from waste treatment facilities is typically less than 5 MW, and a variety of on-site power generation systems are available. Choices range from the traditional engine generators and combustion turbine generators (for larger facilities) to recently developed microturbines and fuel cells. Internal combustion engines are by far the most common generating technology choice. About 80% of the power generation facilities at waste treatment plant use reciprocating engines. Figure 12.7 is a schematic illustration of the use of engine generators for electricity generation.

12.7.2.1 Engine Generators

Engine generators typically consist of a skid-mounted package of a reciprocating engine, generator, and a control panel (Fig. 12.8). The popularity of engine generators is attributed to availability from multiple suppliers and the general familiarity of operations personnel with engines that resemble an automobile engine. The engine is typically a spark-ignited natural gas engine, modified and derated for use with a low-Btu gas.

FIG. 12.8. Engine generator.

Naturally aspirated engines are used by smaller facilities where air emissions are not a concern. The newer turbocharged, lean-burn engines can achieve higher efficiencies and lower air emissions and can operate at gas pressures of only 14–35 kPa (2–5 psi g). The engines typically operate at speeds from 900 to 1,200 rpm. Higher-speed units that reach speeds up to 1,800 rpm are available but, while they have lower initial capital cost, have higher maintenance.

The electrical efficiencies of engine generators operating on biogas can range from 30% for smaller engines to around 35% for larger engines. Approximately 40–45% of the total energy input to the engine generators can be recovered from the engine exhaust and the cooling systems to produce either hot water or steam for combined heat and power applications. The overall efficiency of engine generators can approach 70–80%.

12.7.2.2 Turbine Generators

Turbine generators, available in unit sizes from about 1,000 kW to nearly 50,000 kW, are typically used for larger power generation applications. They consist of combustion turbines connected to generators through speed-reducing gearboxes. The electrical efficiencies of turbine generators are typically lower than engine generators, but they provide a relatively compact package for much greater output capacity. The heat recovered from the turbine exhaust can be used to generate either hot water or steam for heating requirements, including the operation of a steam turbine to produce additional electricity in a combined cycle mode. They are fairly tolerant to contaminants in biogas and have a lower potential for air emissions.

Turbine generators require very little preventive maintenance; however, the units require a periodic (after ~30,000 h of operation) overhaul. In lieu of the overhaul, many facilities prefer to replace the units with new units at the end of the service life.

12.7.2.3 Microturbines

Microturbines are a relatively new technology for on-site power generation for waste treatment facilities, with the first commercial units installed in the late 1990s. They are modular units that consist of a compressor, combustor, turbine, alternator, and a generator (Fig. 12.9). The unit also includes a heat recovery system that captures waste heat from the exhaust gases, improving the overall efficiency of operation. Microturbines are typically rated at less than 300 kW, but multiple units can be installed in parallel for higher capacities.

Microturbines are an attractive option for smaller facilities, making on-site generation available to plants that formerly did not have this choice. They produce minimal noise and have lower emissions compared to engine generators. They only have a single moving part, lowering the routine maintenance requirements. The combustion unit is replaced after about 16,000 h of use and is typically done by the equipment supplier under a maintenance contract.

Initial experience with microturbines on biogas has been mixed. Problems have been reported with siloxane deposition, reliability, and high-frequency noise. Although microturbines offer the promise of lower emissions, their higher costs and lower efficiency have been significant drawbacks. To date, use of microturbines for biogas applications remains relatively limited.

FIG. 12.9. Microturbines.
Source: Capstone Turbine Corp. Reprinted with permission.

Table 12.7. Comparison of biogas power generation equipment.

Parameter	Engine Generator	Turbine Generator	Microturbine	Fuel Cell
System size				
Unit size (kW[a])	150–3,000	>1,000	30–300	200
Appropriate plant size	Small to midsize	Large	Small	Small
Performance				
Efficiency (%)	30–35	25–30	25–30	35–40
Estimated heat recovery potential as a percentage of fuel input (%)	40–45	30–35	30–35	40–45
Overall system efficiency (%)	70–80	55–65	55–65	75–85
Typical costs				
Installed cost (US$/kW[b])	1,200–1,500	1,500–2,000	1,500–2,000	4,000–6,000
Maintenance (US$/kWh[c])	0.010–0.025	0.005–0.018	0.005–0.018	0.005–0.025

Source: Adapted from Pittman et al. (2002).
[a] Capacity is approximate and will vary with supplier.
[b] Installed costs vary with type and amount of auxiliary equipment and the type of structure.
[c] Maintenance costs are dependent on the quality of gas used. Costs are based on supplier service contracts.

12.7.2.4 Fuel Cells

Fuel cells convert hydrogen-rich fuel sources directly to electricity through an electrochemical reaction. Fuel cell power systems have the promise of high efficiencies since they are not limited by Carnot cycle efficiency that limits thermal power systems. Fuel cells can sustain high-efficiency operation even under reduced loads. The construction of fuel cells is modular, making it easy to size plants based on the power requirements. When used with biogas, a fuel pretreatment module packed with activated carbon is required for removal of H_2S.

The use of fuel cells for electricity generation using biogas is still in developmental stages. In addition to the potential for high efficiency, the environmental benefits of fuel cells remain one of the primary reasons for their development. At present, high capital costs, fuel cell stack life, and reliability are the primary disadvantages of fuel cell systems and are the focus of intense research. The cost is expected to drop significantly in the future as development efforts continue, partially spurred by interest by the transportation sector.

Table 12.7 presents a comparison of typical unit sizes, efficiencies, and fuel consumption and heat recovery rates of the various power generation systems.

The electricity generation potential of biogas can be determined knowing the efficiency of the generation equipment. The electricity generation and the heat recovery potential can be estimated using Eqs (12.7) and (12.8), respectively.

$$P = E \times \eta \times 0.000278\,\text{kWh/kJ} \tag{12.7}$$

where P is electricity generation potential (kWh), E is gross energy from digester gas (kJ/day), and η is efficiency of power generation equipment (%).

$$H = E \times \theta \tag{12.8}$$

where H is heat recovery possible from power generation equipment (kJ/day), E is gross energy from digester gas (kJ/day), and θ is heat recovery efficiency from power generation equipment (%).

12.7.3 Biogas as Vehicle Fuel

Biogas also has potential for use as a fuel for vehicles. However, biogas has to be upgraded to natural gas quality for use in vehicles designed to function on natural gas. This calls for removal of particulates, carbon dioxide, hydrogen sulfide, moisture, and any other contaminants present in the gas to increase the methane content to over 95% (by volume). The carbon dioxide present in biogas dilutes the fuel value of the gas and has to be removed. Any hydrogen sulfide present in the gas can increase corrosion in the presence of water. Moisture in the gas can also cause ice clogging at lower temperatures.

Some of the common methods used for carbon dioxide removal include water scrubbing, pressure swing adsorption (PSA), and membrane separation.

(a) Water scrubbing involves introducing pressurized gas to the bottom of a packed column, which receives water from the top. The carbon dioxide and some hydrogen sulfide get absorbed in the water and are removed from the gas stream.
(b) The PSA technology separates carbon dioxide from methane by adsorption and desorption on zeolites or activated carbon at different pressure levels. Any hydrogen sulfide in the gas is typically adsorbed irreversibly by the media. Therefore, a hydrogen sulfide removal step has to be included upstream of the PSA process.
(c) In membrane separation processes, pressurized biogas is passed through acetate-cellulose membranes that separate polar molecules such as carbon dioxide, moisture, and any hydrogen sulfide. Again, hydrogen sulfide removal is recommended upstream of the membranes prior to membrane separation.

Biogas after purification to natural gas quality is typically mixed with natural gas for injection into the low-pressure natural gas grid or into the natural gas filling stations for vehicles. Since clean biogas has the same characteristics as natural gas, the existing natural gas infrastructure can support the use of biogas. Blending with

Table 12.8. Comparison of gaseous emissions from heavy vehicles.

Fuel	CO g/km (lb/mi)	Hydrocarbons	NO_x	CO_2	Particulates
Diesel	0.2 (7.1×10^{-4})	0.4 (1.4×10^{-3})	9.73 (0.035)	1,053 (3.74)	0.1 (3.6×10^{-4})
Natural gas	0.4 (1.4×10^{-3})	0.6 (2.1×10^{-3})	1.1 (3.9×10^{-3})	524 (1.86)	0.022 (7.8×10^{-5})
Biogas	0.08 (2.8×10^{-4})	0.35 (1.2×10^{-3})	5.44 (0.019)	223 (0.79)	0.015 (5.3×10^{-5})

Source: Adapted from Trendsetter (2003).

natural gas can also enhance the distribution of biogas even at relatively small quantities.

The use of natural gas and biogas in vehicles reduces the emissions of NO_x and carbon dioxide. It also eliminates emissions of benzene that is a known human carcinogen. Table 12.8 provides a comparison of gaseous emissions for heavy vehicles on different fuels.

In Europe, there are 14 cities that use biogas as vehicle fuel, with Sweden being the most developed with regard to recycling and reuse of biogas. Sweden has established national standards for biogas as a vehicle fuel, requiring the methane content to be higher than 95%, and has also set limits for dew point, sulfur content, and some other contaminants.

12.7.4 Digester Gas for Cooking

Biogas presents a valuable resource that, if beneficially used, can offset some of the energy requirements even in developing countries. However, the primary consideration in selecting a gas use option for developing countries would be the level of technical expertise required in operating and maintaining the gas utilization system.

Energy for cooking and electricity are two of the basic necessities that constitute a major fraction of the total energy consumption by rural communities. The biogas generated from anaerobic digestion of waste streams can be used directly through conventional low-pressure gas burners for cooking or burned in an engine generator for electricity generation for community use. The use of biogas for cooking is illustrated in Fig. 12.10.

The major components of the biogas system include the following:

Gas metering: Even though a reliable gas flow meter is desirable at a strategic location in the gas collection system to provide an accurate measurement of gas production, in many rural communities, the cost of these units may outweigh its benefits.

Hydrogen sulfide removal: If the gas is burned in an engine, removal of hydrogen sulfide is desirable to prevent corrosion. Hydrogen sulfide concentrations can be reduced by passing the gas through a gas-tight absorption tank packed with iron filings or steel wool.

Gas storage: Biogas produced from an anaerobic system may not be continuously used for cooking or electricity generation. Therefore, it is essential to provide

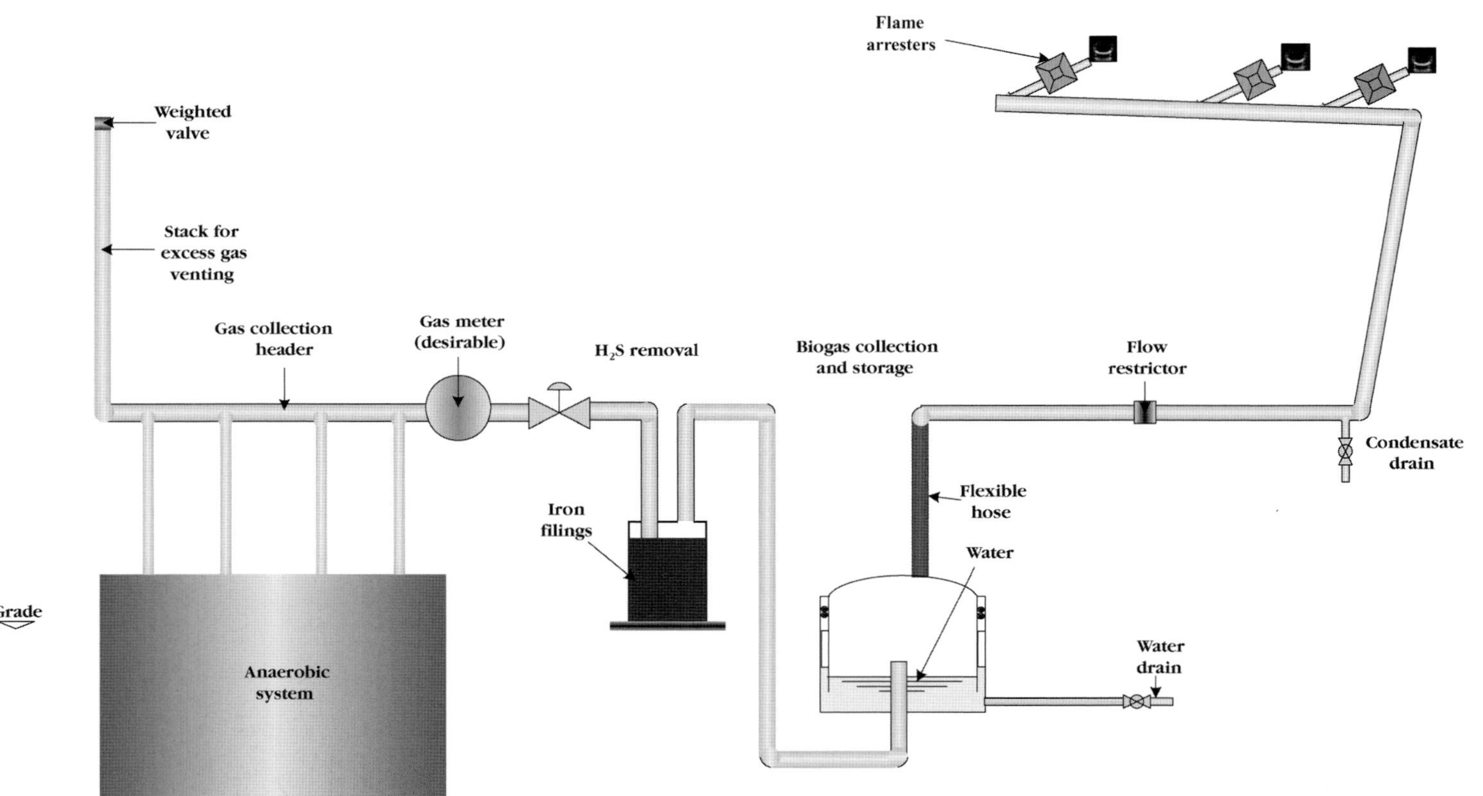

FIG. 12.10. Use of biogas as a fuel for cooking.

adequate storage to minimize gas wastage. The storage capacity requirements can be estimated based on the biogas production rate and the peak demands. The simplest arrangement for gas storage consists of a below-grade reinforced concrete tank, filled with water and equipped with a floating metal cover that provides variable-volume storage. Depending on the volume of gas generated, the cover moves vertically, along a roller and guide mechanism installed along the sidewalls of the tank. The difference in water levels inside the gas holder cover and the annular space between the cover and the tank corresponds to the pressure of gas stored under the cover. The water in the annular space also maintains a permanent seal preventing gas leakage. The bottom edge of the floating cover will rise clear of the water to vent any excess gas. The gas pressure, provided by the weight of the cover, is typically low. The stored gas can be pressurized further by adding weights on the gas holder cover. Any sediment in the gas will also separate out in the storage tank.

Backup pressure relief: As backup to the automatic venting from the gas storage tank, the main collection header has to be provided with a 2.4–3.0 m (8–10 ft) high stack for venting any excess gas. The stack will typically be equipped with a weighted valve that is set to a pressure slightly greater than the maximum storage pressure in the tank. If the tank fails to release excess gas due to a mechanical problem, such as a jammed roller and guide mechanism, the pressure relief valve on the stack will open to vent any excess gas.

Flow restrictor and gas piping: It is essential to include a flow restrictor (only if gas is used as fuel for cooking), which could be an orifice, on the supply line to avoid gas surges to the burners. The size of the orifice and the gas pressure will regulate the amount of gas that is supplied to the burners. To minimize corrosion issues, the gas piping should be made of stainless steel or heavy-gauge black steel. If buried, plastic is also an option, but is more susceptible to damage. A condensate trap should be provided at the low points in the gas line to permit removal of any water resulting from condensation.

Flame arresters or flame checks: Flame arresters have to be installed on all lines that carry methane to the gas stove to avoid the flame traveling back along the supply lines to the main header and to the storage tank. Flame checks prevent the propagation of a flame through heat dissipation.

12.8 Future of Biogas as a Renewable Resource

Biogas is a nonfossil fuel that can be generated from a variety of feedstocks. Around the world, the potential for biogas production is so large that it could replace about 20–30% of the total natural gas consumption. However, if the biogas generated from anaerobic processes is released to the atmosphere, it can offset any reduction in carbon dioxide emissions that could be realized through its effective use. Until recently, the high cost of available biogas utilization technologies and the lack of infrastructure precluded widespread use of this valuable resource. With several government-sponsored financial incentives for the installation and operation of renewable energy technologies in the recent years, the future looks more promising for biogas-generating facilities. In addition, the rising cost and unstable supply of fossil-derived fuels are other economic incentives that could enhance the widespread use of biogas as an energy source.

References

McCarty, P. L. 1964. Anaerobic waste treatment fundamentals. Part four: Toxic process design. *Public Works* 95:95–99.

Pittman, L., Long, D., Brady, R., and Rowan, J. 2002. *Digester Gas Utilization: Is the Time Right for On-site Power Generation.* Proceedings of the 16th Annual Residuals and Biosolids Management Conference, March 3–6, 2002, Austin, TX, USA.

Santha, H., Rowan, J., and Hoener, W. 2007. *Digester Gas Cleaning Requirements Based on End Use.* Proceedings of the WEF/AWWA Joint Residuals and Biosolids Management Conference, April 14–17, 2007, Denver, CO, USA.

Steffen, R., Szolar, O., and Braun, R. 1998. *Feedstocks for Anaerobic Digestion.* Technical paper 1998-09-30, Institute for Agro biotechnology, Tulln University of Agricultural Sciences, Vienna, Austria.

Trendsetter. 2003. *Biogas as Vehicle Fuel: A European Overview.* Trendsetter Report No. 2003:3, Stockholm.

Water Environment Federation. 1998. *Design of Municipal Wastewater Treatment Plants, 4th edn, Manual of Practice 8.* ASCE Manual and Report on Engineering Practice No. 76.

World Health Organization. 2003. *Hydrogen Sulfide: Human Health Aspects.* Concise International Chemical Assessment Document 53, International Programme on Chemical Safety (IPCS), WHO, Geneva.

Index